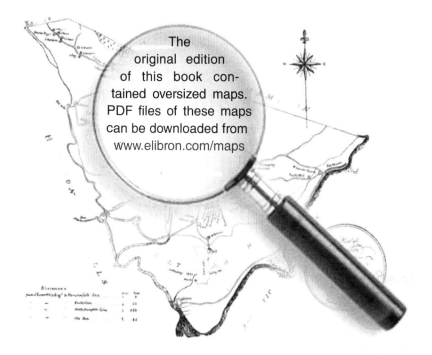

RICHARD FRANCIS BURTON

EXPLORATIONS

OF THE

HIGHLANDS OF THE BRAZIL;

WITH

A FULL ACCOUNT OF THE GOLD
AND DIAMOND MINES

VOLUME II

Elibron Classics
www.elibron.com

This book is an accurate reproduction of the original. Any marks, names, colophons, imprints, logos or other symbols or identifiers that appear on or in this book, except for those of Adamant Media Corporation and BookSurge, LLC, are used only for historical reference and accuracy and are not meant to designate origin or imply any sponsorship by or license from any third party.

THE

HIGHLANDS OF THE BRAZIL.

THE PAULO AFFONSO, KING OF THE RAPIDS, THE NIAGARA OF BRAZIL.

THE

HIGHLANDS OF THE BRAZIL.

By CAPTAIN RICHARD F. BURTON,

F.R.G.S., ETC.

Brazil is usually represented by a Tupy Woman.

VOL II.

LONDON:

TINSLEY BROTHERS, 18, CATHERINE STREET, STRAND.

1869.

[All Rights of Translation and Reproduction reserved.]

EXPLORATIONS

OF THE

HIGHLANDS OF THE BRAZIL;

WITH

A FULL ACCOUNT OF THE GOLD AND DIAMOND MINES.

ALSO,

CANOEING DOWN 1500 MILES OF THE GREAT RIVER SÃO FRANCISCO, FROM SABARÁ TO THE SEA.

BY

CAPTAIN RICHARD F. BURTON,

F.R.G.S., ETC.

VOL. II.

LONDON:

TINSLEY BROTHERS, 18, CATHERINE ST., STRAND.

1869.

LONDON :
BRADBURY, EVANS, AND CO., PRINTERS, WHITEFRIARS.

CONTENTS.

THE

HIGHLANDS OF THE BRAZIL.

CHAPTER I.

SABARÁ TO SANTA LUSÍA.

DEPARTURE.—ADIEUX.—THE RAFT, AND WHAT IS IN IT.—THE "BRIG ELIZA."
—THE STATE OF THE RIVER.

"Messieurs les délicats . . . voulez-vous vous embarquer pour vivre de telle façon ? Comme ie ne vous conseille pas."—*Jean de Léry*.

WEDNESDAY, *August* 7, 1867.—We walked down to the Porto da Ponte Grande,* where the ajôjo or raft lay. I never saw such an old Noah's Ark, with its standing awning, a floating gipsy "pál," some seven feet high and twenty-two long, and pitched like a tent upon two hollowed logs. The river must indeed be safe, if this article can get down without accident.

All the notables of the place witnessed the process of embarkation. Miss Dundas broke the bottle with all possible grace upon the bows, and christened my craft the "Brig Eliza," and two pair of slippers were duly thrown at my head. Many "vivas" were given and returned, and all embarked for a trial-trip—shall I call it, with the Royal Geographical Society, a "tentative expedition"—of a couple of miles. When the fifteen souls came on board, they sunk the article some three palms, and deluged the port platform, making the headman, or pilot, "Manoel de Assumpção Vieira," very nervous—already he began to predict swamping, "going down in a jiffey," and being dashed to pieces by the rapids. We shot past the Pedra Grande, a quartzose rock in mid stream;

* The upper landing-place at the Ponte Pequena Quarter is called "Porto do Gallego," from a stream and an old gold washing hard by it.

the Cámara has threatened for years to remove this obstacle; unfortunately no one here can fire a charge under water.

At the little " church village " of Santo Antonio da Roça Grande, the animals were waiting to carry home the non-voyagers, my wife—who was incapacitated for accompanying me by a bad fall and a serious sprain—included. My hospitable and warm-hearted escort stood—as the setting sun sank behind the mountains—and watched the raft turn the last corner, and float off into the far mysterious unknown. What made me think of the Nile story told by Mr. Curzon, of the white man paddled by dark Amazons adorned with barbaric gold, down the streams unfrequented by the traveller? I confess to having felt an unusual sense of loneliness as the kindly faces faded in the distance, and, by way of " distraction," I applied my brain to the careful examination of my conveyance.

The ajôjo, or, as it is called in other places, the " balsa," here represents the flat boat of the Mississippi, and of the Arkansas " chicken thieves," in the days when, according to Mr. Nolte, men spent a month between the mouth of the Ohio and New Orleans, and then walked back. On the Rio das Velhas, however, it cannot yet be said to have become an institution, and I am the only traveller who has yet passed down from Sabará to the Rapids of Paulo Affonso. As explorers, frontier-men, and other "pioneers of civilization" will have to use it upon the still unknown branches of many a stream, including the Amazons River, a detailed description of the craft may not be without use.

The usual ajôjo* is a bundle of two or three canoes, in the latter case the longest occupying the centre. The best materials are the strong and light Tamboril Vinhatico, and " Cedro," or Brazilian cedar, about one inch thick; mine were of " Peroba,† nearly two inches deep, and consequently too heavy. We drew two palms, approaching a foot and a half (seventeen inches) even without cargo. There is sometimes a helm, always fixed to the longer or the longest boat; if not, the pilot poles or paddles, standing or sitting in the stern. The canoes should be lashed together by hide ropes, with an interval of six to eight inches, not connected as mine were by iron bars joining them at both

---

* Or ajoujo. In Portuguese, as in most of the Latin languages, the circumflex often denotes crasis, or contraction by the omission of a letter whose sound is or is not retained.

† A fine hard wood, formerly reserved by government for ship-building.

stem and stern, and thus destroying all elasticity. Round or squared poles fastened by leather thongs to the gunwales, support the "soalho," or platform, which should fit tight to the sides, otherwise the craft, when "broaching to," may be water-logged. This boarding of ten planks, laid horizontally, projects laterally into coxias, trampways eight to ten inches wide, where the men work.* My canoes, thirty-three feet four inches long, and when joined, six feet broad, formed a solid foundation for the standing awning, a somewhat risky comfort. It was made fast by five wooden stanchions, of which the two pair fore and the one aft, were supported, besides being nailed, by strong iron knees, or stays. The tent was of rough Minas cotton, protected in the forepart, where I slept, by wax-cloth from Morro Velho; and it was a kind of "pál," to throw off the rain. Facing the head, and in the coolest place, was a tall deal writing-desk, which rivalled the awning in catching the wind. Behind this, on each side, stood a Giráo,† or boarded bunk, for sofa and bed, raised on four uprights. Amidships was the table, a locked box of provisions flanked by two stools (tamburetes). In the stern stood the galley, a similar bench, but lined with bricks, and around it the batterie de cuisine, iron kettles and pots, cups and goblets, of course not forgetting the invaluable frying pan.‡ Two large jars of porous earth (talhas or igaçabas),§ carried the supply of water,

* When the Ajôjo carries merchandise, the platform is reduced to the gangway. Coxía also means a stall, a corridor in a hospital, a passage in a warehouse, &c.

† The Giráo or Jiráo, according to the T. D. is properly a hut on piles, used as a granary. Sr. J. de Alencar uses it as the "horse," or small gallows-shaped frame of the Jangada-raft. In the south it is called "Noque." Generally in the Brazil, Giráo is applied to various rude pieces of furniture, shelves of wood or hide, a framework for smoking or sun-drying meat, and so forth.

‡ The provisions were jerked meat (Carne seca), in Pernambuco called Carne de Ceará, in other places Carne do Sertão and Carne do Sol, when simply cut in strips, hung in the air and sun dried, fine coriaceous matter for pulling at with the teeth. Lard (Toucinho) is never wanting in these parts; and rice and beans can generally be found. The men also received a dram of rum (Cachaça) every evening. For my own stores I had a box with a lock: it contained white salt and sugar—

the brown can be found everywhere—mustard and black pepper; here they cannot be bought, while cayenne grows wild. I also had tea—it is no use to carry coffee. The good Mr. Gordon had supplied me with excellent salt beef in rounds, with tongues and with bread, to relieve the monotony of the Brazilian rusk; also, in case of sickness, with a bottle of Cognac and another of gin, which might take the place of Pinga. Finally, a few tins of beef, sardines, and potted meats, for a "treat," were stored in the table-box. Mr. James Smyth, of Morro Velho, gave me a few valuable boxes of excellent Havanahs, which were highly appreciated by my hosts. In Brazilian travel cigars are soon exhausted, and it is the custom to pass round the case.

§ Ygaçába is a Tupy word, generally used in these parts. The first letter had amongst the savages a dubious sound between "i" (or "y") and "ú." Hence the Portuguese wrote it in various ways, as "ira" or "ora," honey, and una for yg, una, a dark stream.

which was renewed every night, and allowed to stand for a day. The President of São Paulo advised me not to drink liquid from the stream, but all on board did so, and so did I. Mr. Gordon had taken care to provide the raft with a stout boat-hook, with an anchor in the bows, a standing wonder to the riverines, who had never heard of Anacharsis the Scythian, and with strong English ropes for "cordelling"*—these are of the greatest consequence when swinging round the rapids.

The crew numbers three,† old Vieira and his sons, who are to receive, besides food, 5$000 per day.‡ Two stand in the bows with poles, which they prefer, as being easier to use than paddles. The former, called varas, and when large, varegões, are stout elastic cuttings of the supple Peroba or Parahybuna wood, fifteen to twenty feet long, by two inches in diameter. They are shod with iron (ferrão), and, when not, the ends must be sharpened before shooting a rapid. The points are of various kinds, the "Ponta de diamante" is a long pyramid, with a ring band; the "Pé de Cabra" is cloven-footed, and the "Gongo" has, in addition, a boat-hook to hold on by; whilst the Forquilha, which rarely comes into use, is a hooked pole, that arrests the course by catching trees. The paddles (remos), used in deeper waters, are artless articles, and vary in shape every few hundred miles; here they are straight and flattened spatulæ. The next set will have handles four feet long, ending in a blunt lozenge one foot broad; its rowlock will be a lashing of hide rove through a hole in the gunwale. This article has no leverage. At the junction of the two streams I found fine elastic paddles of the veined and yellow taipóca wood, which not a little resembled our ash. They were six feet in length, and broadest at the lower end, which was rounded so as to present a clean surface when used as a pole against bank or tree, or ended with trimmed beams of a heavy Cactus, which sinks in water like lead, and which is capable of doing very hard work.

The men were mere land-lubbers, quite unlike those of the S. Francisco. They feel, or affect to feel, nervous at every obstacle. They have been rowing all their lives, and yet they know not how

---

* Locally called "Sirga."

† For the up trip six men are necessary, and the work of one day down stream takes three.

‡ I carried Brazilian bank-notes, taking care that they were new and of small values, between 10$000 and 1$000 : besides these, a small bag of coppers and of silver pieces for especial occasions, was in store. Total, 1:500$000.

to back water; curious to say, this is everywhere the case down stream. They pull with all their might for a few minutes, when the river is rapid, so as to incur all possible risk; and, when the water is almost dead, they lie upon their oars and lazily allow themselves to be floated down. Thus, during the working day, between 7 A.M. and 5 P.M., very little way is made. They have no system, nor will they learn any; it is needless to suggest placing rollers under the canoes or stamping upon the platform when we ground; they never saw such things done, and they don't care to see them. All have the appetites of Abyssinians, and suck sugar-cane like their "Indian" ancestry; they might take for motto,—

> Au boire je prens grant plaisir,
> A viande frieiche et nouvelle:
> Quand à table me voy servir
> Mon esprit se renouvelle.

They are energetic only in performing upon the cow-horn, the bozina de chifre, derived from the ancient savages;* with this they announce arrival, salute those on the banks, and generally enjoy the noise.

My sole attendant is a Morro Velho boy, named "Agostinho," lent to me by Mr. Gordon. He knows something of the river, of gold washing, of diamond digging, and of rough cookery. Despite occasional attacks of dipsomania, he proved very useful, and at Rio de Janeiro he was returned into store with all the honours. "Negra," the mastiff, wild eyed as an ounce, becomes very savage when tied up, and barks as if under a waggon tilt. She is the terror of those who see her for the first time, and she will prove useful—in these parts all men travel with fierce dogs. I have two passengers on board. One is a certain Antonio Casi-

---

* The Tupys called it "Mamiá," and formed it of two pieces of wood joined together with thread and resins. Ferreira, writing in the last century, says of these rude trumpets that, "played in the fore-part of the canoes whilst travelling in the interior, they serve to summon the Indians before starting from the places where the embarcations are moored." According to Prince Max. (ii. 179), the Botucudos (whom he will call "Botocoudys,") termed it countchoun-cocann, and made it out of the tail of the great armadillo (Dasypus gigas, Cuv.). The more civilized Coroados used horns to call one another in the forest. On the Upper Amazons the horn is made of two pieces of thin hollowed wood, joined together by a lashing of twine and coated with wax: they are blunderbuss-shaped, four feet long, with a red mouth-piece, and a deep mellow sound. The Indians use them to frighten away the monsters of the deep, and, like Africans, to show by their noise that they come as friends. My men also enjoy the use of the "bandurra," or small viola, a wire-guitar, and the Marimbáo, a Jew's, or rather Jaw's harp: the name is distinctly Portuguese Angolan.

miro Pinto, popularly called " Onça ; " by profession a fogueteiro, or rocket-maker ; he asked for brandy at once, and the pilot, pointing to his fiery face, exclaimed, " Chupa muito," he sucks (the monkey) much. We presently landed at a breeding estate, where his son, the capataz,* or overseer, looks after some 2000 head. The other was a Southerner-immigrant, Mr. Hock ; this old pilgrim-father had brought with him a party of twenty souls, all had been spirited away by the indefatigable " Sprat " of Sabará, and like Rachel, he declines to be just now comforted. His present idea is to make a railway on condition of receiving alternate sections of sixty square miles, or thirty on both sides of the line. In the United States, where the contractors were satisfied with grants ten times less, the world predicted their ruin ; but the new lots attracted settlers, and paid remarkably well. I would willingly see this system adopted in the Empire, which now suffers from paying seven per cent. interest upon vast sums extravagantly laid out. Mr. Hock accompanied me as far as Jaguára.

Between Sabará and Jaguára the river line is officially twenty leagues, 1,118,490 metres, the breadth is between forty-four and seventy-seven metres, and the average † slope $0^m \cdot 4135$ per kilometre. This distance, about $\frac{1}{6}$th of the whole length, was partially cleared out for 6 : 000 $ 000, and this figure will be useful in estimating the total required. The stream is deeply encased ; the reaches are short, and we seem to run at the bluffs, where high ribs come down to the bed, and cut the bottom into very small bends. As usual in the smaller Brazilian rivers, there is hardly any breadth of valley ; in places it is a mere ledge, hardly to be called " dale " or "level" at the hill-foot. The banks,‡ often perpendicular, are of gravel, sand, or dark puggy clay, and between October and January they are deeply flooded. The pilots speak of 16 to 20 palms rise, and of small bayous, more often flood-lagoons than filtration-lagoons, formed in the flats. The

---

* Formerly called Amo or Vaqueiro ; he receives a certain proportion of the stock as pay, and has complete command over the " Campeiros " or " Moços," who are mostly youngsters.

† Of course the current greatly varies, and in some places the water is almost still. According to M. Liais, the river at Sabará stands in the dry season 695 metres above sea-level, and at the confluence it is 432·3 metres. The distance between the two places is 666,080 metres, or 361·28 miles, or 120·43 geographical leagues, and thus the general declivity is 0·3941 per kilometre. The slope of the Upper São Francisco, between the Paraopéba River and the Rapids of Pirapora averages $0^m \cdot 4890$.

‡ Here called Barrancos or Barreiras do rio, the classical Ribas or Ribeiras not being used.

bottom is of coarse pebbles and finer arenaceous matter, without mud, except where deposited by influents; at this season there are many shoal-islets or sand-bars, and bed-islets in mid stream. We find a few rivers but no "Cachoeiras," or rapids, properly so called. The most troublesome feature is the shallow (raseira);* at places where the bed broadens we ground with unpleasant regularity, and our crew has to tumble in. This part abounds in snags, locally called "tocos," meaning tree trunks; the "sawyer" is unknown, but there are galheiros (pronounced gayyeros), trees with upright and projecting branches. Sometimes they appear like poles, placed to stake the channel. The tortuous bed, never showing a mile ahead, prevents anything like waves, though the wind is in our teeth, and it will long continue so. Where there is much depth, the water boils up† and spreads out, sometimes the effect of a floor uneven with pit holes, and of the mid stream flowing faster than the surface or the bottom, where it is retarded by friction.

At this time we see the worst of the Old Squaws' River. The "Sol de Augusto" is proverbially bad, especially between two and four P.M. Heavy morning mists enforce idleness, and will last till the opening of the wet season, in September to October. There is a minimum of water and a maximum of contrary wind, sometimes, but rarely, chopping round to the south, and blowing with strong flows when the regular current ceases; this is not the case during the rains.‡ On the other hand it is the "Moon of Flowers;" the poor second growth — virgin forest is unknown—teems with the Flôr de Quaresma, with its bunches of purple beauty, and the hill tops are feathered by the tall Licorim and the Guariroba palms.

After about three hours we passed the Pedra do Moinho, the only really bad shoal, made worse by rocks on the left hand; the first sight of human habitation was a little farm near the Lagôa da Fazenda do Baraŏ (de Sabará), a flood-fed pool. Opposite it, on a narrow step of poor ground, was the baronial manor-house with

---

* M. Liais proposes to narrow the stream artificially, between Sabará and Roça Grande especially. But we came down easily in the worst month, drawing, when loaded, at least 20 inches.

† "'Sta fervendo," the men exclaim. This must not be confounded with our popular term "boiling water," that is, when the wind forces the waves one way and the tide checks them the other, thus making them lose their run, rise, dance, and bubble into points.

‡ During the rains there is least wind, and it does not always accompany even thunder and lightning.

a queer green portico, like Mtoni, near Zanzibar City. Then came sundry breeding fazendas and Retiros,* which sell fat and good jerked meat for 3$000 to 3$500 per 32 lbs. The cattle, numerous but degenerate, stand in the water or bask upon the sunny sand, and the horses gathering upon the grassy hill sides, stare snorting at our awning. In rare places there are patches (canaviaes) † of stunted sugar-cane.

Near the house of José Corrêa, where the river forks to east and west, inclosing a hilly island, we found the " Barque Jaguára." She was loaded with the enormous secular logs for Morro Velho. This large flat craft, 105 feet long by 24 feet broad, and 24 inches in depth (pontal), built of the hard Vinhatico and Canella woods, with ribs of Páu d'Arco, and iron-plated bottom, is triangular fore and aft. The weight is 32,000 lbs., of which the greatest part is metal. Unloaded she draws four inches, and increases one inch per four tons ; she carries seventy-two tons down the channel, twenty-two inches deep, between Macahubás and Jaguára, and she makes Sabará in twelve days from the latter place, returning in two or three. Evidently a steam-tug will be a success here, without expending much money upon the river bed.

" You'll never reach Trahiras !" cried the people on board the barque, deriding the " Eliza." And indeed we seemed likely to waste much time. However, if we crept on slowly, it was surely, and the Morro da Cruz of Sabará, which early in the day was a tall bluff to the west, presently gave us a parting look from the south-south-west. As evening approached the weather waxed cool and clear, and the excessive evaporation gave the idea of great dryness ; my books curled up, it was hardly possible to write, and it reminded me of the Persian Gulf, where water-colours cannot be used because the moisture is absorbed from the brush. The first view of Santa Lusía was very pleasing ; a tall ridge about a mile from the stream, was capped with two double-towered churches, divided by fine large whitewashed houses and rich vegetation, with palms straggling down to the water.

---

* The Retiro (dim. Retirozinho) here means a small breeding estate, where the absentee landlord establishes a capataz.

† The desinences "-al" and "-edo," (plural "-aes" and "-edos," as Olival or Olivedo, correspond in Portuguese with the Latin -etum, and the Tupy "-tyba" or "-tuba," e.g. Indaiá-tyba, a place where the Indaiá palm abounds; Uba-tuba, a site where the Uba reed is plentiful. It must not be confounded with -uba, or -uva, a tree.

I landed at the "Porto de Praia de Vicente Rico," above the bridge, and ascended a hill lined by hovels, with torn calico for window glass ; the path showed remnants of a slippery grass-grown Calçada. The "Hotel," kept in the Rua Direita by a "Doctor" Joaquim de Silva Torres, had broken its back, and attendance might be defined as the power of clapping hands and ejaculating "Pst" ad libitum. On the other hand, the bill was a mere trifle.

A walk up town led to two churches, the Rosario and the Matriz, the latter with its steps in ruins. I left my two letters of introduction, and heard no more of them for some time—the recipients, of course, could not call before the next noon. The Baroneza de Santa Lusía, who has a large house in the main street, with a front all windows, was an invalid : the venerable lady is the widow of Sr. Manoel Ribeiro Vianna, who founded the "S. João de Deus de Santa Lusía," a hospital for sick paupers. He died before the work was finished, and his relict magnificently dowered it with a house, furniture, and £3000.

The gold diggings which built Santa Lusia were of two kinds, Cascalho and "Ouro de Barba," Gold of the Beard. The river floods deposited particles upon the bank, the sods were cut* and the grass was shaved off to be panned, hence the picturesque popular term. Hard "Marumbé" iron stone still abounds. The Municipality, which in 1864 contained 22,980 inhabitants, 1915 voters, and 48 electors, might be rich with an improved system of agriculture. The land supplies sugar in quantities, a little coffee and "mantimento," rice and manioc, beans and millet, the Ricinus plant, whose oil is chiefly used for lamps, sweet potatoes (Convolvulus edulis),† and the Cará-tuber, together

---

* After catching the deposit of two years the sods are sliced off one finger thick, and 2 to 3 inches deep are taken up after five years' rest. Lower down stream I saw the cakes heaped on the bank.

† M. Renault, who has made an especial study of the Cará and the Convolvulus edulis, has obliged me with the following information :—

The Carás belong to the family of the Dioscoreaceæ, created from that of the Asparaginæ, and the genus Dioscorea bulbifera. There are six known species, of which all, except No. 5, have a fecula superior to that of the potato. The cultivator opens, in a light soil by preference, large deep holes, to whose proportions the root is supposed to fit itself ; these are filled with dried grass to support the cuttings, which are covered up with a little earth. The root is cooked like the potato, and is eaten with or without sugar or sweetmeats ; its flour enters into cakes and puddings :—

1. The ordinary Cará (D. sativa) produces a spheroidal tuber, at times attaining the weight of 30 lbs.

2. The Cará de dedos, or palmated (D. Dodecaneura), resembles in shape a man's hand.

with small timber; while the river is exceedingly rich in fish, which finds its way to Morro Velho. To judge from the streets, prostitution is the most thriving trade; but all assured me that it was outdone by Cruvello, a city further north, and ten leagues to the west of the main artery. Both of these are "church-towns," visited by the planters on Sundays and holidays.

The little Arraial became on July 8, 1842, the site of the acting Presidency; and here on August 20 of the same year, ended the revolutionary movement. The intrusive President kindly disappeared at night, and the then good genius of the Conservative party, General Barão (now Marquez) de Caxias attacked the insurgents. The fight raged around the bridge, beginning with early morning : the field was still doubtful at 3 P.M., when the 8th Battalion of Regulars occupied the highest point of the village, and put the enemy to hopeless flight. The chiefs, Srs. Ottoni, José Pedro, Padre Brito, Joaquim Gualberto and others, were made prisoners of state, and since that day, to them disastrous, the Ultra-Liberals have ever been called "Lusías."* St. Lucy or Luiz, I may remind you, is the patroness of the blind, and generally holds in her hand an eye apparently gouged.

3. Cará Cobra (D. hyperfolia), supposed to resemble a serpent.

4. Cará Mimoso (D. triloba); its small roots produce a fine fecula.

5. Cará Tinga (D. alba) grows wild in the Capoeiras of Minas, and is the least esteemed. The spheroidal root is a little bigger than an ostrich's egg, the skin is white, and covered with small asperities, and boiling water softens it but little ; it is cooked under ashes, and is eaten when a quill can be thrust into it.

6. Cará do Ar (D. Peperifolia). This species also produces climbers, sometimes 12 to 13 feet long, and as many as 40 fruits, weighing 1 lb., in shape a rhomboidal tetrahedron. The climbers die after fruiting, and reappear next year. This tuber is reproduced from the fruit, and yields within the first twelve months ; whereas the other five kinds are propagated by cuttings of the stalk, to which are attached some of the fibrous roots of the climber. This Cará do Ar has no maladies nor enemies, and it would be a boon to Europe. It requires little care, once planted it lasts for many seasons, it can be crowded without injury, and it wants only a somewhat tall support. A single stem yields ten times more than the potato, and it would save much surface by demanding very little ground.

There is also a "Cará do Mato," the tuberculous roots of a wild Cará much eaten by the Indians.

The Carás, like the true yams and the sweet potato, have often been confounded with the Topinambours (vol. i. chap. 8), because all are tuberous roots, and were imported from America.

The sweet potato belongs to the family Convolvulaceæ, and to the genus Convolvulus edulis. Of this plant there are four well-known species :—

1. Convolvulus edulis.

2. C. tuberosus.

3. C. esculentus.

4. C. varius (Martius).

* "Lusía" was opposed to "Saquaréma," which some travellers call "Sagoarema." It is a village and a water on the seaboard near Rio de Janeiro, and being the head-quarters of the "old Tory" party, especially the families of Torres (Itaborahy) and Soares de Souza (Uruguay), it became a noted name. The term "Cascudo," somewhat similar, is taken from the Rio Cascudo, between Minas and S. Paulo.

CHAPTER II.

MACAHÚBAS OF THE NUNS.—HOSPITABLE RECEPTIONS.

> Que se a abundancia à industria se combina
> Cessando a inercia, que mil lucros tolhe,
> Houverá no algodão, que alli se topa
> Roupa com que vestir-se toda a Europa.
> (*Caramurú*, 7, 48.)

AUGUST 8:—The morning was delicious, and the face of nature was calm as if it could show no other expression. The sword-like rays of the sun, radiating from the unseen centre before it arose in its splendour, soon dispersed the thin mists that slept tranquil upon the cool river-bed. We shot the Ponte Grande de Santa Lusía, leading through Lagôa Santa, distant three leagues, to Cruvello and the "backwoods." It was the usual long crooked affair, with twelve trusses or trestles in the water and many outside, showing that the floods are here extensive: an older erection has disappeared. The girders are rarely raised high enough, and an exceptional inundation sweeps them away, leaving bare poles bristling in the bed, and dangerous piles under water. These must be removed before the stream can be safely navigated.

About two miles below Santa Lusía the water becomes deeper, and the country changes. The right or eastern side is rough and hilly, with heights hugging the bed. Near the other bank the land is more level, and the soil shows a better complexion, by which both sugar-cane and timber profit. On the uplands, extending to ten miles, the superficial formation is of four kinds. The best is the rich ferruginous chocolate-brown alluvium, based upon a mountain limestone, blue streaked with pure snowy lines; the second is the red soil underlaid by the same calcareous matter. The soft black alluvial loam, considered

A 1 in the Mississippi Valley, is here the third; and the worst is the white sun-scorched ground without iron. On both sides are saltpetre caves, and the produce is prepared at the mouths by a simple process which we shall presently see. I heard vague reports of salt-diggings, which probably refer to the Salinas about the Paracatú River described by old travellers.

After the first hour we reached the Fazenda da Carreira Compridar * of the Fonseca family : it supplies provisions and Restilo or rum. The lands extend far up the hills, and the "Engenho" or sugar house is on a ledge near the stream, which loops to the south-east. It was working when we sped by, and the music reminded me pleasantly of certain water-wheels in Sindh, Egypt, Arabia—in these lands of the Future any suggestion of the Past is a god-send. Establishments with water-power motors pay 40$000 per annum, those driven by bullocks half that sum, and upon the produce of both there is, when entering towns, an octroi of 0$320 per barrel of thirty bottles. It will be better for the people when circumstances admit of a much heavier taxation.

This part of the river shows many contrivances for exploiting a far more valuable industry, the vast shoals of fish which haunt the waters. The usual weir (Gamboa or Curral, not Camboa and Coral) is accompanied by the Jequí or Jiquí, a conical crate of wild cane, bound with cipós two feet long, and attached to stakes (estacadas). The Grozeira is a system of thin poles, planted five to six feet apart, and connected by llianas, to which hooks and lines are fastened. The Chiqueiro or hog-stye is a tall roofless closet of cane, some two feet in diameter, and affixed to the bank : it has a perpendicular trap-door, which falls when the fish pulls at a corn-cob. Another self-acting machine, a favourite because a trouble-saver, is the "Linha douradeira," a hollow bamboo with cotton line, hook, and earth-worm (minhoca). The Giráo is a perch on four piles, often planted at the head of a sand bank, and the man who exerts himself upon it with his cana or rod must be hungry indeed. He will, however, find a single take sufficient for the day and its appetite, and the rest of the twenty-three hours and fifty-five minutes may

---

* "Of the long quarry;" it is said that white lime is here found. I shall mention only the principal Fazendas which struck my attention ; a complete list is given by M. Liais.

be expended in doing nothing. I can hardly persuade my crew to throw a hand-line overboard when we anchor; the pretence is that they have brought no hoe for digging out earthworms. But they can catch half-a-dozen sprat-like "piábas" or "piaus"* by heaving up a calabash full of water, and by throwing it upon the bank; or they can shoot a bird or rob a nest, which will do equally well for bait. A fish-gullet best fits the hook, and will not come off, but they do not approve of this "new-fangled fashion." Salt is here wanting, but sunshine is not, and two days will extract all moisture from the fish-meat when cut thin and hung in the air. For long journeys these can be fried and potted with vinegar and spices. The flavour is preserved by frying the game when quite fresh from the water; it can be "warmed up" when wanted; fish-soup is invaluable, but it requires too many ingredients for a traveller to succeed in making it enjoyable. As a rule the people reject the scaly fish, because they say the spines are dangerous.

Those who visit these streams should be provided with fishing tackle, with the largest fresh-water hooks, and with the stoutest running gear, or the "cats," sometimes weighing upwards of a hundred pounds, will surprise them. On the other hand guns are useless. The crew generally carry their shooting irons, the locks guarded as in Africa by a sheath of monkey's skin; but little game appears upon the banks; it was confined to a water-hog, a single small deer, doves, and at rare intervals, a few Penelopes. Wild fowl, especially ducks (Marecas, called by the aborigines Jerere or Ierêrê), were sometimes seen, and cranes were heard screaming from the bayous within the River Valley; to get at these places, however, requires much marsh-walking and nothing else to do. In the Brazil those streams which, like the Tiété and the Paranápanéma of the São Paulo Province, ignore the white man, even the squatter, and can be reached only after a week of much travelling from the coast, afford magnificent sport; not so those where the gun is well known. Sportsmen

---

* The Piau is a small fish, which has given its name to the vast Province of Piauhy. Gardner mentions the Piau branco, one of the Salmonidæ, one to two feet long, with large scales. It is taken with the hook, and is held to be good eating. On the Rio das Velhas the bait is a bola of manioc flour. By night the Piau used to jump into the tender canoe; the light slate-coloured back and white belly reminded my companion of the "silverside." We heard of the Piau certia, a large species, some white, others dark, and of the Piau de Capim, a sea-fish which feeds on grass.

visiting the Brazil will do well to bear this in mind; tapirs, ounces, and anacondas are still found near the sea-board, but they are exceedingly wild and troublesome to seek out, whilst the climate is bad and the walking is detestable.

Another hour carried us to the Port and Fazenda of the Capitão Frederico Dolabella, where we sighted the first cotton-plantation, and right well it looked. It is mostly herbaceous, the seed having lately been introduced; but still lingers the Brazilian "kidney-cotton." This, after some years, becomes a tree fifteen feet high, and thick as a man's leg, with large luxuriant foliage, red yellow blossoms, and bearing a strong medium-staple lint, that covers moderate-sized and naked black seeds. This is the "Gossypium arboreum," of which travellers in this Empire speak—the more exact limit the term to the "purple-blossomed, green-seeded, short-stapled, small cotton tree of India."* There is a mine of neglected wealth in cotton and fish, and the more we see of it the richer we shall find it. The hills were clothed with thin brown-grey grass, looking, in places, as if they were frosty with hoar, and they were profusely tasselled with noble Macahúbas or Coqueiro palms.

The snags and "branchers" were bad as those of yesterday, and we lost an hour by grounding at the Volta dos Pinhões, a "broad" and a bend in the river. Then we ran at the "Penedo," a tall fronting mass of bare stone, protruding from the trees which straggled over it from base to summit; a little below it was another hill, all forest, and between the two a pile of wood awaited the "barque." On the right was the Rio Vermelho, a little stream coming from the Arraial da Lapa, east of Sabará, and allowing unloaded canoes to ascend it for a league.† Presently another bend showed certain white lines between the river fringe of trees, and a hill fronting west; this was the "Macahúbas das Freiras"—of the Friaresses.

Before making fast to a "porto" or gap in the clay bank, here called a Port, I gave a passage across to a traveller from Lagôa Santa. He wore a cow-skin hat, shaped like the Petasos of

---

* So says Major R. Trevor Clarke. Here the cotton has more lint than usual; 1200 lbs. will give 500 lbs. of cleaned fibre, whereas in Alabama 1500 would be required. The people usually replant the shrub in its fourth year.

† Thus all my informants. M. Liais calls it "Rio de Macahúbas," and makes it a stream of some consequence, with a contingent of 20 metres per second, which makes the Rio das Velhas of "great importance," and gives it a debit of 62 metres.

Mercury, a white shirt streaked with indigo—an old style still lingering—a paletot of Minas cotton, and deer-skin riding-boots built to reach the thighs, but falling below the calf as if he stood in his carpet bags. An impure path, winding past cascalho-heaps, by a dirty pond, and through offals of pig-sties, leads to the high site of the Recolhimento or Recluse House. On both sides of, and attached to, the church, are long double-storied wings of whitewashed pisé, based upon the usual fine blue lime-stone, and all the windows are jealously latticed and barred. To the left is the Vicar's house, and at a lower level rise clay and thatch huts, inhabited by slaves and porkers, fowls and turkeys. All appears exceedingly foul, but the people declare that with godliness, but without cleanliness, they live to a great age.

As there was no Venda we went to the Tropeiro's Ranch, and were surlily received by the housekeeper. This chattel of the " Recolhimento " was making pots, of course without wheel, out of a grey, iron-coloured clay ; she refused to give coffee before we declared our names. Such is the effect of a single party of highly Protestant emigrants visiting so highly Catholic a place. I at once sent my card and letter to the Rev. Padre Lana, whose first cousin had been so kind to me at Itacolumi of Ouro Preto. This amiable Mineiro, educated at the Caráça, at once called upon us, ordered dinner, and carried us off to see the lions.

The " Madre Regente," or Reverend Mother, rather a pretty person, received us at the door, kissed the Padre's hand, and led the way to the little college-chapel, white and gold with frescoed ceiling. We visited the dormitories, which had nothing new, and from the windows we could see the inner square, which may not be visited without an order from the Bishop and his coadjutors. The galleries are long; the rooms, large and airy, reminded me, in their roughness of unhewed beams, of a Goanese establishment which I described nearly a score of years ago. The lecture " sála " showed a black board for " cyphering," some old maps, and creditable specimens of caligraphy, em-broidery, and artificial flowers. The Infirmary contained one sister and four invalid girls. The thirty-six reverend women are dressed in white veils, and petticoats with black scapulars in front, and over all a blue capa or cloak. The twenty-five educandas or pupils followed giggling in the steps of Galatea, concerning whom it is written,

Et fugit ad salices, sed se cupit ante videri.

The grounds consist of six acres walled in, and producing an abundance of well-watered "green meat;" here, however, the brown scummy river, ugly to look at but tasteless, is generally used; indeed, below Jaguára the people prefer it to that of the Córregos. The vegetables, especially the salad, are excellent; the vine, which at Sabará as at Barbacena bears fruit twice a year, is a failure. For the first time in the Brazil I saw the Coqueiro palm (Cocos butyracea) not wholly neglected; the fruit-pulp makes good tallow for lamps, and the kernel gives a medicinal oil;* besides which the "cabbage" is by no means despicable.

We then visited the church Nª Sª da Conceiçaõ, and found the Santissimo exposed and the nuns singing behind the grated choir-cage, which, as usual, fronts the Seat of Honour or High Altar. At the "Speak-House," where a grille allowed us to address the unseen inmates, and where an upright barrel with a stave or two knocked out, pivots in and out their humble wants, we were allowed to take the Livro das Entradas; it begins with an interesting paper dated July 18, 173–. After collating it with the Claustro Franciscano (Frei Apollinario, Lisboa Occidental, MDCCXL.), and lastly with the Relatorio of the Vice-Director General, the Chantre José Ribeiro Bhering (Ouro Preto, 1852), I compiled the following account of the oldest religious house in Minas.

About 1710 two brothers, Manoel and Felis da Costa Soares, "godly men and of a goodly house"—in those days the "vulgar" colonist would hardly have dared to be better than his neighbours —came here from Pernambuco, in search of lands, bringing sisters, nieces, and a widowed daughter. On August 12, 1714, they began to build a secular house, which "had no meum and tuum." This "Convento Velho" lay south of the present site, and its ruins still show in the thin palmetum. Felis met on

---

* St. Hil. (I. ii. 378), says that this palm tree is very remarkable. "Car, s'il existe une foule de sémences oléagineuses, l'olivier est, à ma connaissance, le seul arbre dont le péricarpe ait été signalé jusqu'ici comme fournissant de l'huile." Yet he must often have seen the Elæis guineensis, the Dendé of the Brazil, and perhaps he had eaten "palm-oil-chop."

the banks of the Rio das Velhas a hermit, habited in a garb then strange to him, but which he presently found to be that of "N^a S^a da Conceição de Monte Alegre;" the recluse mysteriously disappeared—perhaps, said Padre Lana, it was a vision—and the laic, being unmarried, resumed the garb minus only the hat. Thus arose in the "Sitio de Mocaubas," the first convent of the Recolhidas, dedicated to the "Immaculate Mother of God." The "Seraphic Order," then in lusty youth, came to its aid, and soon raised for it by alms 60,000 crusados,—say £60,000 of this our day.

The Sister Catharina de Jesus became the first Reverend Mother—a fact about which there is some confusion in the Livro das Entradas—and died in 1717. She was followed by Felis on Oct. 11, 1737. The old convent suffered from a torrent, and the present building was completed Dec. 25, 1745. D. Fr. Manoel da Cruz made it a branch Third Order of St. Francis, and it became a Mosteiro on Sept. 23, 1789. According to the "Relatorio," a rule was given to it by Padre Antonio Affonso de Moraes Torres, Superior of the Caráça.*

The Recolhimento receives nothing from the Government, but, as will appear, much land has been left to it; it lives by agriculture and cattle breeding, and it no longer works the once rich mining estate. Of late years the revenues have been simplified by conversion into Government Bonds. Its object is to give the "usual instruction required by the mother of a family," and in 1851 a sister and a pupil were sent to learn, from the Sœurs of Marianna, a better system of instruction and house management. The hypercritical declare it to be a kind of "bush"-school, and the confessor had never heard of the Bull Unigenitus. The name of Professor Agassiz, who had been repeatedly quoted by every journal in the Empire, was utterly unknown to him. How many millions of men ignore, we may ask, such persons as Alexander, Cæsar, and Napoleon, the great Triad, the mighty Avatars of humanity?

Padre Lana accompanied us to the Venda, where we sat down to a long conversation. Here we found a weak old woman,

---

* Even until very lately, throughout the Brazil pious women have collected together in houses, and have cohabited for devotional purposes. The foreign ultra-montane priests, who are here flocking like eagles to the battle-field, reprehended the harmless and often beneficial practice, and forced upon these sisterhoods the "rules" of Europe, which are often nothing else but a mere system of old Asiatic asceticism.

who had worked at the Morro Velho mine—the sisters will let, but will not sell their slaves. I asked her how she had been treated : "nunca apanhei"—"I never catched it" said the poor nanny-goat voice. We bade an unwilling adieu to the excellent Padre, who complained that I was paying him a "visita de Medico," in the Brazil not so complimentary as our "angels' visits." Mr. Hock, who complained that he had been stiffly treated by a former vicar, that found him to be a "herege," asked me, with Ay-merican gravity, if I really thought that the "sisters" were chaste ; it is curious to see how these men, so jealous of their countrywomen's honour, find "libertinism" everywhere. "What a sad (triste) race they seem to be," quoth Padre Lana on his side, as he looked at the old man champing in melancholy silence, behind his thin drawn-down lips, a huge quid.

The moon and stars were unusually bright, and the night was delightfully clear and cool. Before dawn in the next morning I was aroused by the moan of the dove and the small piping of the Saracúra—commonly called the Saracúla (Mr. Bates Serracúra, Gallinula Cayennensis)—crane, that useful enemy of cockroaches ; the cry of the Siriéma or serpent bird, which resembles the whining of pups, and the gabbling of bubbly-jocks mingled curiously. Land and water were obscured by a thick white fog,[*] but the Eliza was not a Rhine steamer to be stopped by it. The pilots consider it a sign of a still day, and presently it lifted, showing a wondrously high vault, stretched with cirrus in long curved brushes. [†]

Friday, August 9.—We set out at 7 A.M., and presently ran down to "Coqueiros," a fine site for a house, a dwarf level at the mouth of a gap between two hills, one grassy, the other feathered and forested with palms. To-day the effect of a large influent appears in reaches somewhat longer, there is less of dead drift-wood lining the banks, and the bed now begins to show "Remansos," still places in deep pools. We grounded but three times, and only once our men were obliged to "tumble in." The stream is admirably embanked, the bottoms are more extensive, while the lands, higher and drier, are of superior

---

[*] Popularly known as Neblina or Noroega; this latter is probably an imported word, often applied to a dark place where the sun is little seen, e. g. "Catas Altas de Noroega."

[†] Generally known as Rabo de Gallo—cock's tail.

quality and less desert. Women washing upon the margin no longer ran away unless we disembarked, and some asked with a scream if we were making a "planta" (map). The negroes were loading corn-cobs upon carts with plank floors, fenced round the top with square wattles four feet high; sometimes this woven work sloped backwards from a high front, like the classical biga and the car of triumph. There is a scarcely perceptible rapid called "das Alprecatas,"* near the mouth of the Upper Ribeirão de Taquarussú, whose yellow and shallow waters head some eight leagues away. Near this place are settled a Mr. and Mrs. John Wood, whom I failed to find.

Near the Taquarussú influent the bed, which has formed a neck, narrows, leaving a broad sandbank to the west; this increases the swiftness of the stream from two to four knots,† and the sharp turn and shallow water make the boatmen rejoice when they have passed it. Huge blocks of stratified sandstone (lapa) are tilted up at a shallow angle towards the river, forming gloomy caverns, recesses and natural piers, which continue till near the ruinous "Fazenda do Mandim"—of the Mandim or Snorter.‡ The last time that I heard the song of the fish was in the port of S. Paulo de Loanda.

Then the hills fall, and the low cultivable sides are those of an English water, whilst Campo-ground appears in the distance ahead. Fields of the liveliest colour, telling the richness of the sugar-cane, contrast with the darker greens and wintry browns; the Ubá§ or arrow-reed, with lanceolate fan-shaped leaves and whitish flowers, here grows twenty feet high, and forms impene-

---

* The Alparcátas or Alpargátas sandals.

† M. Liais calls the large sand-bar above the Taquarassu "Proa-Grande," doubtless a misprint for Corôa-Grande.

‡ The Mandim (M. Liais writes Mandim), called Roncador or Snorter, from its grunting noise, especially in the hot afternoons before rain, was known to the Tupys as Mandué or Mandubé. Some of the pilots declare that the noise is produced by friction of the head upon the canoe bottom. It is one of the Siluridæ, and resembles the Mississippi "cat." The usual length is from 18 inches to 2 feet, the yellow-brown skin, with dark round spots, is scaleless, the long barbacels give it the Anglo-American name, and the three dorsal fins are dangerous. It keeps near the bottom, bites voraciously, and, as it has few bones,

the white meat is tolerable eating, at least the otters find it so. There are many varieties: Mandim-assu; M. Amarello; M. Armado; M. Capadelho; M. Esquentado, &c., and M. Halfeld remarks (Rel. 215) that "all these qualities are diminishing." "Roncador" is the name given to several fish, especially on the south of the Villa da Vittoria. (Prince Max. ii. 157.)

§ Gynerium parvifolium, Mart., Vubá or Arundo sagittaria (because the Indians used it) of the System, and Saccharum Ubá of St. Hil., who (III. i. 18) says that Luccock is wrong to write "Uva." Yet Uvá is preferred by old authors. In S. Paulo it is called Ubá, from the Tupy uy'bá, an arrow. The Mineiros know it as "canna brava," or wild sugar-cane.

trable thickets. This Calamus seems almost independent of climate, and enjoys the coast-levels as well as the Highlands of the Brazil. Another narrow, where the drift-sticks hanging to the trees mark a flood rise of at least fourteen feet, leads to the first of the curious formations called "Lapa de Stalactite." Here the limestone rocks on the left were hung in front with long tongue-shaped lappets of thin stone, which have a strange effect.

The next interesting point is the Ponte de Dona Ignacia. Since M. Liais wrote, the tall weed-grown bridge has opened a central gap of 30 feet, and people cross by the normal ferry, an " ajôjo " of four canoes, with railed platform, worked by a chain and pulley. Opposite the large white Fazenda and distillery, now belonging to Lieut.-Col. Luiz Nogueira Barbosa da Silva, was wrecked the first steamer that appeared upon these waters, or indeed upon any of the island lines of the Brazil. M. William Kopke,* who came out as interpreter to the Cocaes Gold Mining Company, and who obtained a concession to navigate by steam the Rio de São Francisco, had the energy and enterprise to build her at Sabará in 1833–4. Like Captain Fitzgerald, of Larkhana in Sindh—who, by-the-bye, blew himself up—M. Kopke was obliged to make the greater part of his own engine, and some-times to use wood where metal was wanted. The experiment was so far successful, but no farther—the steamer here went down " snagged."

On the right bank, a little below this place, is an Olho de Agua, or pool, which they say communicates by a " sinker,"† with a lake on the other side of the river. Bits of wood have been thrown in and have been recognized on re-appearance ; of course these natural tunnels are possible in a limestone country. Presently

---

* M. Kopke (or Kopque?) whom the decree calls "negociante Hamburghese," losing his steamer, rigged up a boat and visited the Paracatú River. His brother, Dr. Henry Kopke, is still at Petropolis. After the first concessionist, whose permission to navigate the Rio des Velhas was decreed Aug. 26, 1834, and was extended to the São Francisco November 14, 1834, M. Tarte, a Belgian engineer, applied for the same exclusive privilege, but did not obtain it.

The first steam-ship that ever plied in the Brazil was built in 1819 at Bahia, by Sr. Felisberto Gomes Caldeira Brant Pontes,

afterwards Marquess of Barbacena. She ran to the then Villa of Cachoeira, and was wrecked by a storm upon the Monserrato beach. In 1822 a steamer was sent from Rio de Janeiro to Santos, carrying a deputation of distinguished men, and the Desembargador João Evangelista de Faria Souza Lobato. They persuaded the patriotic José Bonifacio de Andrada e Silva to accompany them, and returned to the capital on January 16, 1822, a week after the Prince Regent had declared that he would not leave the Brazil.

† Popularly called the "Sumidouro."

the sun set, the cold made us gather round the galley-fire, and the moon rose with low, uncertain light. The crew, not having seen the bed during the last four years, became very nervous as we swung round the Cachoeira de Jacú, with its swift deep current impinging upon the right bank of the narrow bed. I felt that a stick or a stone might spoil my whole journey, and I allowed them to make fast at the "Porto do Bebedor."* We scrambled up the steep bank to the house of Sr. Antonio Lourenço, and were admitted to the strangers' room, as soon as the key would turn, by the daughter of the house. D. Conrada, still in her teens, was the mother of three children and the widow of a tropeiro : she made coffee, warmed our beef, and sat chatting with us till we slept—a rare and recordable incident of hodiernal Brazilian travel in the Far West.

August 10.—The morning was mistless, and we set off early. After nearly two hours we saw on the left bank a large and much decayed square of white-washed and red-tiled building, backed by a neat church—the Fazenda de Jaguára.† At the "port"

---

* The "drinker;" a drain, not a drainer.

† Some explain Jaguára to be the name of the well-known ounce—puma or S. American lion. Others explain it by Jahú or Jaú-guára. The "Jahu-fish (is here) abundant."

Jaguára, corrupted Jaguar, Iagoar, and so forth, is properly "Ja," we, us, and "guara," an eater, a devourer (of us), and was applied by the indigenes to all man-eating beasts. Doubtless in the early days of colonisation, when these large cats knew nothing of the gun, they were dangerous enough. At present their courage seems to have cooled, and the Matador de Onças—tueur d'onces—once so celebrated in the Brazil, finds a large slice of his occupation gone. Many travellers have seen nothing of this king of the cats, except the places where it sharpens its claws. I have had experience of one live specimen, and that too by night. The people still fear them, especially at night, and have many traditional tales of their misdeeds. They are still very dangerous to dogs, monkeys, after which they climb, to the Capyvara, an especial favourite, and to the young of black cattle. There are four large varieties of these Felidæ :

1. The Onça çuçuaranna, or çuçurana, (Mr. Bates "Sassú-arána, or the false deer"), whence the barbarously corrupted "Cougouar," derived through the "Gua-

zouara" of Azara. It is variously termed Felis Onça, or brasiliensis, or concolor, the last term being the best name. It is one of the biggest. I have seen a brown-red skin 5ft. 8in. long, not including the tail, yet it is the least dangerous. The range of this puma, or red lion, appears to extend throughout the tropical and temperate zones of the New World. It is evidently the "painter" (panther) of the United States.

2. Cangouassú or Cangussú, the largest variety, with smaller rounded spots of a lighter colour, on a dark brown-red skin. Prince Max. informs us (iii. 188) that in Bahia it is applied to a small animal whose pelage is marked with small blacker spots.

3. The Onça pintada (painted ounce), also called the Jaguarété (true or great eater). This "Felis discolor" is a very beautiful animal, especially when the white field of its maculæ has a light pink blush, in shape much resembling the "cheetah," or hunting leopard of Hindostan, it is the most dreaded ; it does great damage to cattle ; it worries and destroys far more than it needs, and after gorging itself with blood, it returns at leisure to eat the flesh.

4. The "Tigre," or Onça Preta, is the black Jaguar, a rare animal now in the Brazil, but still found, I am told, on the banks of the Upper Paraguay River. As a variety it probably resembles the black

where the Ribeirão de Jaguára falls in, I was met by Dr. Quintiliano José da Silva, ex-President of Minas, and now here officially as Treasury Judge (Juiz dos Feitos da Fazenda Nacional).
He led me up to the house, introduced me to the mistress, D.
Francisca dos Santos Dumont, the daughter of our host at Ouro
Preto, showed me to the strangers' room, and lavished all the
hospitable attentions in which his countrymen are such adepts.

leopard of the Niger Valley ; and the dark
spots upon a sable skin render it peculiarly
interesting.

I have seen good collections of these skins
on the Rio das Velhas. Here, however, as
elsewhere, they are expensive, and are soon
bought up for local use. All classes covet
them for saddle-cloths, pistol holsters, travelling bags, and even hunting caps. Of
course the spotted ounce is preferred ; and,
as a rule, the skins are as thoroughly spoiled
as if they had been handled by negroes.
They are ruthlessly deprived of head, legs,
and often of tail. *En revanche* the leather is
well and carefully tanned.

CHAPTER III.

AT JAGUÁRA.

RIDES ABOUT THE PLACE.—THE VEGETATION.—EXCURSION TO LAGÔA SANTA.
DR. LUND.—M. FOURREAU.—WHAT THE WORD "CACHOEIRA" MEANS.

> A distant dearness in the hill,
> A secret sweetness in the stream.
> *Canning.*

AT this hospitable house I spent five pleasant days, whilst another crew was being engaged, and arrangements for my reaching Diamantina were being completed. "Jaguára" has, in its day, caused no little sensation in the Province, and the following are the heads of information touching the "extincto vinculo"— the "cut-off entail."

Half a century ago, a certain Colonel Antonio de Abrèu Guimarães amassed a large fortune with 750 slaves, and still more by forgetting to pay the Government dues on diamonds exported from Diamantina and other places. He held an enormous property of 36 square leagues (427,504 acres), which was afterwards divided into seven great estates. The first was Jaguára, containing 1000 alqueires, (each 6 × 2 square acres): this was lately bought, without the 200 slaves, by M. Dumont's father-in-law, for 12 contos, 1200*l.* The next was the Mocámbo, actually belonging to Colonel Francisco de Paulo Fonseca Vianna. Then came the Bebída, including Casa Branca, Saco das Egoas, and Saco da Vida. It once contained four square leagues, now it is reduced to 1300 or 1350 alqueires, and it is to be sold for 3000*l.*— 30,000*l.* with a total of 170 slaves:—we shall visit it down stream. Number 4 was the Riacho of João Paulo Cotta ; then ranked the Pindahyba, now Ponte Nova, including the Tabóca, formerly the property of Antonio José Lobo and Domingo José Lobo, nephews of the Abrèu, and afterwards purchased by Colonel Domingo Diniz Couto. No. 6 was the Brejo of Francisco Fernandez

Machado and his brother; and lastly, the "Mello" was the nucleus of the estate.

The old contrabandist, who had also farmed with exceptional success the ruinous royal tithes, presently went to Lisbon, repented him of his sins, and was ordered by his confessor to build a church to Nᵃ Sᵃ da Conceição; furthermore, by way of fire-escape, he was directed to tie up (vincular) the greater part of his enormous estate for the benefit of religious houses. He wrote from Portugal to his brother, Francisco Martins de Abrêu, with all directions to carry out his orders, and the latter, much against his will, was compelled to sign all necessary documents by the authorities of Sabará, who met him, they say, on the road, and led him into an adjoining cave. The old man died in the Convento da Cartuxa at Lisbon, some declare miserably poor, others represent in miserly wealth, of which he had dropped but a small portion.

The revenue of this vast estate was divided into five portions, of which three were made over to the Misericordia of Sabará, one was given to the Recolhimento of Macahúbas, and the fifth part was distributed amongst the relations of the mortgager, the families of Abrêu and Lobo. The Governmental administration was placed under a Junta, or Commission, who levied the rents, and paid them through the Juiz dos Feitos Provincial, into the Provincial Treasury. It is needless to say that the revenue declined; it gradually fell to 4$800 per annum. Decree No. 306, of Oct. 14, 1843, "extinguished" the mortgage, and permitted the sale of the property. Since that time it has fetched, they tell me, some 40,000l. The seventh estate, called the Mello, is still being surveyed for sale,* and this accounts for the presence of the high officials at Jaguára.

Dr. Quintiliano kindly rode with me about the estate. There is a garden close to the stream, on a fine ledge of rich, red-brown clay (maçapé), which might be extended for many acres. My companion was emphatic upon the immense fertility and salubrity of the place,† and truly, as the spring was setting in,

---

* The Mello contained 63 sesmarias (here generally half a square league). Of these 10 were measured in 1865; 38 in 1866; and 15 in 1867; leaving 63 for survey. It has been bought since I left the river by the Provincial Government for the benefit of the American settlers.

† Another estate, Páo de Cheiro, some three leagues down the river, and belonging to 7 or 8 proprietors, is held to be a sanitarium.

and the birds were making love, and the trees were weaving their new coats of many colours, the microcosm looked enchanting. He showed me some dry sticks, which a few days before he had planted in the ground with ashes of decayed wood, and upon which he had turned a tiny stream : all had budded; the effect of the subjacent limestone, the finest natural manure. The tenements are in poor condition : the low, long walls, and the hollow squares suggest the "Hishán" of the Arabs; these, however, are white-washed and tiled. The out-houses are in a still more tattered state ; the owner cares more for the exploitation of the Rio das Velhas * than for agriculture or horticulture. The only part tolerably well preserved is a detached building, the Casa da Junta,† where the Commissioners met; the little church had been lately repaired, but its congregation was mainly the "Sanharó," ‡ a fierce species of wasp, dangerous to other honey-makers.

Our next visit was to the lakelets and to the vast limestone formations on the north-west of the estate. We passed a red digging, an open cut from which much gold had been taken by the ancients. Thence we issued upon a prairie of "spotty soil," here rich and red, there white with gravel. No lack of good grazing ground, and the cattle on the estate had, I was told, been worth 4000*l.* The vegetation was that of the Campos about Barbacena, the trees were hard gnarled Barbatimão, Patáro, Geáo de Gallo, Piquí, Tinguí,§ and Sicupíra. Besides these, I remarked the Sambahyba (Curatella Sambaiba, also written Sambaüva), with valueless fruit, a rough leaf used for brushing cloth, and astringent bark, good for tanning and for dressing wounds ; it has the effect of iodine in resolving chronic inflammations. Another common tree was the Cagaitéra (Eugenia dysenterica), an ugly name, but a pretty growth, with white flowers and milk-producing leafage : the Cagaita, or berry, is a strong drastic. Here grows

---

* I obtained a copy of a map survey of the Rio das Velhas by M. Henrique Dumont, dated October 1864. It agreed well with the labours of M. Liais.

† I found the Casa da Junta (B. P. 208°·80, therm. 72°) = 1807 feet above sea-level. Pelissher's aneroid gave (29·46, therm. 64°) = 543 feet. Mr. Gordon's observation (29·44, therm. 74°) = 553 feet. All these observations are curiously under-estimated. The river is here about 646 metres above sea-level (2120 feet), or 49 metres lower than at Sabará.

‡ It resembles the Pelopæus lunatus described by Azara and Prince Max. (i. 139). The latter makes it attach its pyriform nest to trees as well as houses.

§ This must not be confounded with the Tingi, Tingy, Tinguí, or Tiniury da Praya, a kind of Iliana (Jacquinia obovata), which, like the Paullinias, is used for intoxicating fish. The branches are cut, bruised, tied in bundles, and thrown into water whose course has been arrested by a dam.

in abundance the stunted Acajú or Cajú, which we call Cashew (Anacardium occidentale, Linn.; Cassuvium, Jussieu): amongst the aborigines it was a growth of great importance,* they numbered their years by it, they kept the nuts to remind them of their age, and they made of it their most valued Cauim or wine. The Goanese extract from it a neat brandy; here it is mostly made into sherbets, and strangers have burnt their lips by eating the dark reniform kernel that grows outside: the bitter gum called by the Tupys Acajú-Cica (for "icica," resin), is used by book-binders, and keeps off worms. In the lower sites there is a kind of salsaparilla (Salsa do Campo and do Matto), which appears on ant-hills under the trees. The root is large and white; the yellow being preferred in Europe and the United States; the people declare that it should be drunk with milk, to disguise its acridity, and use it much, but with care, avoiding it for instance in the middle of the day. The garden-grown salsaparilla is all cut at this season, and the shops here ask 2$000 per lb. of the dry old twigs sent from Rio de Janeiro.

The only birds were the Siriéma,† that hunted the serpents from our path; its favourite "big brother" the Ema (ostrich) which never gave a shot under 200 yards, and the pretty little Tiribá paroquet, with cuneiform tail (Psittacus cruentatus, Mart.),‡ which shrieked as it passed us like an arrow. The "Campeiros," or herdsmen, wild as the Somal, were picturesque in their leather wide-awakes, sitting loosely upon ragged nags with wild equipments; huge spurs armed their naked heels, and the wooden box stirrups which the cistus renders necessary in Portuguese Algarves, defended their toes. They were wiry and well-grown men; here it is remarked that even the slave-boys

---

* They called "Acajú acai piracóbá" what the Brazilians term Chuvas de Cajú, which fall in August to September, and which injure the inflorescence of the Anacardium. Southey (i. 181) confounds the "Caju" with the "Auati" (Olfi moquiiia, a Chrysobalan), a "Madeira reservada," or hard-wood forest tree, of which there are many species, some bearing a fruit that yields an intoxicating drink.

The aborigines began their years with the heliacal rising of the Pleiades. Their months were called, like the moon, "Jacy," from "ya," we, or our, and "cy," mother. Like most savages, they had not learned to convert the quarters into weeks.

† The Cariama of Marcgraf. Prince Max. (iii. 115) describes it as an "oiseau défiant," but I have seen it tame enough, especially as the people do not molest it. It is easily domesticated. My friend Sr. Antonio da Lacerda, jun., of Bahia, has or had a specimen. It flies for short distances, the wings being feeble, the body heavy, and it may be run down where there are no trees.

‡ Described by Prince Max. (i. 103), who was reminded of the "Croupion" (P. erythrogaster) of the Berlin Museum.

who are mounted in early life, are much taller and stronger than those bred in the house. This may partly be owing to their abundant diet of milk and cheese, farinha, and sun-dried meat. Here and there were scattered the huts of "aggregados," squatters who are permitted to live upon the Fazenda, but who do not acquire by residence any right to the soil.

The lakelets are of little importance: they are the Lagôa Seca, then dry; the Lagôa dos Porcos, where porkers are bred and cut up; the Lagôa de Dentro, which overflows, and leaves after retreat a thick, short-piled carpet of soft sweet grass, and the Lagôa de Aldêa, so called from an Indian settlement, which has now disappeared. These pools, fed by rain-drainage, and sometimes by springs, are scattered everywhere over the country: they are natural vivaria, producing in abundance the "Trahira" fish.*

Presently crossing a wave of ground, we entered a small Mata or patch of dwarf forest in the Bebida estate. The low-lying soil is fine, as we are told by the Mutámba or Motámba tree (Guaxuma ulmifolia),† which bears an emollient gelatino-saccharine fruit, and whose gum refines sugar. The leguminous Angico (Acacia Angico), delicately feathered, whose bark abounds in tannin, is also a good sign. My attention was called to the Maçela do Campo, whose yellow flowers, resembling immortelles, are used to stuff pillows ; to the Fruta Cheirosa (one of the Anacardiaceæ), with a large "baga" or berry, now green and milky ; and to the Almecegueira (Icica or Icicariba Amyris, Aublet), with sweet-smelling wood, and perfumed resin used for a variety of technologic purposes.‡

I could not but observe how abundant was the antefibrile element: the Formulary quotes 15 species, several of them resembling those of Peru. In the denser growths was the Quina

---

* Gardner writes Traíra (Prince Max. Traïra), and describes it as "rather slender." I found it short and thick, like a doubled John Dory. It extends all down the river, and has several varieties, Trahira-assú, T.-mirim, and so forth. The flesh is good, but too spiny to be eaten with pleasure. Its dark back, ugly mouth, and rat's teeth make the people call it Páu de Negro—negro wood—and refuse to touch it. The Trahira, like the Piabanha and the Piau, is commonly met with in the rivers that fall into the Atlantic Ocean.

† Mutámba is an Angola word ; the Tupys knew it as Ibixuma.

‡ In Portuguese Almecega is gum mastic (Amyris) ; hence the Brazilian tree is named.

" A almecega que se usa no quebranto."

"The gum of mastic used for inner hurts," says the Caramurú (7, 51). On the coast it acts like pitch ; and the aromatic balm is everywhere applied externally for internal injuries, as hernias, ruptures, and so forth. The word "Quebranto" classically means " fascinatio," the evil eye.

do Mato (Chinchona Remigiana); and with it the "Poor man's Quinine," a tree with bitter bark and sweet fruit, called by many names, Páu Pereira (Geissospermum Vellozii), Ubá-assú, Páu Forquilha, Páu de pente (comb-wood), Camará de bibro (for bobbins), Camará do Mato, Canudo Amargoso or Pinguaciba.* There is also an abundance of the Chá de Pedreste, or de frade (Lantana Pseudo-thea). The giants of the forest are there, especially the Jatobá† (Hymenæa, whose leaves are in pairs), which in August yields a wine, said to be very pectoral; it bears gum animé (Jutay Cica), a good pottery varnish, and a copal used by the Indians in making their labrets and other ornaments; the flowers are enjoyed by the deer, especially that called Mateiro, and the long chestnut-coloured pods that strew the ground supply a flour of insipid taste, which serves, however, in times of famine. The most beautiful growth is the Ipé Amarello, or Páu d'Arco, "bowdarque" (Bois d'arc, a Bignonia), a tall thin trunk, as yet without leaves, which will appear after inflorescence; its trumpet-shaped blossoms, in tufts of yellow gold, would make the laburnum look dull and pale.‡

Presently we came to the foot of the Pedreiras, where the land wants water, a fatal objection in the present state of things. This is a lump of naked, fine black-blue and stratified limestone, weathered so as to resemble basalt from afar: it runs from north to south, when it joins the forested Serra d'Aldêa, also a calcareous formation, large enough to supply the Province for centuries. The outcrop is marked with striæ and holes of dull, dead white, from which spring trees, and especially Cactus, whose figs

---

* System (p. 95-97). In the Campos are the several Chinchonaceæ, Quina do Campo (C. Vellozii?) with dark and spotted leaves, and a sweet fruit upon which birds feed. St. Hil. (III. i. 229) mentions a Quina do Campo or de Mendanha, which he found to be a Strychnos Pseudo-quina. The other common species is the Quina da Serra (C. ferruginea). Camará is the local name of a plant called in Portugal "Malmequer;" bibro (from "volvere") is "fusus."

† This fine feathery forest tree, which prefers the dry woodlands, has many other Tupy names, for instance, Jataby (Jutahí and Jutahí-Síca (Mr. Bates, i. 83), Jetaby, Jetaíba, Abati-timbaby, Jatai-uvá (or ubá). According to Sr. J. de Alencar "Jatobá" is derived from Jetahi, the tree,

"oba," a leaf, and "a," augmentative, alluding to the dense and beautifully domed foliage. The bark was used to make the native "ubá," or coracle. The wine must be drawn before the young leaves appear.

‡ Of the Bignonias there are many kinds, e.g., Ipeúna, whose heart supplied the hardest and best material for bows; Ipéroxo with mauve and purple blossoms; Ipé-tabaco, so called because the heart contains a fine powder of light green; the Bignonia cordacea (Sellow), with blossoms of tender yellow; Ipé-branco, with large white blossoms. On the coast the young foliage of brown and burnished tinge, curiously contrasting with its neighbours, is put forth in early spring, at the end of August. In these Highlands it is later.

are here appreciated. To the west of these "Bald Knobs," I was told, flows a broad stream, arising near the hill-summit, a common feature in Kentucky and other limestone countries. After running 300 yards it disappears into an underground passage, from which it presently emerges. My "American"* informant told me that it could work any amount of machinery. Hereabouts are caves which yield saltpetre, and where Dr. Lund made some of his greatest discoveries.

On our way back we passed by the Lagôa Grande, the largest of the pools; around it was a Campo Novo—a "new," that is a newly fired prairie; the bright green grass started up from between the stones, which are supposed to defend it by preserving the moisture. Here also were fair slopes of graceful rounded forms, where the plough can act perfectly. From the rising ground we saw to the north the long line of the Cipó Range, limestone forested with Mato Dentro. To the northeast was the box-like apex of the Serra do Baldim (pronounced Bardim), and to the south-south-east the quoin-shaped and cloud-crowned head of our old friend the Piedade near Cuiabá.

My next excursion was to the Lagôa Santa, in company with Sr. José Rodriguez Duarte, whose amiable family we had met at Ouro Preto. The path was southerly, hugging the left bank of the Old Squaws' River. From the uplands before 8 A.M., the Valley appeared a serpentine of dense white mist, clinging to and curling up along the wooded bed: a suggestive spectacle, which never loses its interest. Presently we passed the rich fish-pool, Lagôa do Córrego Seco; its village of four houses boasted of an Inspeitor de Quarterão, the humblest of police authorities, facetiously called Juiz de Paz. After a total of an hour we crossed the southern limit of the Jaguára estate, and at six miles for head-quarters we sighted the "Sumidouro" or Sinker.† This pool is said to be connected by a tunnel with the Olho de Agua on the right bank. To the west lay the village, lazily creeping up the wild slope, and much resembling a scatter of termitaria.

---

* Americano in the Brazil always means a citizen of the United States.

† The place alluded to by Southey, iii. 48. "From his (Fernando Diaz) head quarters at the Sumidouro (or Swallow, as those places are called where a river sinks into a subterraneous channel) he explored the Serra of Sabara Bussu." The feature reminds us of the subterranean river which is supposed to run under the good city of Tours.

The next feature was the "Quinta do Sumidouro," a one-streeted village with a brand-new chapel, Nª Sª do Rosario; it is mainly the work of an Italian, the Rev. Padre Rafaelle Speranza, who, if half the tales told about him are half true, has been left to live by a kind of miracle. Here men still remember a tragical episode in the eventful career of Fernando Dias Paes Leme, one of the most adventurous of the Paulista explorers. He was then seeking for "green stones" or emeralds, near a pestilential water known as the Vepabussú or Great Lake, and the hardships caused many of his Red-skin auxiliaries to revolt. They were prompted by one of his illegitimate sons, to whom he was greatly attached. When the mutiny was quelled, the father took the first opportunity of asking the youth what penalty was deserved by a man who had dared to rebel against the king's majesty.

"He should be hanged," said the son.

"Thou hast pronounced thy own doom!" replied the father, who, stern as the first consul of Rome, ordered the sentence at once to be carried into effect.* The old man died a few days afterwards, "Vnhouzzled, disappointed, vnnaneld," on his way from the Lagôa Santa to Sabará.

Sr. Leite, an intelligent store-keeper at the Quinta, which is about half a mile from the River, assured me that the ground had lately been subject to shocks, which were most frequent about full moon; he seemed to fear for it the fate of Mendonça. In this limestone region I could detect no sign of igneous action, plutonic or volcanic; but the earthquake at Alexandria, and another which I witnessed at Accra on the Gold Coast, prove that sedimentary formations are by no means exempt from the visitations of Ennosigæus.

The rest of the road was over wild and picturesque Campo, where the bright little Ribeirão Jacques will some day be valuable. Presently, after 3^{hrs} 30^m = 12 miles, topping a long hill, we saw below us a shallow basin, with a church and a scatter of white and brown houses—the town of Lagôa Santa. The streets were formed by the "compound" walls: tile-coped, and protected by a few inches of taipá or pisé, resting on a layer

---

* Southey (iii. 49) recounts the story nearly in the same words. St. Hil. (I. ii. 189) places the scene of the "Octagenarian's" adventure in the Province of Porto Seguro, and declares that the "Vupabussú" was afterwards called "Lagôa Encantada," because it could not be found.

of rushes, which projects on both sides and defends the lower part of the perpendicular mud. We rode up to the square, "Praça de Nª Sª da Saude, so called from the Matriz, to the east of which is a fine fig-tree being rapidly devoured by the "Bird Herb" (a Polygonea ?). The place, now so quiet and sleepy, has seen wild times. Successful at Queluz (July 27, 1842), the insurgents retired to the Capão de Lana, and, after a week, when the "Oligarchy" rendered this position untenable, they retreated and entrenched themselves in the Arraial da Lagôa Santa. An ambuscade of forty men wounded the loyalist colonel, Manuel Antonio Pacheco, afterwards Barão de Sabará, and repulsed his 750 men. The attack was renewed, the Revolutionists fought stoutly, and an aunt of Adrianno José de Moura assisted them by serving out ammunition; on the 6th August, however, they were obliged to take to the bush. The conduct of the late Baron was praised, even by his enemies; he was one of few who treated the captured with kindness.

We rode up to some horse-posts (estácas) opposite a door, over which was inscribed F. F., and, having heard of a French hotel, we knocked. The house was opened by a very English-looking dame, who proved to have been born at Malta; we asked to see M. François Fourreau, and we were told to dismount. After shaking hands and exchanging salutations in the "language of Racine and Corneille," we ordered breakfast unceremoniously enough; the host joined us, and we enjoyed an excellent soupe and bouilli, not often eaten outside French walls. An old sous-officier of the 16me Léger, he had been taken prisoner in the Russian Campaign, and the result was that he, a très joli garçon, set up a circus, and had travelled all about Western Asia. His three stalwart sons, including "Bibi," were still conducting the business at Diamantina; his daughter, a pretty ecuyère and married, as "Pedrinho" proved, lived with her parents. The good old soldier had bought considerable property at Lagôa Santa, he lusted to escape from it, but he did not see the way out. He was by no means one of that wretched race, which belongs to France or to England, not to the world. We passed the night with wine and jollity, and when I suggested the "addition," M. Fourreau laughed in my face. I am sorry to say that Madame did likewise; yet I left them with regret.

On arrival we sent our cards to Dr. Lund, the illustrious

Dane, the hermit of science, who had spent a portion of his life in the bone-caves of Minas Geraes. I was most anxious to ask him about the "fossil man," or "sub-fossil man," as opposed to the "primeval" or "prehistoric man." The term has been prematurely decided to be "a misnomer, since the thing so designated is of all things the most desired, the most sought after, but perhaps the least likely to be found." Still the influence of Cuvier! I was also desirous to know if the incisor teeth of the fossils had naturally oval upper surfaces (not worn down), and of longer antero-posterior diameter than transverse. Dr. Lund has for years been prevented by consumptive tendencies from living out of the Brazil; he has bought a house in the square of Lagôa Santa, and, as might be expected, he has become bed-ridden by rheumatism. He is said to live chiefly on Caparosa-ptisane,* which combines theine with caffeine. We perforce accredit others with our own feelings, and I felt sad when pic-turing to myself the fate of so great a traveller, doomed to end his days without a relation by his side, in the social gloom of this gorgeous wilderness. M. Fred. Wm. Behrens, the savan's obliging secretary, came over with many excuses and prayers that we would wait till the next morning. We did so, but with-out success. I suspect that our failure was caused by the nervous fear of strangers, which often affects even strong men after a long residence in the Brazil, and indeed in the Tropics generally.

Having heard many curious lake tales † about what proved to be on inspection a vulgar feature, I spoke to M. Behrens, who led me to his employer's lust-haus on the holy lake, launched

---

* "Caparosa" is primarily our copperas (sulphate of iron), also applied to verdigris, and the shrubby tree got the name on ac-count of the tender blue-green leaf. It is known at once by the cut or torn part of the twig turning dark and tarnished. Ac-cording to the System it contains tannic acid with a solution of iron, which may be made into ink, and which supplies a black dye. The abuse of its ptisane has, I was told, been already fatal to some who have followed the example of Dr. Lund. The celebrated Paullinia Sorbites, better known as Guaraná (from the Tupy Guarana-uva) also combines theine and caffeine.

† These lake superstitions are common in the Brazil. La Condamine, Humboldt, and others speak of the Lagôa Dourada. Henderson mentions that of the Lagôa Feia.

Prince Max. records the fables of the Taípe, and heard of other traditions on the banks of the Rio dos Ilheos and the Mucury. The Parimá or Parimó Lake of Guiana is equally rich in legends. Connected with lakes of golden sands was the city of Beni, Grão Pará, Grão Pairiri or Paititi, alias El Dorado, whose streets were paved with the precious metal, and where the Emperor of the Musus, the great Paititi or gilded king of the Spa-niards, was smeared with oil as he rose in the morning, and covered with gold dust blown at him by his courtiers through long reeds. Castelnau (vol. vi. 41) relates those of the Boldivian "Opabusú." This word, like Southey's Vepabussú, is a cor-ruption of Ypabussú, ypaba in the Lingua Brasilica meaning a lake.

the boat, and struck out with the paddle. The piles and poles
which have been said to denote pfalbauten or crannoges, were
probably an old palisading now flooded. The length is about
one and a half miles from south-west to north-east bending east,
where a sangrador or drain, some eight to nine miles long, discharges it into the Rio das Velhas, near the Fazenda called of
Dona Ignacia.* The southern side had greatly shrunk, and we
saw at once what causes the "bubbling surface." Here, during
the rains, is a Cabeceira or head stream, one of the many feeders
from the basin-sides, which gently rise to grassy Campo ground.
On the opposite margin of the little reservoir rises a pretty bit
of cockney forest, which has been pierced with toy paths. The
lake is said to be filling up, and the greatest depth in the centre
is three fathoms. The sides are overgrown with a fine pithy
rush (junco), of which mats are made; this is one of the local
industries; the others are fishing and rude pottery, glazed with
yellow and green. The poor almost live upon the Trahira, the
Curumatão,† and the dreadful Piranha.‡ The vegetation around
is stunted; we are still in the lands of the plantain and the pine,
but the Araucaria is short and ricketty, evidently finding the air
too hot to breathe.§

The Holy Lake was originally called Ypabussú (Vupubussu),
or Lagôa Grande; it owes its pretentious name to superstitions

---

* Mr. Gerber's map makes it heart-
shaped, lying north and south, with the
apex to the south, and he drains it by a
greatly exaggerated "Rio Fidalgo." The
latter is the name of an estate belonging to
the heirs of the late Cirurgião Mór, Serafim
Moreira de Carvalho.

† The name of this fish, one of the Salmonidæ, is variously written by authors. Prince
Max., Crumatam; Pizarro, Corimata and Curmatan; St. Hil., also Curmatan; Gardner,
Curumatám; Halfeld, Cumatá or Curimatá,
and the Almanak, Curumatá—the latter
two neglecting the nasal sound which it
certainly has. I hesitate whether to write
Curumatão or Gurumatão, the first conso-
nant being doubtfully sounded. This fish
is about two feet long; it leaps like our
salmon, with its silvery scales glancing in
the sun, and it must be caught in drag-nets,
as it will not bite at bait. There is also a
salt-water fish of this name, soft and full
of spines. The savages shoot it with arrows (Prince Max. ii. 137), Mr. Bates (ii.
140), "Caught with hook and line, baited
with pieces of banana, several Curimatá

(Anodus amazonum), a most delicious fish,
which, next to the Tucumaré and the Pes-
cada, is most esteemed by the natives."

‡ The well-known Scissar fish, Piranha
in Tupy meaning scissars. Our authors
call it "devil fish." Cuvier named it
Serra Salmo Piraya, and unconsciously sanc-
tioned the vulgar Mineiro and Paulista corruption of Piranha to Pirayya (so Canúyya
for Canalha). The fish is common in the
Upper Uruguay and the Paraguay, as well
as in the São Francisco. Those that I saw
were from one foot to eighteen inches long
by about ten inches deep, flat but short and
thick. The carnivorous fish swims vertically, and is supposed to turn on one side
when it bites; the serrated teeth bend
backwards; they easily tear off the flesh,
and a shoal will, they say, in ten minutes
reduce a bullock to a skeleton. I found
the meat dry, full of spines, and with poor
flavour. On the Lower São Francisco the
people refuse to eat it.

§ The Lagôa Santa proved to be 2228
feet above sea-level (B. P. 208°·1, therm.
76°).

which have now died out. In ancient days people made of it a
Pool of Bethesda, and a Dr. Cialli, in 1749, found that its waters
contained medicinal properties. The tale which Henderson recounts about its surface being filmed over with a silvery pellicle
like mercury, was unknown to all. They preserved, however,
the tradition that, "once upon a time," a woman used to be seen
hovering over the centre, whilst a silver cross arose from the
depths. Many a hardy fellow, doubtless in a pitiable state of
nervousness, paddled to make a prize of a precious metal, and
was sunk by a mysterious whirlpool, when, as the Arabs say, he
passed without loss of time from water to fire. The spirit was
exorcised—a common process in Hibernian legends—by some
holy man, whose name has fallen into unmerited forgetfulness.
Similarly in the Manitoulin Islands of Lake Huron, the Manitou
(popularly and erroneously translated "Great Spirit") forbade
his children to seek for gold ; the ore was supposed to be found in
heaps, but no canoe could reach the spot before being over-
whelmed by a tempest. All these have vanished :—

> The intelligible forms of ancient poets.
> Die alten Fabel-wesen sind nicht mehr.

and humanity is no longer sorely tempted upon the Holy Lake.

Disappointed, to Jaguára we returned, and I found it difficult
to tear myself away from the pleasant society of my new friends,
Dr. Quintiliano and Sr. Duarto. I little thought at the time that
the latter was so near his end : he had been treated for ulceration
of the leg ; the wound was healed, but when he returned to Ouro
Preto he died suddenly. Hospitality is the greatest delay in
Brazilian travel. It is the old style of Colonial greeting ; you
may do what you like, you may stay for a month, but not for a
day, and the churlish precepts and practices of Europe are unknown.* At length, however, I found a pilot, Chico (i. e., Francisco) Diniz de Amorim, who had a farm near the Retiro das
Freiras : he was described to me as very "fearful" (medroso),
meaning skilful and prudent. The others were Joaquim, the son
of Antonio Corrêa, overseer of Casa Branca: a useless shock-

---

* The Basque proverb says,—

> Arraina eta arroza
> Heren egunac carazes, campora deragoza.

" Fish and guests after the third day stink, and must be cast out of the house."

head, unable to work. I presently bought for 40$000, a kind of "Igára," * a tender-canoe, and used to send him ahead to explore the Rapids. The third was João Pereira, of the Rio de Jaboticatúba, a freedman of the late Padre Antonio : he was the hardest worker of my five crews, but as fierce and full of fight as a thorough-bred mastiff. We got on well together; I did not, however, engage him for the Rio de São Francisco, lest his readiness with his shooting-iron might get me into trouble. These men were to receive 5$000 per diem, and 2$000 whilst returning to their homes : they asked a couple of days to prepare, and they caused no unnecessary delay. Usually, every excuse is offered, the favourite one, both here, on the Rio de São Francisco, and on the Amazons, being that the wife is about to grow another olive branch.

As far as Jaguára, the River has shown us mere broken waters (Quebradas), tide-rips (Maretas), and runs, properly called "Correntezas," "Corradeiras," or "Corredeiras," and "pontas d'agua," when the stream swings swift around the points. The traveller, however, will hear them denominated Cachoeira,† a generic term, equally applied to the smallest ripple or Stromschnelle, caused by a sunken tree, and to the Paulo Affonso, King of Rapids. The word, therefore, will be used for convenience, without attaching to it any importance. To a certain extent it is correct, the difference of levels in most of the rapids is unimportant, and we shall not find a fall or drop (Salto) till we reach the São Francisco. The little perpendicular steps in the Cachoeira, or Correnteza, are called Corridas and Corredóças,‡ and especially occur in the scatters of rocks, known as 'Taipaba, a corruption of Itaipába. § On the other hand, the "Canal" is the fair-way through the Cachoeira.

---

* This is the Tupy word from "yg," water, and "jara," lord. My "Lord of the Water" was of Mandim or Peroba wood, twenty-five feet long, with average breadth of nineteen inches. As usual here, and the same is the case on the Mississippi, in defiance of all the rules of displacement, the dug-out was made leek-shaped, bulging at the bows, with a head larger than the body, under the raftsmanlike idea that this facilitates progress. We can only compare them with the "plough bows" and the "short bluff ships" which are now figuring in our naval estimates.

† Also written Caxoeira, which has the same pronunciation. The word corresponds in part with the Raudales of the Orinoco. In Tupy it is "aba-nheendaba," which means equally a rapid (Cachoeira) or a cataract (Cascata or Catadupa). In parts of the Brazil, especially the Province of São Paulo, Cachoeira means a rivulet, without conveying the idea of rapids. Cachoeira is a classical Portuguese word, scil. Cachãoeira, a place abounding in "cachões," plural of cachão, derived by Constantio from Coctio, boiling (water).

‡ This is also a classical word, locally used in a limited sense.

§ The word is Tupy, "Ita-ipa" mean-

The Cachoeira proper is a place where the river skirts a hill, or breaks through a range which projects into it rocks that cause rapids. Generally it extends from one side to the other : its diminutive form is the "Camboinha," a "Carreira," or a "Corredor." The upper strata in the Rio das Velhas being mostly limestone, the obstruction is often a narrow wall of loose stuff (pedras movediças) through which a few Irishmen with picks would open a way in twenty-four hours ; once opened, the water laden with sand and gravel would not allow it to close. Before this is attempted, I should advise, however, the use of the diving-bell, or helmet, in each deep pool (fundão) which precedes the break. These basins where the water slackens (remansos, poços, aguas paradas), and which lie close above the rapids, are in fact huge flumes and cradles where the gold * and diamonds washed down to the river-bed will be found to have settled, whilst the rock-bars crossing the stream must preserve the deposited matter from being swept away during the floods. In the Rio de São Francisco the Cachoeira is much more serious, because formed either of the hardest sandstone or of lumpy granite, whose crest numbers feet when here we have inches.

The Cachoeira, like the "Pongo," or "Mal Paso" of the Upper Amazon, is nearly always found at the mouth of a tributary, a river, or a córrego or stream which brings down mud, "creek-sand," and gravel. It causes inundations by arresting the flow, and these floods would be easily remedied, whilst the stream would not be injured by additional velocity. In rare places it may be necessary to canalize across a neck of ground, but the Brazil is not yet prepared for such expenses.† On the Rio das Velhas there are generally houses near the Cachoeiras, but, as a rule, in the dangerous parts the people know nothing of the river a league above or below their doors : they use canoes for fishing, crossing, and paying short visits, but they travel by the roads along the banks.‡

---

ing a stony reef. It is translated "Gurgulho" or "Pedragulho," coarse gravel. Castelnau (i. 424 and elsewhere) mentions upon the Tocantins River the "Eutaípava," probably a peculiar way of spelling.

* Two attempts have been made to turn the bed of the Rio das Velhas ; one was below Santa Lusia, and the other was above Jaguára. The success was partial, the precious metal was found in quantities, but after an enormous expenditure of human labour, the floods came down and the stream returned violently to its old course. At present the people cannot reach the bottom of the bed, and coffer-dams, dragging machines, and diving bells are equally unknown to them.

† Upon this subject I shall offer some observations in Chapter 15.

‡ In the Brazil, as in British India,

The bad "Cachoeiras" on the Rio das Velhas number ten, and all will require more or less work before a tug can be employed upon the river. They are "wild rapids," Cachoeiras brabas (bravas), the others being "meia braba" and "mansa," or tame. There is no rule for passing them. Sometimes the raft must creep down the sides; at other times the pilot must make for the apex of the triangle, whose base is up-stream, and whose arms are formed by jumping water. In many of the tide-rips there is a double broken line, containing a space smooth as oil, which shows the deep bed. The rock or snag, on the other hand, is known by the triangular ripple, with the base down stream. The paddles should be taken in, and the raft must be pointed down with poles (sobre vara): if the men are lazy they will spare themselves this trouble, and they will probably come to grief. Where the current is very rapid, it is advisable to diminish the pace by dropping down stern foremost.[*] "Cordelling," stern foremost by a rope from the bows, is mostly confined to the tail-end of islands, where there is a gate in the rocks through which the raft that would otherwise be swept down by the current, must pass. Of course, the seasons make the greatest difference in the rapids;[†] some of them which are formidable during the floods, are safe when the dries set in. Generally they are most dreaded in the winter weather, when I passed them: during the inundations between December and March, a small steamer might pass over many of them without knowing that they are there. The boatmen swim like ducks, despite which many are drowned. A stranger without a life-belt would have little chance of escape; it is therefore advisable to prepare for accidents by attacking dangerous places *en chemise*.

water communication, which should have been first undertaken, has been left to the last. I shall have more to say upon this subject.

[*] Commonly called (virar or descida) "de bunda," more prettily "de poppa;" opposed to the normal way "de bica" or "de corrida."

[†] M. Liais was on the Rio das Velhas between April 10, 1862, and July 3, 1862. His head pilot was one Clemente Pereira of Tabatinga, in the Vinculo do Mello. Hence the names of the Cachoeiras, and other features, which are not all correct.

CHAPTER IV.

TO CASA BRANCA AND THE CACHOEIRA DA ONÇA.

THE COUNTRY HOUSE.—THE COUNTRY GENTLEMAN.—VISIT TO JEQUITIBA.— UGLY RAPIDS.

O echo do Rio que o trovão simula,
E lento se prolonga reboando.
Domingos José Gonçalves de Magalhães.

FRIDAY, *August* 16, 1867.—After a week at Jaguára, I packed up my chattels by an effort of the will, and, accompanied to the " Porto " by my kind hosts, embarked. We parted with many hopes to meet again, and with long wavings of the hat: presently I found myself, once more, like Violante in the pantry— alone.

M. Liais records in May from Jaguára downwards, a constant depth of two metres, and no danger of grounding except from carelessness: this, however, was not my experience. During the ten miles of to-day there was little to observe. We passed the bar of the Rio Jaboticatúba,* and we shot through a broken bridge and by a ferry raft with chain and pulley, belonging to the Fazenda de Santa Anna of Sr. Antonio Martins de Almeida. After another bend we sighted on the left a square of white and brown houses with turreted entrance and private chapel. This place, the head-quarters of the Casa Branca estate, lies below a plantain-covered hill rising above the wild growth of the banks. The land is of rich limestone, with a wealth of water ; is rich in cotton and sugar, maize and rice, haricots and the castor-plant; it breeds horses and mules, black cattle, and pigs ; and on the river's banks large granular gold in rusty quartz looking like iron is still

---

* The name is that of a fruit, somewhat like the common Jaboticaba (Eugenia Cauliflora) ; but the tree is taller, the bark has a different appearance, and the berries do not grow so low along the trunk. Canoes ascend the stream for five leagues ; it heads in the Serra do Cipó, and is navigable for the smallest craft to the Ribeirão de Abaixo, distant some twelve leagues. Further down is the Córrego da Palma, whose bend, a little below the mouth, is called the Roto da Palma.

washed. The four square leagues may be bought for 300:000$000, or less. ,

A small party of Anglo-Americans met me on the bank and introduced me to the owner, Sr. Manoel Francisco (de Abrêu Guimarães). He was a fine, handsome, middle-aged man, Portuguese
by birth ; about eighteen years ago he inherited half the estate of
his uncle, Major João Lopes de Abrêu. The manor house was
in the normal style, fronted by a deep verandah, from which the
owner can prospect the distillery, the mill, whose wheel informs
us that sugar is the staple growth ; and the other offices. At the
end of the verandah is the Chapel of N^a S^a do Carmo, with her
escutcheon of three gilt stars upon a wooden shield painted blue ;
here there is chaunting on Sunday evenings. The Senzallas or
negro quarters are, as usual, ground-floor lodgings within the
square, which is generally provided with a tall central wooden
cross and a raised wooden stage for drying sugar and maize ; the
tenements are locked at night, and, in order to prevent disputes,
the celibataires are separated from those of the married blacks.
These Fazendas are isolated villages on a small scale. They
supply the neighbourhood with its simple wants, dry beef, pork,
and lard, flour of manioc and of maize,* sugar and spirits, tobacco
and oil; coarse cloth and cotton thread; coffee, and various teas
of Caparosa and orange-leaf. They import only iron to be turned
into horse-shoes ; salt, wine, and beer, cigars, butter, porcelain,
drugs, and other "notions." There is generally a smithy, a
carpenter's shed, a shoemaker's shop, a piggery, where during
the last month the beasts are taken from the foulest food, and an
ample poultry yard.

The life of the planter is easily told. He rises at dawn, and his

---

* "Farinha de Milho" should be steeped
(molhado) for 24 hours ; the manipulation
is delicate, and especially the water must be
flowing, or the flour turns sour, and acquires a nauseous taste (farinha podre). It
is then pounded (socado) in the stamps
(pilão) and sifted (penerado) ; the dough
(massa) is toasted by slow degrees, otherwise it will be injured, in large pans of
tile stone or metal (Furnos de cobre, &c.)
fixed in masonry over the fire. Travellers
have used these articles for drying skins
and plants. This farinha is best when
eaten with milk. The people ignore the
corn bread of the United States. In making
manioc-farinha, the bag in which the raw
raspings were strained (tapety or tapiti, in
French colonies "la Couleuvre ") is supplied by placing palm-leaves above and
below the massa when in the press (prensa) ;
the sediment of the juice that comes from
the mass is called tipióca (our tapioca), and
the liquid is thrown away. The Indians,
like the Dahomans, prepared a much roasted
and hard meal, which they called ouy-entan,
and the Portuguese know as "Farinha de
guerra" (Prince Max. i. 116). In the
Brazilian forests there is a poisonous species
called mandioca brava ; in Europeans it
produces fatal vomitings, but the wild
people are said to eat it after keeping it
for a day.

slave-valet brings him coffee and wash-hand basin with ewer, both of solid silver.* After strolling about the mill, which often begins work at 2 A.M., and riding over the estate to see that the hands are not idling, he returns between 9 and 11 with his family, and if a bachelor with his head men, to breakfast. The sunny hours are passed either in a siesta, aided by a glass of English ale —there is often nothing English in it but the name—in reading the newspapers, or in receiving visits. The dinner is between 3 P.M. and 4 P.M.—sometimes later; it is invariably followed by coffee and tobacco. Often there is another relay of coffee before sitting down to tea, biscuits and butter or conserves, and the day ends with chat in some cool place. The monotony of this Vida de Frade—Friar's life—is broken by an occasional visit to a neighbour, or to the nearest country town. Almost all are excellent sportsmen, good riders, and very fond of shooting and fishing. They are also doctors, great at administering salsaparilla and other simples, and at prescribing diet. In Gardner's time Buchan's "Domestic Medicine," translated into Portuguese, was the book; now the Formulary of Chernoviz must have made a little fortune; it is part of the furniture, as was " Guillim " in the country-houses of our grandfathers. Homœopathy † throughout the Brazil is in high favour, and generally preferred to the " old school " and the "regular mode of practice." The choice is the result, I presume, of easy action upon the high nervous temperament of the race, and the chemist who deals in the *similia similibus*, makes more money than his brother the allopath.

We will now visit the Engenho, or sugar-house, the simplest expression of a mill. In the more civilized establishments a light wheel works by a cogged axle, the two iron or iron-banded cylinders placed horizontally.‡ The old three perpendicular

---

* This is still the custom of Turkey, Egypt, and Persia. On the Rio das Velhas metal is preferred to the more frangible material ; for everywhere in the Brazil negroes break whatever they handle.

† The establisher of homœopathy in the Brazil, who corresponds with Dr. Samuel Gregg in New England, was Dr. B. Mure, a Frenchman, a most active and energetic proselytiser, who worked the press with unwearied energy. " You and I are the only men who love homœopathy for its own

sake," Hahnemann said to him. He died I believe on the Red Sea, riding his favourite hobby-horse towards and for the benefit of India. The "Instituto Homeopatico do Brasil" published his "Pratica Elementar," and it has reached several editions.

‡ "Engenho de ferro deitado," opposed to the ancient system of upright cylinders called "Engenho de páu em pé." When not worked by a water-wheel, a long lever is carried round the walk by cattle.

rollers are waxing obsolete; and a hopper sometimes protects in these days the slaves' hands from mutilation. There is an utter absence of European chemical science and of modern machinery: the vacuum-pan, the "subsider," and the "steam evaporator," are equally unknown. Even the simple use of bone black and lime, to remove the albumen and the acetic acid of the sugar, have not been adopted. The ripe stalk should be ground as soon as cut: it is often piled in the yard for days, and the accidental rents in the outside skin, hacked by the awkward black, acidify the juice by admitting the air. The caldo or garápa* is run right into the pans, which often are not thoroughly cleaned; it is slowly boiled down in coppers exposed to atmospheric action, and the laziness of the boilerman prevents his skimming the juice with care. Hence, in this Land of the Cane, those who prefer loaf sugar must send for it to Europe.†

The "American" party consisted of nine souls, including a wife and three young children, white-headed, blue-eyed, red-cheeked rogues, always blessed with health, restlessness, and accidents; they are extreme contrasts to the slow, dull, whity-browns of the land, and here the southern "cross" is uncommonly strong. They had been living for some four weeks in a house assigned by the host, and during that time their united

---

* The word is, I believe, Indian: it properly means Caouy, or wine of sugar-cane, and wild honey: and it extends far. It is simply the fresh cane-juice, which the people are fond of drinking after the Indian fashion, warm; to me it is much more agreeable cold. Garápa is a favourite beverage with Tropeiros, and it stands on the shelves of every Venda, together with Capillaire and other mixtures. For cattle, and especially for horses, it is an admirable fattening food.

† The following is the rude system in this part of the Brazil. The canes are ground by the rollers, and the juice (caldo or garápa, the Spanish huarapo) flows into the boiling coppers (caldeiras): of these there are usually three, worked by a single negro. The trash (bagaço, in French bagasse) is still rich: it is good for animals, especially pigs, it would supply fuel for a steam engine, and it is excellent manure, returning silex and saccharine matter to the soil. Now it is generally piled in a heap and left to decay. From the coppers the juice passes to the cooler (resfriadeira), where feculences and impurities subside,

and the "caldo" crystallizes. It is then placed in vats (formas) whose bottoms, half the breadth of the tops, are pierced with holes, and are provided with bungs. These troughs are covered with brick-clay, never animal charcoal: when the molasses (melado) has drained out, the sugar is dried in the open air, raked about by negro boys, and allowed to become thoroughly impure. Finally, it is stored in the sugar chamber (Caixaõ de assucar).

For distillation the molasses from the troughs is led by channels (bicas) to a large canoe-shaped wooden cistern (coche). It is then mixed with the scum from the boilers, and reduced to about 11° Reaumur, in the tank, for alcoholic fermentation (tanque de azedar). It is then carried to the still (alambique), an old-fashioned and rarely cleaned machine like a retort. This usually receives three feeds (alambicadas) in the twelve hours: work being rarely done at night. Finally the spirit is poured into a square wooden bin containing some 500 kegs: this "tanque de Restilo" when hollowed out of a single log is called "Paiol."

"venda bill," food included, had been only 26$000—say, 30*l.* per annum. A wonderful infirmity of purpose seemed to affect them; the only reasonable cause of delay was a wish to try the effect of a rainy season before squatting in the new Alabama. Some liked the place, because it is above the difficult rapids, and it is connected by land and water with Sabará the market, a sine quâ non here. Others abused it; they held 'it unfit for the plough, and objected to the Brazilian style of spontaneous production, where the land is uncleaned, where the only implement is a bill-hook at the end of a long handle used to lop off the sprouts of the young wood springing from undecayed roots, and where gathering is the only work and care. They naturally enough objected to plant in the same field cotton and corn, beans and Palma-Christi, the sole rude succedaneum for a rotation of crops now known in the Brazil. The best lands are here sold at 15$000—40$000 per alqueire of 6 × 2 square acres, and large tracts may be purchased for much less. To work profitably, however, they require stock and fifteen black hands—the latter at present a very expensive article, ranging from 50*l.* to 100*l.* The production per acre is of cleaned cotton, one bale of 500 lbs., worth a minimum of 200$000; 40 bushels of corn fetch from 40$000 to 80$000, and the same is the price of an equal quantity of beans and ricinus seed.* The acre also produces 100 lbs. of tobacco, worth 60$000, and the price will be raised by proper treatment. Not being over-burdened with money, the colonists must rely mainly upon time-purchases. I heard afterwards that they had bought a raft, and descended the river to Trahiras. One of them, Mr. Davidson of Tennessee, volunteered to accompany me as adjutant-general; I liked the man, and gave him a passage to the Rapids of Paulo Affonso.

The host was a bachelor, and the evening of my arrival was ushered in by music and dancing; a "pagoda," however, not a "fandango," nor the peculiar Congo style of saltation known as the "batúque." † I could not enjoy it, the sun had been over-powerful, and the breeze had been too cool: my principal sufferings were from cramps in the fingers, here, apparently, a common

---

* This mamona-oil sells at 1$200 per alqueire—nearly 8 imperial gallons. The Southerners are familiar with the plant, but they rarely burn the oil, as is done in this part of the Brazil.

† Not batucca, as it is written by Prince Max.

complaint. I had arrived on Friday, but the host would not give me leave to depart before Monday, and then, also, not till after breakfast. My raft was plentifully supplied by him with fine "Restilo," or rather "Lavado," whose exceeding strength provoked the wonder and admiration of the river. A single wine-glass of this spirit before turning-in, especially when the wind and rain rushed under the raft-awning, was a protection against ague. Thus, Peter Pindar :—

> " Would you, my friend, the power of death defy ?
> Pray keep your inside wet, your outside dry."

I found also a six-months' provision of fine, white, clayed Rapaduras sugar bricks, 9 × 6 × 2 inches. Sr. Manoel Francisco accompanied me to the "Eliza," embraced me, and wished me the best of voyages ; I parted from him with regret.

August 19.—After two hours we passed on the right bank the Paracátu influent,* a buttress of caverned rock ending a hill ; it was the first of three picturesque cliffs composed of calcareous blocks, tufted with trees, and separated by shallow green hollows. In front the distances were charmingly painted by the pink-blue air of the Brazilian spring, which lasted us twenty-three days longer, till we reached the Rio Pardo ; the gauzy, filmy sky blurred the outlines of the vegetation and rendered mirrory the surface of the stream. The timber was small, the tallest growths being the Jatobá and the Angico Acacia ; the most spreading was the Gamelleira or wild fig, that kindly gift of Nature, with dense, cool, dark-green foliage, and "beard of wood"† garnishing its widely-extending boughs. Clearings extended from the water to the hill-sides, making brown patches of dead vegetation ; and oranges and bananas showed where the dwelling places lurk. There was the usual beautiful variety of hue and form, so attractive to all who have an "eye for trees." The mauve Quaresma, the chrysoprase of the young sugar, and the fan-shaped Arrow-cane (ubá),‡ here 14 to 15 feet high, tasselling the long, smooth reaches, and a hundred tints of leek-green, gold-green, dark-green, spinach-green, brown-green, pink-green, and red-green,

---

* This must not be confounded with the Paracatú influent of the true São Francisco. The word thus written means good (catu) stream (pará) ; others hold it to be a corruption of Pira-catu, good fish (pira or pyra).

† Barba de Páu or Tillandsia.

‡ "Uira" is also in Tupy a shaft or arrow, and Uiraçaba, a quiver.

contrasted with the white flowerets of the Assa-peixe branco, with the silver-lined leaves of the Sloth-tree, and with the coppery foliage of the Copahyba.* Here rose a tall skeleton, blasted by lightning, or slain by the annual fires; there a nude form enjoyed the disrobing of the dry season before assuming the impermeable of the rains; there a panachéd palm rose bending and rustling in the wind. Now the trees shot boughs horizontally over the stream and curled up or put forth secondary branches towards the light; orchids were rare, but the llianas were as usual rampant, and pendulous birds'-nests occupied the best places. There half-cut trunks bent their heads into the water, whilst others, inclining down the river in the teeth of the wind, showed the force of the floods. Masses of vegetation rolled bulging down the bank. We especially remark the massive digitations of the Castor-plant, and the Tabóca Cabelluda (hairy bamboo), a graceful, maidenly shape, but armed with angry thorns cockspur-shaped, and disposed in threes. The Hibiscus, 10 to 12 feet high, here known as the Mangui or Mangue,† will long attract the eye by its yellow cotton-like blossoms, by the young cordiform leaves with velvety lustrous green, and by the dead infoliation washed with faint vermilion, looking from afar like spangles of red.

Below Paracatú is the Poço feio, or "ugly well," where a rock projecting from the left bank caused the little whirls and regurgitations here called, from their shapes, "panellas de agua," or water-pipkins. Three hours carried us down to Páu de Cheiro,‡ thus long had it taken to coast this part of our friend's grounds. The estate, belonging to half-a-dozen owners, is estimated at 200 alqueires, and may, they told us, be bought for 8 to 10 contos of reis. A Californian who lately visited it, declared that he could make 2$000 per diem by panning the gold which lies unworked in the banks. Then we came to the Lapa, the longest and tallest limestone bluff on the river. This "rupes præcelsa sub auras" is broken into a thousand cracks and holes, whilst the cavern is fronted by the most corpulent of stalactites. Here the Calcaire is based upon an iron-stone grit, which stains the banks with

---

* The Copahyba, also written Copaïva, Copauba (Copaifera officinalis, copaier, "capivi" tree) will be mentioned in Chap. 6.

† Arruda calls this Malvacea Guachuma

(Guaxuma) do Mangue (Hibiscus pernambucensis).

‡ Literally "perfumed wood," a Laurinea.

iridescent water and rests upon sand, evidently the old bed. In many parts the slopes are frosted over with a curious incrustation, which lasted to the mouth of the Rio das Velhas. The crew declared that it was the efflorescence of arsenical pyrites from Morro Velho. We dissolved it in boiling water, strained it through flannel, and made a hardish cake of uncrystallized matter like impure sugar; the taste was that of alum and salt-petre. The latter, as in Kentucky, often overlies a whitish-yellow, arenaceous soil, whose pores act as strainers. The rest of the surface was a rich soil some six feet deep, or double what satisfies the farmer on the fertile Mississippi.

Now the currents are becoming rapids, and the bed is studded with islets of calcareous stone, dangerous during half-flood. At the Porto da Palma * M. Dumont's navigation at present ends. Four huts stand at the Barra de Páu Grosso, justly so called from the huge timber of its banks. It is said to head near the Rotulo † estate, which was bought from a certain Marquez (P. N.) of Sabará by the English Company at Cocaes, who intended it to supply their miners with provisions. The survey of this Fazenda extended over a year, and cost some 1400*l.* The overseer under the General Manager, Mr. J. Pennycook Brown, is a Mr. Broadhurst, whose father, together with a son-in-law, Manuel Simplicio, bought from Sr. Bonifacio Torres part of the estate called "Cana do Reino." Mr. Broadhurst the elder brought out English machinery for carding, spinning, and weaving cotton; he was afterwards drowned in the Cipó River, which runs out of a dip in the mountain. The same happened to two or three other Englishmen—an accident charitably attributed to the superior excellence of the rum. The Fazenda do Rotulo has fine red and black soils, based on limestones, and in two places saltpetre has been worked. It is to be sold for 50 : 000 $ 000, but it has the disadvantage of being far from water carriage. On the other hand, it is some six leagues long by two broad, and it would support a little settlement of forty families.

At 5.10 P.M. we idly came to anchor off a sand-bank, the Praia da Cançanção : ‡ it is backed by land bare of grass, and a few huts

---

* Or Porto das Palmas.

† Rotulo means a roll or label; it is generally corrupted to "Rochelo."

‡ "Of the nettle" (Jatropha urens). The wild men, who were well acquainted with and had given names to the medicinal growths of their forests, used this plant in local phlebotomy. They switched with it

are on the other side. We slept on board the " Coffin," and
were pleasantly surprised to find no insects. The night was
still as the grave, and at times curious sounds from water, earth,
and air reminded me of those described by wanderers in the
Amazonian forests—the work of some night-bird or beast, or fall
of heavy fruit, or the plashing of hungry fish. At midnight, tall
distinct pillars of white mist, silvered by the moon, formed a
majestic colonnade slowly progressing down stream. At 4 A.M.
the hot humid air of the River Valley was clear ; before sunrise,
however, a cold draught swept from the Serra Grande or do
Espinhaço on the east,* and condensed the vapour into a thick
fog. During the day the breeze chops round to the north,
forming a head-wind which refrigerates the surface stream ; the
fish will bite at midnight but not at noon.   The evenings are
mild, serene, delightful.

August 20.—We resolved to set out betimes, but the vapours
kept us at anchor till long after sun-rise, and we had reckoned
without (including) our host. The country now assumes a type
which will last. In the offing is a grassy table-land or ridge
either with one or two distances, bristled with a few trees, and
rising high above the avenue of bush and forest, through which
the stream flows. After a couple of hours we paddled under a
split bridge which had been carried away in 1858 : like that of
Casa Branca it should have been raised at least 50 instead of
32 feet (10.30 metres), a fair allowance for extraordinary floods.
The site is, as usual, badly chosen ; instead of being divided into
two a little lower down, it runs like a causeway right across a
branch channel formed by inundations on the left bank.   The
original cost had been 2:800$000, and the holes made for plant-
ing the piers had yielded 4:400$000 of gold. An engineer
offered to repair it for 600*l.*—instead of 60*l.*—and the owner
therefore prefers a raft.

Just below it, to " larboard," is the pretty little village of
Jequitiba ; † here is a lakelet draining into the main channel

---

the part affected, and when sufficient inflammation was produced, they made a
great number of incisions with a stone or a
knife, a style of cupping more barbarous
even than the African.

* This corresponds with the south-east
wind that blows at sunrise on the lower
Mississippi.

† Or Gequitibá, a magnificent forest
tree (Couratari legalis, Mart. ; Pyxidaria
macrocarpa, Schott.). The colossus is often
180 feet high, and its spreading shade
would shelter a small caravan.

further down. Opposite we sighted the Fazenda do Jequitibá, a sugar estate belonging to Colonel, better known to the people as Capitão Domingos Diniz Couto. It was impossible to pass him, and the visit led to the expected result; a room was shown, breakfast was ordered, and with difficulty I extracted a promise for dismissal on the next day—after the early meal. One cannot sympathise with the Northron's estimate of Brazilian hospitality. Besides the fact that the guest has obligations as well as the host, I always find in the Fazenda sufficient intelligence, especially on local matters, to make up for lost time. At Jequitibá I was asked about the murder of the Baron von der Decken; at Jaguára my name was shown to me in the "Revue des Deux Mondes," a publication which, not having been salaried, persistently abuses the Brazil, and consoles the Brazilians by its gross ignorance of the subject which it maltreats.[*]

Colonel Domingos has a fine taste for good soil; people wonder that he still works at adding acres to acres, but the process has now become part of his existence. He has some forty square leagues of land, and travelling down-stream for three days we shall pass his estate. Besides this Fazenda he owns the Ponte Nova on the Barra de Jequitiba, about six miles distant, the Paiol with 100 head of negroes, the Bom Successo with upwards of 300, and the Laranjeiras. He will sell any or all of them, and from 1 pair to 500 pair of hands; he begged me to publish this sporting offer, which I accordingly do.

We spent a pleasant day, and were visited by M. Bruno von Sperling, a German engineer, married, and settled near Ouro Preto; he is now surveying the Mello estate. A small Portuguese landholder told me that he had heard of coal in the neighbourhood, but exact information was not to be obtained. As the Colonel was suffering from cataract he sent Sr. Antonio Justino de Oliveira, his kind and civil administrador, to show us his fine grounds. The place would be a Paradise with a steamer passing by it once a month. The gardens, sloping down to the stream, give a pretty view of the little Arraial on the opposite side, with its chapel, backed by pink-blue hills in the far distance. The many acres were planted with a few roses, cockscombs, and

---

[*] I refer especially to the "Review's" articles upon the Brazilo-Paraguayan war. Either it knows the truth, and conceals, or rather travesties it; or it ignores facts, and should seek information.

other flowers; the fruit trees were mangos, figs, Avocado pears (Abacutis, Persea gratissima), and large Cuyétes or gourd trees (Crescentia Cujetè); the rest was sugar* and bananas. There was a noble row of Jaboticabeiras (the well-known Eugenia cauliflora) with cupped or rounded summits, dense foliage, and smooth myrtaceous bark, everywhere studded along bole and bough with small yellow-white flower-tassels and young berries, little larger than a pin's head. In São Paulo the tree bears fruit only once a year in early summer, October and November: here it is continually productive. I had looked forward to the myrtle season as one does to the strawberries in England and the cherries in France; the tree, however, is not found on the Lower São Francisco—a great disappointment. Its fruit is one of the most delicate, in size a little larger than the biggest gooseberry, with a tough coriaceous skin like that of the Brazilian grape. The flavour is lost when the Jaboticaba is brought to market; the proper thing is to eat it off the trunk; a tree may be hired at São Paulo for 10$000 per annum, and " andar à Jaboticaba " † *en famille* is a very pleasant picnic.

August 21.—Having offered some parting advice to our host touching a visit to some ophthalmist at Rio de Janeiro, before couching became too late, we set out at 7 A.M., much condoled with. The river was beautiful; its grassy bluff seemed to bar the course, and the irregular lay of the heights told us what was coming. At 1.40 P.M. our troubles began, they were to last for five days. Our awning nearly came to grief at a sharp volta or bend ‡ a little below the Barra do Diamante. Twenty minutes afterwards we came to the Saco da Anta or d'Anta. The Saco or Reviravolta here corresponds with the " Horseshoe bend " of the North American rivers; the stream makes a sharp turn, at times running almost parallel with itself, and the land on its convexity becomes a quasi-peninsula with a narrow neck.

---

* I have rarely seen finer sugar-cane, certainly none in the Brazil. It is the Cayana quality, and the stalks when cut are 10 feet long by 1½ inch in diameter. Such is the effect of the Maçape soil.

† "To go to Jaboticaba."

‡ Usually pronounced in Minas and São Paulo "Vorta." The confusion of the *r* and *l* are as common as in China, and

I have heard a "University man" and a Provincial Deputy call the Estrella da Alva (the morning star) Estrera da Arva. As has been said, many of the "Indians" cannot articulate the *l*. Moreover, in Tupy it is popularly asserted that *f*, *r*, and *l* are wanting. This, however, certainly does not appear in the Lingoa Geral, which ignores *d*, *f*, *h*, *l*, and *z*.

Here a grassy bluff on the right bank fends off and loops the stream; the tall rock falls into the bed, throwing over a ridge which causes the water to break nearly right across; the material is lamellar shale, porous, and full of holes; it might easily be removed by a small steam-hammer. The current, as we can see, swings to the left, having a large sand-bank to the right, bends in the latter direction under a tall bank and disappears;* the course is from west to east. Chico Diniz went down in the tender carrying our damageable goods, and cut away some branches dangerous to the awning. We then floated along the bank to port under pole, and stern foremost, an occupation which cost us eight minutes, and the preparations for it half-an-hour.

After the " Tapir's bend" we at once came to the Funil—here, as in São Paulo, pronounced Funí. This name—entonnoir, or funnel—in land formations means a defile; on the Brazilian rivers it is usually applied to rapids breaking across the head of a long, straight reach that ends in a vanishing point. Here an eyot or sand-bank, covered with gravel and pebbles, bisected the upper entrance, and the course was from west to east. We bumped down the island's right side, hugging it to escape bad rocks on the river's bank to starboard; then we poled over, always a delicate operation, to the proper left side, avoided the " brush," and made fast. Bag and box were sent down the left branch in the tender, which ascertained that the rock-bed was now too much exposed for the raft. Perforce we again bumped across the stream below the heavy central break to the right bank, where canoes, plantains, and a hut denoted the Fazenda do Funil.†

At 5 P.M. we prepared our dormida (bivouac) on the Praia do Funil, a dry sand-bank to the left. The first step was building the hearth, and this did not take long, fuel being found everywhere. I observed that, contrary to the African habit, my people preferred the small fire, which was the practice with the " Indians," who, to warm their naked bodies even in the wigwam, and to defend themselves against wild beasts, used to make their

---

* The total windings are south-south-east, south-east, east, north-east, north, and at last the general direction, north-west. M. Liais, who descended the Rio das Velhas, where the river must have been somewhat fuller, shows nine detached rocks, five on the right and four on the left. This obstacle would severely try the engines of a tug going up-stream.

† M. Liais shows a clear way between the sand-bank on the left and three lumps of rock dotted along the right bed.

women keep wood burning all night.* Carne Seca and fish, when any is caught, are skewered and planted by the blaze. The next operation is to make Angú, that almost universal dish; porridge, hasty-pudding, stirabout, polenta, mush, and the ugáli of Unyamwezi. Fubá or maize meal is thrown little by little into boiling water and moved with a stick, or it will be lumpy : it should be eaten as soon as the whole is wetted.† The favourite national dish, feijão floating in lard,‡ is kept upon the fire all night so as to be ready for the dawn-breakfast. The men pass the evening chatting and smoking till ready for sleep, when they spread their mats and hide well in the smoke-drift, and no wonder that they so often suffer from Cadeira or lumbago.

The air was delightfully pure, and I sat for some time listening to the voice of an old friend. "Pst"—the blow—"Whip-poor-Will." This Caprimulgus begins to be vocal with the crepuscule, somewhat like certain owls, especially the Strix Aluco of Europe,§ and his loud and remarkable cry will extend, with certain intervals, all down the Rio de São Francisco. His man-

---

* Like Africans, they used to light fires by the side of newly made graves, not to frighten away evil spirits or the devil (according to travellers), but for the personal comfort of the defunct.

† Another form is called Mingáu (not Mingant, as Prince Max. i. 116) ; it is made of manioc, farinha in water, and sometimes with a little cinnamon. A third preparation is termed Cariman, derived from Caric to run, and Mani Manioc, "running manioc." In old authors we find "mingan" or "Ionker," potage or thick "bouillie," made with salt, pepper, and manioc-meal. Yves d'Evreux mentions a Norman interpreter named David Mingan. The Pirão is farinha mixed with hot water, or better still, with broth of fish or fowl ; it is a favourite accompaniment with fish.

‡ Popular writers inform us that fatty and carbon-producing substances, so necessary to the inhabitants of the Arctic regions, lose their use as we approach the Equator, and are supplanted by fruits, rice, and similar light food. This is by no means the case. The Italian consumes a quantity of oil which would make an Englishman sick. The Hindu swallows at a meal nearly a tumbler full of Ghi or melted butter, and few, if any, Northrons can eat his greasy sweetmeats with impunity. The naked negro, panting near the Line, saturates his food with palm oil, and even at Bahia in

the Brazil, where the "coloured cuss from Africay" is comfortably clothed, where he can buy meat in abundance and obtain any quantity of ardent spirits, the oily and spicy caruru and vatapá (palm-oil chop, &c.) are eaten by all classes. Near the Equator, the damp heat has much the same effect upon diet as the cold of high latitudes ; strong diffusible stimulants, port, sherry, and stout are better than thin claret and French wines, and meat is much more digestible than vegetables. Practice is worth all the theories or rather the hypotheses of pseudo-theorists, and the habit of one writer copying from the other without an attempt at independent inquiry traditionalises a variety of error.

§ Prince Max. mentions sundry other Engoulevens. There is a larger species than the common Whip-poor-will, which Marcgraf calls Ibiyaou, and he (i. 267) Bacouraou. Another (described i. 370) is the Caprimulgus æthereus, which soars high in the air like a bird of prey. A third is the Mandalua (C. grandis), white mixed with brown : and its sharp whistlings fill the forest. The German ornithologist described for the first time (iii. 91) the Curyangú, a day-bird which flies during the light, and mixes with horses and black cattle in the pasture ; and the Caprimulgus leucopterus (iii. 178), whose beak is like that of C. grandis.

ners, as far as we observed, resemble those of the N. American species, and we often saw by day a pair nestling in the sand. The Portuguese call the bird " John cut Wood," and it is a curious commentary upon the "ding-dong" theory that one race hears " Pst—Whip-poor-Will " and the other "João Corta Páu." By mentally repeating the words I could produce either sound, but the Latin version seems preferable.

August 22.—We were aroused at an early hour by the Coryangú or Curyangú (not Criango) bird (Caprimulgus diurnus, the Na-cunda of Azara), which seemed to say, "How well ye woke!" This goat-sucker has a musk-coloured coat, with white spots and bars on the wings. I often disturbed a quiet pair nestling by day in the shade of rock crevices; the flight was that of our night-jars, and it was always short and low. We set out at 6 A.M. somewhat prematurely, and the "smokes" obscuring the river-surface, nearly caused an accident; a tree on the left bank, which could have been cut in ten minutes, drove us amongst the stones of a "rush."

At 8 A.M. we shot the Saco do Barreiro (de Gado)* the Bend of the Salt-Lick (of Cattle). These places abound on the Rios das Velhas and de São Francisco; the banks of red, grey, yellow, or dull brown clay are burrowed with lines of holes by the tongues of beasts and the beaks of birds, which usually visit them in early morning. As in the United States, the lick is often saline only by name, and the practice must be compared with the earth-eating disease of Africans in the New World. In parts the breeders mix salt with the clay and throw it upon the stream-side to produce an artificial glaisière, but as a rule it is not considered sufficient to lay down salt, as the lick requires a peculiar sort

---

* M. Liais calls it "Cachoeira do Paó Seco." Here the stream runs from south to north, and is faced by three low blue hills. We easily descended in four minutes, crossing from right to left, and thus avoiding the breaks on both sides.

According to Azara (i. 55) the "Indians," who ignored the use of pure salt, supplied it by the saline "barro," which they devoured in abundance. Prince Max. remarks (ii. 257) : " La glaise du Brésil n'a pas le goût salin, et je n'ai rencontré chez les habitans indigènes de ce pays aucun mets salé." A curious commentary upon the supposed necessity of the condiment. It must, however, be observed that the Tupys were eminently carnivorous, and thus they found their salt in their meat. This of course would not be the case with "vegetarians." Earth eating is not unknown to the Brazilians. I have shown that in Africa, as amongst the Ottomac Indians, whom Humboldt describes as intrepid geophagi, it is eaten in large quantities without doing injury. I cannot, therefore, with St. Hil., hold that the Ottomacs are the sole exception to the fatality of geophagism. He declares that the Brazilians prefer the clay of the termitaria; this is also the case in Unyamwezi, where it is called " sweet earth."

E 2

of clay. After two unimportant features,* we drew near the Maquiné Rapids, which have a very bad name. No one could explain the word; our pilot "guessed" that it was that of a huge "kraken" like the "worms" of "strange dragons of vast magnitude" which haunted England in the "good old" days. It is called the "Maquiné Pequena," to distinguish it from a creek lower down the river.

The first symptom was a fragmentary ledge on both banks, dark friable limestone tilted up at an angle of 40°; this is called the Cabeceiras do Maquiné. We made fast to the left bank near a fine cotton-field that runs up a gently sloping hill. Here we could look down the straight reach, some 400 yards long; about 600 feet of smooth water separate the Upper from the Lower Rapids, which are considered to be the worse. They are formed by the bluff end of a short range, whose general course is to the north-east, but which bends to the north-north-east, throwing the stream from its main direction to north-east 25°. The limestone base forms an oblique ridge from north-west to south-east, where the water breaks right across, and even at this season only one rock appeared well above the surface. The friable limestone, split and stratified, is easily broken with the hand; before approaching the narrow wall there is a fundão or hollow at least ten feet deep, and thus nature would keep open the narrowest road.

After reconnoitering, we embarked with the "trem" or luggage in the tender, which now drew 4—5 inches. Apparently there was a fair way on the right, but it is not shown in the Plan, and the pilots always prefer the left. We went to port of a central rock-knob, and, safely crossing the broken water, we made for the half-way house, a sand-bank on the starboard side fronting the smooth that divides the Rapids. Hence we watched the "great ole barque" take her lumbering way; after two or three chancy swings and half broachings-to she obeyed the pole, and came down gallantly.†

Having rested till noon, we prepared to attack the Maquiné

---

* The Cachoeiras das duas Barras and das Cabras.

† M. Liais's plan shows a clear way in the midstream, and two main obstructions. The upper break is of two blocks of stones, with the thalweg in the centre. Then after the smooth, come three detached rock-piers on the right, and opposite them a corresponding formation, but smaller and more broken. In this section there are two stones, which must be removed from the thalweg.

Abaixo or Lower Rapids. Fortunately, I left my books on board the "Eliza." We went to the left, grounded on the rock ridge, which slants like the upper formation, and were whirled round against the trees; I could save only my journals, somewhat like, to institute a modest and uninvidious comparison, Cæsar, Camoens and Mad. André (de la Mediocrité). Reaching the left bank we viewed from the feathery shade of a charming Jatobá the doings of the ark. A second portage had been made, each occupying some two hours, and, thus relieved, she slid safely down in her usual playful elephantine way. But she was assisted by certain moradores of the neighbouring hamlet of Maquiné Pequeno, José Luiz de Oliveira, who, accompanied by his two cousins, stripped, and lent a hand in lifting the "Eliza" at a critical moment. They would take no reward, but a glass of our fine cohobated Lavado and a few cigars seemed to content them.

After shooting this "Long Sault," the line, "barring" the easily avoided scatter of sunken rocks (pedras mortas), should have been safe, but it was not so. My men had worked well, but they had drunk still better. They dashed upon a limestone rib near the left bank. They then bumped heavily and unnecessarily in two places; the tender was almost lost, and I felt by no means comfortable as we approached the Cachoeira da Onça. Probably from these adventitious circumstances, the Ounce Rapids have left with me a more unpleasant impression than all the other combined difficulties of the Rio das Velhas. *

After about two miles we turned to south-south-east and entered a gorge already gloomy at 4 P.M. " 'Sta gritando," it is crying! said the men, giving anxious ear to the roar. Advancing swiftly for a few yards we saw the Cachoeira, breaking across with dangerous projecting rocks. We poled down the left side, and by opening too much to starboard we struck heavily upon the stones, and the water spouted up between the planks of the platform. Having escaped this shock, we crossed the stream to a smooth on the right and prospected it. The result was a stern-

---

I went down entirely by the left; the stream, however, evidently runs at the middle of the bed, and this, when opened, will give a clear passage. In the Upper Maquiné the detached rock or rocks must be knocked away, and in the lower the wall must be pierced. It would, I think, be easy here to work a large hammer, not by steam but by water power.

* Yet the Plan shows only a stone pier, and two hard heads on the right, faced by five scattered rocks which may easily be removed. The danger is from the detached stone upon which the current breaks immediately below the upper "gate."

presentation, and we slipped down in eight minutes, narrowly shaving to port a dark laminated stone, dipping 50°, which was angrily throwing off the waters, and upon which the current broke.

The crew was tired and out of condition; I resolved to avoid, by an early halt, the risk of a bad accident. We found on the left, opposite a clump of five huts called Jatobá, a few yards of sand under a precipitous bank of yellow clay; the men termed the place Praia da Cachoeira da Onça. The day had been wearying work, we had nearly boxed the compass.

An angry mass of purple-brown cloud hung in the western sky; my men, hoping that the stream would be swollen, prayed for rain, which at this season sometimes lasts three or four days. At night the view was suggestive. On our right was the ominous growl and the lurid flashing of the Ounce Rapid: from the left or down-stream came the rattling babble of the Corôa braba, the "Fierce Sandbar," whilst the sky was red with the last gleam of day, and flashed with the frequent prairie-fire.* In front flowed the stream, dark steely blue; the further waters were scolloped with the black reflections of the trees, which rose high where the Little Bear should have been.

And this desert stream will presently become a highway of nations, an artery supplying the life-blood of commerce to the world. The sand-bank upon which we lay may be the landing-place of some wealthy town. The "Ounce Rapid" and the "Fierce Sandbar" will be silenced for ever. And the busy hum of man will deaden the only sounds which now fall upon our ears, the baying of the Guára wolf, † and the tiny bark of the little brown bush rabbit.

---

* St. Hil. (III. i. 202) declares that in Western Minas cultivators fire the grass only during the moon's wane (dans son decours).

† The word is G-u-ára-ă, a great eater, very voracious. "G" is the relative, "u" is to eat, and -ára (in Hindostani "wálá") is the verbal desinence. Guara (an eater) is intensified by the post positive ă. The name is of the animal here called wild dog or Brazilian wolf (lobo), the old Ursus carnivorus being very well calculated to mislead; the Aguara-guazu of Azara, and the Canis mexicanus of Cuvier. I have seen closely but a single specimen, which much resembled the French wolf, except that the coat was redder. This carnivor especially favours the lands where forest and prairie meet or mix. I have never heard of its attacking man; but, on the other hand, there are no snows to make it ravenous.

There is also a swamp-bird called Guará or Gara (an Ibis), a word derived from ig, water, and ará, a parrot or parroquet: "water-parrot," from its fine pink-red colour.

As a desinence, guará means lord or master; e.g. pyguara, a guide, literally lord of the path or foot (py).

We are taught to dwell far too much upon what has been; upon the ἀρχή, the proem, the first canto of the grand Epos of Humanity; we are too indifferent about what is to be, in the days when the whole poem shall be unfolded. Rightly understood, there is nothing more interesting than travel in these New Worlds. They are emphatically the Lands of Promise, the " expression of the Infinite," and the scenes where the dead Past shall be buried in the presence of that nobler state to which we must now look in the far Future.

CHAPTER V.

THE CORÔA, OR SAND-BAR.—PREPARATIONS TO VISIT DIAMANTINA.—THE
PLEASURES OF SOLITUDE.

> The day is placid in its going
> To a ling'ring sweetness bound,
> Like a river in its flowing—
> > *Wordsworth.*

BEFORE setting out it will be necessary to describe the " Corôa "*
feature, of which a neat specimen awaits us.

The " crown " is the " sand-bar " of North American rivers,
an island in the stream, but very unlike our " holm," inch or
eyot. It is mostly, as we have observed of the Cachoeira,
at the mouth of some little stream where the influx of fresh water
slackens the flow, and it is often built upon detached stones or
upon rock-ridges. The current swings to either side, leaving in the
centre a bald convexity like the shaven pole of a Corôado Indian,
and of all sizes, from yards to acres. The water is shallow above
it, deep below, on both flanks, and in the baylets and concavities
where fish live to plunge and cattle to bask. When the formation
is very low the drift wood floats over it; otherwise, tree-trunks
are mostly found at the sides, and snags must be feared, especially about the head or up-stream. Often the Corôa is double, or
even treble; it is always elongated down stream by the current;
never circular as in lake formations, and the edges are either flat
with the water, or stand up in dwarf precipices.

The surface is pebbly and gravelly—of all sizes, from an inch

---

* Pronounced C'rôa ; it is the Latin
Corona, certainly not to be written with
Prince Max. "Corroa." The feature is
opposed to Praia, a "sandbank," attached
to the side. The Tupys called the former
Iby cui' oçu, "Corôa de Areia," opposed
to the latter Iby cui' praia ; with them Cua
was the river plain (varzea) where liable to
inundation, and "Coára," literally a hole,
was a little bay (enseada) or river port,
where canoes can be made fast.

to a foot; these scatters come from the banks, and are brought down by the floods. The material is mostly of quartz in its protean forms, jasper, touchstone, pingas d'agua (Quartzum nobile), crystallized, stratified, and almost always red or rusty with iron. There is also an abundance of sandstone, limestone, and chlorite, which may or may not contain gold,* together with bits of "cánga" or ferruginous conglomerate, the gift of the upper country. In places the sand is very loose, admitting the foot to the ankle. In the hollows where rain sinks there are large flakes of mud three to four inches deep, and wherever the waters extend, the pebbles in the dry season show a coat of indurated slime, whose base may be either cascalho (rolled gravel), soft sand, or hard mud. These Corôas pure and simple are haunted by gulls and terns, hawks and kingfishers, ducks and herons, plovers, sandpipers, and other birds which will be mentioned.

A scattered vegetation of stunted trees and verdigris-coloured grasses and shrubs, forms, generally beginning with the end down stream, and thus the sand-bar becomes wooded.

The typical growth is the Araça guava, with comparatively small thin leaves, and an exaggerated strength of wood, self-adapted to its locality. Another common shrub is the Ariuda, also called "Alecrim da Corôa;" the leafage is smaller than that of the Psidium, the stem and branches are as stout and tough, and it is bent down stream by the force of the inundations; this plant also appears upon the sands. In places the water-sides are edged with a sedgy grass, whose blades average a finger and a half in breadth. It is used for stuffing pack-saddles. Upon the Rio das Velhas we shall not find the osier-like and broom-like growths which we first observed in the Rio de São Francisco below Remanso.

The sand-bar first forms under water, when it is called Areão, or "big sand;" it rises by degrees, and where the annual floods are not too violent it presently becomes an "Ilhota" or islet; a "Carapùça" if cap-shaped; and if large, an "Ilha" or island. Many of them, like the Mississippi tow-head, are partly timbered, the wooded portion up-stream, the sandy below, or *vice versâ.* The feature is then permanent, and the figs and mimosas bind the

---

* From some, for instance the Corôa da gallinha, gold has been taken; the people dig deep into the sand.

soil like the "cuttun woods" of the United States. Passing the Rio Pardo we shall see another complication, where blocks of blue limestone, thinly grown with sturdy shrubbery, cumber the surface, and lower down on the São Francisco, a combination of tall rock, timber, and sandflat.

M. Liais advises these obstructions to be removed by "draguage." With diffidence I differ from him; but would not the obstacles upon which they are formed themselves require dragging? A single rock will, like a stick in the sandy desert, produce an accumulation of matter; the same causes continue to be in operation, and doubtless every flood would renew the effects.

August 23, 1867.—The warm morning tempted us again to set out at 6.30 A.M., half-an-hour too early. The course was from east to west, and we found our babbling friend the Corôa braba a complicated affair of stone and sand-bar. On the left was a rock, then gravel, then another rock; to starboard rose the sand-bar, upon whose dexter side we lost no time in grounding heavily. We poled off with difficulty, and I did not like the look of things. Luckily we met a ragged youth, punting a dug-out towards the village, and, for a consideration, Herculano Teixeira de Queiroz was persuaded to accompany us. He landed, and presently returned a smart young waterman, in white shirt and pants, with straw hat, and the inevitable bone-handled sheath-knife strapped round his waist.

After about three-quarters of an hour the "Eliza's" head was turned to the north-east, thus describing a long horse-shoe with a very narrow heel. In places the river is to the land route, 3 : 1, not an unusual ratio; in others, 5 : 1. Before us rose the tall blue broken wall of the Serra do Baldim, the "Baldoino" of M. Liais, which bore north-east of Jaguára; it is said to contain deposits of alum, like those which we found on the São Francisco. Half-an-hour afterwards we passed the Cachoeira dos Paulistas, whose ledge does not run right across; the Plan makes it part of the "banks of Cafundó." * It became evident that the rapids were now waxing less laborious and far more dangerous, with deeper water and

---

* This is apparently Ca fundão—here (is) a deep place—fundão—where the pole does not reach. Near the right bank there is a sand-bar; on the left, and a little up-stream, is a hard gravelly sand-bank flanked by two rock-piers, one above, the other below it.

narrow channels, likely to jam the raft. We hugged the right point and then made the mid-stream, steering for the apex of a smooth equilateral triangle strongly defined by borders of foam or ripple broken against stocks or stones—here the usual guide to the clear way.

Then came a complicated obstacle—a bold bluff of ferruginous stone to port deflected the steamer to starboard, almost from north-east to south-east. Avoiding two sand-bars and two rock ledges, we went to the right, and nearly rounded the Corôa— going with the sun—from south-east viâ east and north-east to north-west. A couple of rock-piers in our way made us cross to the left, and bending to the north we found a break formed by detached lumps of limestone. This "Cachoeira da Barra do Engenho de Manuel da Paixão" was an affair of eight minutes; the deviations are risky, and, before a steamer can ply, the bed must be cleared of rocks, after which the current will dispose of the sand and gravel.

After winding some four miles, where a voice could be heard across the neck of the loop, we saw ahead fine cotton-fields in full bloom, and a tier-like succession of gently swelling hills in far perspective. A field of plantains on the left bank, and four huts, of which at least one was a Venda, told us that, contrary to prophecy, we had reached Sta Anna de Trahiras. This place is on the highway of the Tropas, travelling between the Provincial Capital and Diamantina;* it became a parish in 1859, and it is now under the vast municipality of Curvello. In 1864 the population was computed at 4298. I was told 12,000, which, as usual, doubles the probable number.

Here were two ferries, one with a chain and belonging to a kind of company, the other with a civilised wire-rope, procured from Morro Velho; the latter was the property of Sr. João Gonçalvez Moreira, to whom I had an introductory letter. He met us on the bank, and showed me a tree marked by the water ten to twelve years ago, 40 feet above the present stream level. On this occasion the floods swept the riverine valley to the foot of the Campo hills, and people were taken by canoes out of their thatched roofs. In average years the inundation rises for a few

---

* The distances are by land, 25 miles from Diamantina, 21 from Sabará, 24 from Morro Velho, 9½ leagues (24 by water) from Casa Branca, 6 leagues (by the river 20) from Jequitibá, and 4½ to our present destination, "Bom Successo."

days to the ferry chain. If foreign railway engineers in the Brazil, which is everywhere subject more or less to these exceptional deluges, recurring with a somewhat vague periodicity, had taken the advice of the natives, and had built their bridges and drains accordingly, they would have saved themselves much trouble and their employers more expense.

We walked to the village on the right bank; the ground was somewhat stony, pebbly and poor. It was rich in the low shrub with a leaf like the Mimosa, known to the Tupys as Tarcroqui, to the Brazilians as Fidegoso (Cassia occidentalis, sericea, etc). The "stinkard's" root is a powerful drastic, homœopaths infuse it in spirits of wine and employ it as quinine; the beans are sometimes made into coffee, as maize is in the United States. The village main square on the highest ground has two chapels— Santa Anna and the Rosario, a few young palms and some Vendas, especially the double store of Sr. Tóttó (i. e., Antonhico or Antonio) Rodrigues Lima, and the apothecary's shop of the Professor of First Letters, who, though his father was named Custodio Amancio, has preferred to term himself "Emmanuel Confucius of Zoroaster."

The houses may number 200 or 300 within church-bell sound; all are one-storied, and mostly of the meanest. The only thing that seems to flourish is the goat; the "Cabrito" is here, unusually in Minas and São Paulo, favourite food. Our kind guide led us about to the several Prud'hommes, who invited us to pass the day. Sr. Antonio Gomez de Oliveira, a relation of Colonel Domingos, asked us to breakfast, and gave us some good English stout. His house was the neatest in the place, a long building fronted by a bit of shrubbery; of course it contained a shop.

Our temporary pilot had done work enough, and we sent to invite two others, but without the least chance of an answer for three days. Chico Diniz politely intimated his utter despair, and we returned to the ferry. Sr. Moreira enticed us to his home on the other side, and whilst he despatched a peremptory message, introduced us to his wife, and showed us the garden, in whose oranges and cabbages he took no little pride; here the soil is an improvement upon that where the village lies. He spoke warmly, evidently not believing a word, about the coming Steam Navigation; to him the Cachoeiras were insuperable, and when we spoke of cutting away the obstacles we talked manifest Greek.

In 1853 a Government engineer had spent six months at the rapids above Trahiras; the people remembered his fusées and mule-loads of tin cylinders for mine-charges; all agreed, however, that he had not removed a single difficulty, and most men opined that he had left the place worse than when he found it. At last, worn out by delay, we bade a friendly *au revoir* to our host, and we quitted Trahiras, satisfied that if the opening of the Rio das Velhas be abandoned to men who receive public pay, and to those who live upon passing mule-troops, the splendid stream will remain long closed.

We set out shortly after noon, and the day was a succession of sand-bars and rapids, with rocks on the right, on the left, and in the central thalweg.* The first serious feature was the Ribeirão da Onça, a rapid on the left of a triple " Corôa; " it is so called from a little green-set rivulet up which canoes go for several miles. Presently we came to a place where four men were loitering; we offered to pay for pilotage, but they refused. They did not object, however, to assist us in cordelling down the Cachoeira da Barra do Ribeirão dos Geraes, alias Cachoeira dos Geraes (do Lamego).† Whilst they held the tow-rope we hugged the left bank, a drop of loose sand; the broken ledges of horizontally stratified dark stone project from the right shore above the rivulet-mouth, and deflect the stream to the left, thus doing engineer's work.‡ Above the rapids much gold has been dug.

A couple of hours carried us down to the Cachoeira do Lagedo,§ a small rapid formed by a porpoise nose of wooded bluff on the right; from its summit, they say, the Piedade of Sabará may be sighted. After sundry unimportant features,‖ and passing the

---

* It began with two bluffs of rock, flooring the hill to our north. At the Córrego da Tabaquinha (the little Tabóca, Taquara, or bamboo), a rock-outcrop from the left bank intrudes upon and deepens the stream.

† The first name would mean the rapids at the embouchure of the stream of the General Lands, an influent from the right. Geraes are mostly lands out of the reach of the river, either Pasto, Campo or Mato, and bearing general produce, cotton, tobacco, cereals, as well as breeding cattle. St. Hil. (I. ii. 99) confines the use of the word to pastures, and says that "Matos" must be expressed when forests are meant. I did not find this difference, nor did the people ever employ " As Geraes " to mean " As Minas Geraes."

‡ In order to drive the stream to the left, M. Liais proposes a "tunage avec enrochment" on the right with a passage through it for the streamlet; a gigantic work.

§ In the Plan rocks are placed on the right bank; in the description (p. 8) on the left.

‖ The Fazenda do Jardim belongs to the

Corôa do Jardim, almost an islet, and to us a new spectacle, we anchored at the usual hour, shortly before 5·30 P.M., at the Praia da Ponte.* Below was a Corôa of the same name, which made music for us all night. Behind the hole-riddled bank were a few hovels with patches of sugar-cane growing poorly in rough, scrubby soil, good only for ticks. A few boors came up and stared at the menagerie; they would neither eat with us nor take anything but fire for their cigarettes, and we were as formal as they were. I had been warned to treat them with "agrado e gravidade"—civility and gravity—otherwise that they may become quarrelsome or lose respect. They spoke of a pilot, and we sent for him; but, as usual, he was ill. Two women bringing fowls for sale, squatted near us with feet wide apart like Africans, and chuckled their remarks to each other; nothing could be less like certain Buffalo girls. At sunset all disappeared, touching their hats in the deepest and gloomiest silence.

I felt saddened by this contact with my kind. It was the Present in its baldest, most prosaic form; the bright kaleidoscope of cultivated life here becomes the dullest affair of unvarying shape and changeless colour. There is no poverty, much less want; nor is there competency, much less wealth. There is no purpose; no progress, where progress might so easily be; no collision of opinion amongst a people who are yet abundant in intelligence. Existence is, in fact, a sort of Nihil Album, of which the black variety is Death. I prefer real, hearty barbarism to such torpid semi-civilization.

August 24.—The cold night made the fog hang long over the water, and we did not set out till 7 A.M. Two Corôas, neither of them in the plan, gave us some trouble. Thence the river entered a gorge, each side alternately being high ground, —wooded above and stony below. Before the hour was finished we were at the Cachoeira das Viólas; † but, instead of going down

---

widow of the Capitão Herculano; a streamlet comes in from the right bank, and below it there are two sand-bars: the first with a clear way to starboard, the second on the other side. Then came the Saco de Pindahyba, where the river loops to the south-west, and the Ribeirão de Luiz Pereira on the left.

* A Ponte is the name of a córrego

which does not appear in the Plan.

† Or, da Vióla; probably some one lost his fiddle here. The stream runs north to south; and the obstructions are two rock-walls from the right; then one from the left, and lastly detached rocks on the right. I include this feature amongst the bad ones, as it has done much harm in its day.

the mid-stream, we took the left to avoid driftwood, and we bumped like the bucking of a mule. A charming reach, with beautiful woods, appeared ahead, and the material of the latest clearings strewed the land ; here the direction of the limestone (?) is north-east, and the dip 12°—15°. After sundry unimportant features* we left on the west a fine bit of land, the Fazenda do Boi, belonging to Sr. Delfino dos Santos Ferreira. The people crowded down the yellow bank to stare and to frighten us about the Cachoeira Grande, a place of which we had already heard ugly accounts. The dialogue was in this style :—

" Do you know the Rapids ? " we inquired.

" We know them ! "

" Will you pilot us ? "

" We will not pilot you ! "

" For money ? "

" Not for money ! "

" And why ? "

" Why ? Because we are afraid of them ! "

This was spoken as the juniors ran along the bank like ostriches or the natives of Ugogo ; they are beginning to lose the use of their un-Latin " yes " and " no," and to answer by re-echoing half your question—the true old Portuguese style.

Shortly before noon we landed on the right side and examined a thick layer of Cánga or pudding-stone, probably auriferous, and possibly diamantine. The almonds were dark, rusty quartz, in the usual iron clay paste, and from this point downwards we shall see large deposits of it. Further on, sandstone lay facing the south-east in nearly horizontal courses, ready for quarrying. The men crossed to an orchard on the left bank, and brought back baskets of fruit and sugar-cane, which they tore and chewed like Botocudos. They sounded the horn, but as no one came they put off. Alas ! they had robbed the church ; the ground belonged to Padre Leonil, and worst of all, his oranges

---

* As the general course of the river is from north to south, I shall call the right bank east, even when it is not, and so forth. The obstacles here are a rock (os Pandeiros) in the centre, which causes a break, and allows passage on the left. Then to star-board enters the Ribeirão de São Pedro amongst rocks and sand-banks. Followed the easy Cachoeira da Agua Doce near the Fazenda of the Sr. Nicolao de Almeida Barbosa. We swang with the stream to the left, avoiding the thick shrubbery clothing the bank, and finding a clear way between it and the three detached rocks of the Plan. Then a larger sand-bar than usual led to broken water, and sent us down by the right.

were not worth eating.* This, however, is here a venial offence. You may freely take from a plantation—a Roça cannot be robbed, is the saying—but you must not touch, for instance, a little plot of onions or other vegetables upon which the proprietor bestows pains, such as entering it at dawn. For the former are as feræ naturæ; the latter is a park or a poultry-yard.

Luckily for us—the Cachoeira Grande was no joke—we found at the Saco Grande, on the right bank, a small crowd preparing for a "Sámba," or to keep "Saint Saturday," and perhaps "Saint Monday" with dance and drink. The men carried guns in hand, and pistols and daggers under their open jackets—evidences that they did not intend to be recruited. The women were in full dress —brilliant as rainbows—with blood-red flowers in the glossy crows-wing hair; but of the dozen not one was fairly white. After a few words with Chico Diniz, the bow pole was taken by a certain "Felicissimo Soares de Fonseca," the stern was occupied by a "yaller"-skinned elder with curly white beard, "Manuel Alves Pinto," and his son Joaquim. This looked like business. The new comers were men of few words; they saluted us civilly, and they pushed off.

The beginning of the end was the little Rapid of the Saco Grande or "Big Bend," where the river bed turning sharply from south-east to north-west makes parallel reaches. To avoid the rock-pier on the left we floated stern foremost down along the right bank, here a mass of ferruginous sandstone, striking to the south-east and nearly plane (3°—4°). After bringing the boat round, we left, on the right, two sand-bars and as many de-tached rocks; upon the opposite side also was a mass of blue stone,† which must not be approached. This elbow is too sharp for a tug-boat, and the obstructions absolutely require removal.

---

* They were very like the Laranja da Terra, the "indigenous orange," alias the imported orange run wild. The taste is a fade "mawkish" sweetness ending in an unpleasant bitter; I have, however, seen the juice beneficially used in ptisane for one of the severe catarrhs (constipações or defluxos) which abound in Minas and São Paulo. I presume that, like the Laranja Secca or juiceless orange, the "bitter orange" is the effect of a high country, rough soil, and other local conditions. St.

Hil. (i. 280) quotes Pizarro, who enumerates three subvarieties, one sweet, another agro-dolce, and a third very sour, and believes that the Laranja da terra is a return to the primitive type of the sweet fruit. "Personne," he says, "n'aurait probablement songé à nommer un arbre qu'on aurait fait venir d'Europe, oranger indigène." This ver-bal argument is worthless; many productions imported in ancient times are now called by the Brazilians "da terra."

† In parts of the Rio das Velhas it is

Presently we turned to the east-south-east, and faced the dreaded Cachoeira Grande, which is formed by another sharp bend in the bed, winding to the north-east. The obstacles are six several flat ledge-like projections of dark stone on the right bank, and four on the left, mostly awash, and cunning is required to spiral down between them. We began by passing the port of No. 1, then we made straight for No. 2, to the left; here, by pushing furiously up-stream—had a stick broken we should have been nowhere—the "Eliza" was forced over to the right, was swung round by main force of arm, and was allowed to descend, well in hand, till within a few feet of No. 4, which rises right in front. Finally leaving this wrecker to starboard, we hit the usual triangle-head, with plenty of water breaking off both arms. A single bump upon a sunken boulder (pedra morta) was the only event. The descent occupied sixteen minutes. The Great Rapid is more dangerous, but not so serious an impediment to navigation as the "Maquiné." Any form of ram would easily knock off the heads of the rock-piers, and open a way in mid-stream— all that is wanted.

After many congratulations our friends made a show of taking leave; all had some important business, which proved on inquiry to mean "doing compliments." As the dangers were not over, the keg of Restilo was produced, it was tasted and pronounced "muito brabo" (very hot in the mouth); the Ma-a-jor (myself) became so irresistible that all would accompany me to the Rio de São Francisco or—elsewhere, anywhere. The poles were twirled and wielded with a will. We left to port broken water and an ugly stone, a hogsback, known as the Capivára (Hydrochærus), and then we crossed to scrape acquaintance with a sunken mass in front. This place is called the Rapadura; it is a mere "correnteza," but the many "dead stones" would render it dangerous for a steamer.

The end was the Cachoeira das Gallinhas,* to which we presently came. We gave a wide berth to a rocky wall on the right bank, and stuck to the left side of the Corôa, till we had reached its tail down stream. Here is a narrow gate formed by two rock

impossible, without testing the rock, to determine whether it be sand, clay, or lime.

* M. Liais treats it as a matter of little moment; we thought it quite the reverse, and evidently so did the pilots.

piers, projecting from the shores, and in such places " cordelling" is always advisable. The men sprang into the water with loud cries of "Hé Rapasiáda,"* and pulled at the hawser till the current had put us in proper position; they then cast off and sprang on board before we could make much way. We left to starboard two blocks and one sunken rock of fine blue limestone, brushing them as we passed. The "Rapid of the Hens" occupied us nine minutes, chiefly spent in shouting. The right channel may easily be cleaned: a mass of drift wood is all that obstructs the left, and knocking away the rock walls would soon start the "Corôa."

A second dram of the "wild stuff," and all our friends in need ruled. They blessed us fervently but stammeringly: they prayed for us somewhat the wrong way, and they unintelligibly invoked for us the protection of the Virgin and all the saints. They landed with abundant tripping and stumbling, carrying 1$000 and a bottle of the much prized restilo. I had every reason to be grateful to them, for they had most civilly saved me an immense amount of trouble; but, shortly afterwards, reports of certain "little deaths" in which they had been actively concerned, showed that they were not exactly lambs, except after the fashion of Nottingham.

By this time my men were "pretty well dead beat." I anchored a little above the Barra da Cerquinha,† opposite the Córrego do Paiol. The ground was sandy and unusually clean, whilst the valley soil, apparently arenaceous only, produced cotton in quantities. To-day the river, except where disturbed by rapids, has been a vista of beautiful amenity. Mr. Davidson was in ecstacies, and began to talk of the Yazoo and to sing something about "Down the O-hi′-o!" The grandly moving stream, hardly broad enough to suffer from winds, is not too narrow for vessels to thread their way up, while steamers could easily turn in the fine reaches. At nightfall the sugar-wheel of the "Paiol" ‡ creaked and sang in curious contrast with the accompaniment of nature; the distant hum and the nearer

---

* "Now, my lads!"

† The "Embouchure of the small hedge or paling (stream);" it is not named in the Plan.

‡ Properly a "bread-room," but often applied to places where coffee, sugar, and even rum are stored. This Paiol has been mentioned as one of the estates belonging to Colonel Domingos. I afterwards visited it; the soil is fine, the water abundant, and there is a large house, with the usual chapel and sugar mill.

cries of birds and beasts, frogs and toads,* and a noisy little rapid fretting and snorting down stream.

We were now approaching a place of rest, and I contemplated with satisfaction a fortnight of land-march, even on mules. Rapids resemble in one point earthquakes—the more you see of them the less you like them, and the stranger at first is disposed to look contemptuously upon the prudence and precaution of the " old soldier." Shortly after dawn we went down the small but ugly Cachoeira da Cerquinha, between a bad rock on the right and a stone ridge on the left, to which we inclined. It was followed by another little break.

After two hours' work we turned from the main stream up the Córrego do Bom Successo. Here we made fast the "Ajôjo," and the crew agreed to keep guard in it at night. As a rule the riverines avoid sleeping in these places between the days of the new year and of the St. John. The waters bring down much earthy, decomposed matter : it is easy to smell the difference of the branches and of the main line, and especially during the Vasantes, or annual retreat of the waters; they dread the dangerous marsh fevers, remittent and intermittent, called the Malétas. At Jaguára I had been warned that the Rio das Velhas below Bom Successo required certain precautions, such as to eat much pepper, to avoid the cold night damp after the day heats, not to wash or bathe when perspiring, and not to drink coffee in the open air.† I could not, however, be troubled with so much " coddling," and we both found the climate perfectly healthy.

After making the necessary arrangements we walked up to the Manor House; the air was crisp and dry, and the soil gravelly but rich. The stunted Cashew everywhere grew wild, and there was an abundance of the Jaboticabeira myrtle, justly called cauliflora, the aspect of the dark leafage being exactly that of an enormous cauliflower. The other fruits were the Mango, Plantains in a fine patch on the hill to the left : the Gabiróba ‡ and

---

* Humboldt, on the Orinoco, heard by night the sounds of the sloth, the monkey, and the day-bird. This is not the case here, at any rate at this season.

† The two latter somewhat whimsical precautions are general on the São Francisco River, where the people, seeing an old hydropathist bathe in a state of violent perspiration, quietly remarked, "You are calling upon Death !" I have often known

Paulistas, even in the healthiest part of the Province, refuse coffee out of doors.

‡ In the System, "Guaviróba" is the name of sundry Eugenias. The Tupy Dict. writes the word Guabiraba. St. Hil. (III. ii. 270) tells us that the small species of Psidium "à baies arrondies" are called Gabiroba, opposed to Araça, those with pear-shaped fruits. I believe this to be correct.

the Araticum,* of which all are so fond. At the tall gate we found
a fine fig-tree planted only fourteen years ago. The garden to
the north-east of the house contains vines, as usual trained to
lath tunnels; here Bacchus apparently refuses to live without
support. The flowers were, as usual, few. The Brazil has many
more of the wild than the tame.

I remarked the pretty white Beijo de Frade, or Friar's Kiss,
and the Poinsettia bracts, brilliant as the "flame tree," and
generally known as Papagaio, the parrot. There is also a
graceful tobacco (N. ruralis or Langsdorffii), with thin leaves
and pink flower: it is, I believe, the "Aromatic Brazilian,"
much admired in the United States, and there found to lose its
aroma after the second year. The Tropeiros learned from the
Indians, who used it for smoking, and in medicine, to clean with
its infusion their mules of the Berne-maggots. The traveller will
do well to remember that a leaf rubbed over his hands and face
will compel the greediest mosquitos to buzz harmlessly about
him. According to the System this Nicotiana grows spontaneously, and is a Brazilian indigen, local as the Missouri variety:
I have always found it a companion of man, and flourishing un-
planted about the houses and villages. The Coqueiro palms
were peculiarly fine, although here as elsewhere the reticulum
pendent about the throat, a kind of vegetable goître, is never
removed. The Jenipapeiro † (Genipa americana, L.; Jenipa brasiliensis), whose fruit is compared by strangers with the medlar,
but which appears to me even more nauseous, is a noble tree; its
fine white flowers had already fallen. Wheat will grow at Bom
Successo, but it is subject to rust, and the flour, which is made
into bread, is of a dirty-brown tinge.

I introduced myself to Dr. Alexandre Severo Soarez Diniz,
nephew and son-in-law of Colonel Domingos; his family occupied
the Sítio, now the Fazenda of Andréqueicé, mentioned in 1801
by Dr. Couto. There is nothing to describe in the establish-

---

* Also written Araticú, and pronounced
"Articum." The name is given to many
Anonaceæ (A. muricata, A. spinescens, &c.).
Thus the fruits are distinguished from the
Anona squamosa, the custard apple of India,
here called pinha, fructa do Conde, and at
Rio de Janeiro by its Hindostani name,
Atta (for Ata).

† This is the tree, le Genipayer, well

known to the "Indians," who painted their
bodies with its juice, yielding a dark
blue dye. The fruit is called Jenipapo,
Jenipabo, or Genipapo. Such is the general rule in Portuguese, as Cajú, the Cashew-
apple: Cajueiro, the Cashew-apple-tree.
At times, however, the former is used by
synecdoche, as grammarians call it, for the
latter.

ment, which was the Casa Branca on a large scale. Here, for the first time, Friday appeared honoured by fish and eggs. After meals all stood up with clasped hands and prayed, ending with crossing themselves. As is the custom of old Minas, the slaves in waiting did the same. I do not know why St. Hilaire was so much scandalised by the anticipatory process. During the evening the household and the field-hands sang a long, loud hymn, and recited the " Christian Doctrine." On Sunday the prayers were more elaborate.

At Bom Successo, until four years ago, globules of free quicksilver were found adhering to the cross-battens of the "bica" or race of raised troughs which feeds the overshot wheel. Several bottles were filled, when suddenly the yield stopped. Mercury is reported to have been discovered on the Jequitinhonha River, and in other parts of the Minas Province; but a suspicion arose that it came from ancient gold washings. Here, however, all agreed that this could not be the case; we therefore resolved to inspect the formation. We followed the course of the Rego or leat which supplies the race. These water channels, sometimes 12—13 feet deep, are of vital import- ance to an estate, and are levelled by the eye, like the Kariz of Belochistan, to great distances. An Irish ditcher, if he could be kept sober, would soon make his fortune. The banks were green with grama (Triticum repens) pricking up from between the stones; the Herva do Bicho,* held sovereign for headaches; the bamboos were the Tabóca de Liceo, and the Cambahúba, which resembles the tasselled Criciúma. These gigantic reeds fatten cattle well, but it is believed that the food affects the wind of horses and mules. We were on the left of the Bom Successo stream, which heads three leagues to the north-east, and in it we found argillaceous shale, unelastic sandstone, slaty, talcose, and laminated,† fine blue limestone in bits and boulders, and quartz of many colours—white and yellow, rusty and black, and especially black and white—passing into one another. In the small creeks feeding this main line scattered fragments of cinnabar appeared, and a bit about the size of a nut was found in the leat.

---

* This well-known term is usually ap- plied to the Polygonum anti-hæmorrhoidale, the Tupy "Cataiá" or "Cataya." This Polygonea supplies a bitter peppery decoc- tion, used to cure the disease known as "O largo."

† In fact, diamantine Itacolumite. There are several diamond diggings about Bom Successo.

After about four miles we reached the dam at the head of the leat; here stakes were bent down-stream, and weighted with stones, so that the floods might pass over them with as little damage as possible. Evidently the metal came from below this point; if not, it would have been deposited beyond the possibility of being washed down, in the deep water above the weir. We therefore thought it probable that, as has happened in Spain and Austria, in Peru and California, the water or the pick had struck the gangue of native mercury, and had set free the disseminated globules. The deposit in the earthy water would be washed out and exhausted, and thus the ore would not appear until another cavity may be laid bare.

Intending to visit Diamantina city, I had engaged at Jaguára an old Camaráda and employé of Casa Branca, named Francisco Ferreira. He had preceded me for eight days, acting as guide to Trooper Manuel and to the four mules obligingly sent for my use by Mr. Gordon of Morro Velho. Matters did not look pleasant; the "talkeey" elder reported with a hiccup and a stagger, that it was "aw right;" and landsmen and watermen at once engaged in a general "drunk." It was in vain to take away the keg; in these Fazendas liquor is always to be had gratis. Mr. Davidson's health did not allow him to accompany me; and my three Calibans — Agostinho was to act page-cuisinier — would, without the strictest supervision, be in a normal state of disguise.

On the other hand my old longing for the pleasures of life in the backwoods—for solitude—was strong upon me as in Bubé-land. I sighed unamiably to be again out of the reach of my kind, so to speak—once more to meet Nature face to face. This food of the soul, as the Arabs call it, or diet of the spirit, as Vauverna-gens preferred—has been the subject of fine sayings, from the days of Scipio to those of J. G. Zimmermann; it is the true antidote to one's entourage, to the damaging effects of one's epoch and one's race; it is like absence, which, says the proverb, extinguishes the little "passions" and inflames the great; from those who think with others it takes all power of thought, but the "totus quis" comes out in it, and it largely gives to him who wishes to think for himself. "Homo solus aut deus aut dæmon," is almost half true; Væ soli! is evidently professional, and "O Solitude, where are thy charms?" is a poetical study.

How unhappy is the traveller who, like St. Hilaire, is ever

bemoaning the want of "society," of conversation, and who, "reduced to the society of his plants," consoles himself only by hoping to see the end of his journey ! " Une monotonie sans égale, une solitude profonde ; rien qui pût me distraire un instant de mon ennui." This, too, from a naturalist, " * * * Je finis par me désespérer à force d'ennui, et je ne pus m'empêcher de maudire les voyages." One understands the portrait which he draws of himself, veiled, with parasol to ward off the sun, and a twig to switch away ticks. It suggests a scientific Mr. Ledbury.

CHAPTER VI.

TO THE CIDADE DIAMANTINA.*

PARAÚNA RIVER AND VILLAGE OF THE CABOCLOS.—THE WINDY RIVULET.—THE
SERRA DA CONTAGEM.—COMPLETE CHANGE OF COUNTRY AND VEGETA-
TION.—CAMILLINHO VEGETATION.—BIRDS.—GOUVÊA.—DONA CHIQUINHA.—
SOLAR ECLIPSE.—BANDEIRINHA.—ARRIVAL.

> Hæc Boreas . . .
> Pulvereamque trahens per summa cacumina pallam,
> Verrit humum, pavidamque metu, caligine tectus,
> Orithyian amans fulvis amplectitur alis.
> *Ovid, Met.* vi.

I SECURED a sober start from Bom Successo by sending
forward my Calibans to bivouac at a place beyond the reach of
liquor, and I followed them on the morning of Tuesday, August 27,
1867.

The cold windy night had hung the north with heavy blue
fleece-pack, outlying an arch of lighter and more scattered

---

* Itinerary from Bom Successo to São João viâ Diamantina (approximately).

				hours		miles	
				1.15′		6	1st day,
1. Bom Successo	to Burá	time			distance		23 miles.
2.	,,	to Paraúna R.	,,	3.0	,,	9	
3.	,,	to Riacho do Vento	,,	2.10	,,	8	
4.	,,	to Contagem	,,	2.15	,,	8	2nd day,
5.	,,	to Camillinho	,,	1.15	,,	4	28 miles.
6.	,,	to Gouvêa	,,	4.15	,,	16	
7.	,,	to Bandeirinha	,,	3.45	,,	14	3rd day,
8.	,,	to Diamantina City	,,	3.0	,,	10	24 miles.
9.	,,	to S. João Mine	,,	4.30	,,	18	(Generally held to be 16.
		Totals		25.25′		93 miles.	

The Guides reckon ten leagues or forty miles between Bom Successo and Camillinho.
They place Diamantina sixteen leagues (forty-eight miles) from the Rio das Velhas, and
half that distance from the highest navigable point on the Paraúna River. From Bandeirinha to the Datas Mines they lay down three leagues, and I rode from the São João
Mine to Bandeirinha (twenty miles) in four hours thirty minutes.

Diamantina is usually held to be fifty-six leagues (224 miles) from the Provincial
Capital, a distance which greatly requires shortening. The Mine of São João is placed at
thirty-two leagues (128 miles) from the Villa de Guacuhy, at the mouth of the Rio das
Velhas.

vapour—signs of galey weather. Whilst the wind blows from the north or east we shall find the road dusty, not muddy; *vice versâ*, if it shift to south. Here the rains open in early October, either with or without thunder-storms (trovoadas); if the 15th be still dry, people fear for their crops. The grass-burnings (queimadas), began about 9th—10th August, and will last through September: the patches are fired in alternate years, so that forage may never be wanting, and we shall sometimes see half a dozen blazings in different directions. The custom is old and poetical.

> ———— to fell the virgin wood,
> To fire the second growths while young they grow,
> To feed with fattening ashes all the field,
>   The grain in holes to hide.*

There is no doubt of the real injury, independent of the loss in timber, which such romantic and picturesque practice entails upon the woodlands. It must greatly affect the vegetation, and kill out all but the strongest species. In these rugged Campos, however, there is less to say against it; the grass sprouts at once, and the potash is believed to be wholesome for cattle.

I fell at once into the Caminho do Campo, the western high road to Diamantina City, on the occidental skirt of the Serra Grande or do Espinhaço. It is separated by an interval of ten to twenty leagues from the Caminho do Mato Dentro, on the eastern flank, and viâ the Serra da Lapa: this latter is the shorter, the more trodden, and the better, but still very bad; and both are equally detestable during the rains.

The path runs over the crests and round the flanks of familiar Campos ground, whose surface is sandy, gravelly, or pebbly, with scatters of loose stones, bearing stunted vegetation, Cerrados, Capoẽs † and "Matas," or dwarf woods, clear of underwood, like the charming forests of France. The ground, strewed by the fierce north winds with dry leaves, was over-rich in ticks. Water gushes everywhere from a white or red clay, now compact, then a silty dust; and the vile bridges are logs loosely laid over a

---

* ... derrubar os virgens matos;
Queimar as Capoeiras ainda novas;
Servir de adubo á terra a fertil cinza
    Lançar os grãos nas covas.
        (Gonzaga, Lyras, part 1, 26).

† There are two principal Capoẽs, separated by two miles, the Capão das Moendas (of the Mills), to which it supplies hard wood, and do Padre (Antonio). Both are near waters flowing to the Bom Successo and thence to the Rio das Velhas. The usual desvios mark the worst places.

pair of sleepers. There is very little of human life in view; on the left is the "Rissacáda," * a Retiro, or shooting box, consisting of a few poor huts, belonging to Colonel Domingos, and after an hour's sharp riding I reached a similar place, the Retiro do Burá—of the Burá bee. Here my Calibans and animals had passed the night, and I was most civilly received by the honest, burly feitor, Sr. Paulino.

The inevitable coffee duly drunk, we pushed on merrily over broken ground at the foot of the hills, thick with copse, and showing green grass sprouting from the ashes of the dead. Where clearing was in process, the people worked off the reed-like vegetation with a bill-hook at the end of a long handle. Crossing the limpid streams,† and passing the Tapéra (da Maria) do Nascimento, the ex-home of a defunct widow, where the vultures were enjoying a dead bullock, we reached the Serra do Burá, which divides the basins of the Bom Successo and the Paraúna‡ streams. Up this buttress, which is partly grassy, and partly white and stony, with boulders of blue limestone striking south, there are two steep windings divided by a step or level.

From the summit we have a perfect command of the country around us. We see in front the tall blue wall through which the Paraúna breaks : in places the summit appears level, in others there is a feature locally known as Tapinhoacanga, § or Nigger-head, a porcupine-like lump, with out-cropping ledges of dark bare rock. Behind us the Campos roll as usual in flattened waves to the blue horizon, a smooth ring except where fretted with some solitary peak or notch of darker hue which suggests the Koranic "W'al Jibalu autádän,"—a peg to pin down earth. Everywhere in the Brazil the idea of immensity suggests itself, and nowhere more than on the Campos.

Beyond the Burá Crest begins a yellow descent, rough with gravel, soft laminated clay-slate, and porous iron-stone, like slag

---

* Translated "Bosque." In the dictionaries Ressaca or Resaca is the French ressac, the back drag of the tide.

† The first is the Córrego da Rissacáda, which at times swells and is dangerous ; the second, an unimportant feature, is known as the Córreginho—the streamlet.

‡ The Blackwater River, from "Para" and "una."

§ St. Hil. (III. ii. 103) derives the word from Tapanhúna, which he says in the Lingoa Geral means black ; the latter, however, is Pixuna, Pituna contracted to Una. The dictionaries give Abá (man) tapŷŷn-húna or tapŷŷiuna contracted to Tapan-húna or Tapanho, meaning a negro, and "acánga," a head.

or laterite. This leads to the " Cerradão," a taboleiro or plateau, about four miles in length; at first something sterile, but presently becoming a rich red soil with fair vegetation. The grass is the Capim-Assú, whose grain, often compared with rice, keeps cattle always fat, and amongst the dwarf woods are Palms in abundance, the Licorim, delicate, with ragged leaves,* the Indaiá,† and the Coqueirínho do Campo, which rises but little above the ground. The plateau ends at the Olhos de Agua, where a few huts gather near a Córrego that supplies pure water. Below us, to the right, lies the Paraúna, a dull dark (turvo) stream, running in snowy sand, with banks of white clay.

After three hours we reached the wretched little Aldêa de Paraúna, on the left bank of its river. It has a single straggling street of some seventy mud hovels, including one large open Rancho and eight Vendas: most of the tenements are tiled, few are whitewashed, and many are in ruins. On the right bank are six huts and a tilery. This old Indian settlement was once rich in gold, it flourished in the days of the " Diamantine Demarcation," which here began: in 1801 it was an Arraial, with most of its houses shut or fallen, and tenanted by a guard to prevent precious stones being smuggled. It lives now upon its excellent-stapled cotton, which fetches 2$500 to 2$800 per arroba, and by supplying travellers. The people are famous for their churlishness, possibly the effect of the moody Indian blood, and a curious contrast to those further on. As we found no civility at the house of a Caboclo shopkeeper, by name Sr. Tóttó, we rode up-stream to the little Fazenda do Brejo, an Engenhoca (small sugar-house) belonging to Manuel Ribeiro dos Santos, better known as " Manuel do Brejo," Emmanuel of the Marsh. When unable to visit it, I heard of a place called the Brejinho, where there is a salt stream that might be utilized.

The Paraúna, whose mouth we shall presently pass, drains the

---

* The Licorim palm must not be confounded with the Aricuri (Cocos coronata), which is common along the coast latitudes. It grows twenty-five to thirty-five feet high, with foliage like the true Cocoa-palm; the fruit hangs in bunches, and each nut is covered with a deep yellow and sweetish pericarp. The Macaws are fond of these Cocos de Licorim, and break the kernels with their powerful beaks.

† Also written Indaja, and in places pronounced Andaiá. Prince Max. calls it Coco Ndaiá assú, and describes it (ii. 30). On the coast range and shore we may truly say of this Attalea compta, "l'arbre est majestueux; c'est un des plus beau palmiers dans ce pays." On the Campos it is a stunted growth, almost without bole. The leaves are not eaten except by the hungriest of cattle; the nut is small and exceedingly hard, with an almond resembling that of the Cocos nucifera.

western slopes of the Serra Grande: it is a useless shallow
stream, here about 200 feet broad, full of rapids and choked by
drift wood: the banks are of hard, white, rain-guttered clay.
The valley, a flat of red and grey silt, edged by gravel and stones,
is narrow, and the lower vegetation at this season is browned by
the burning sun. The hill tops preserve their black verdure,
whilst the flanks are yellow, and dark clumps are scattered about
them. The ferry is six leagues by water, or four to four and a
half by land, from the Barra or Embouchure into the Rio das
Velhas. In opposition to the map-makers,* all assured me that
the Cipó stream, which is fed by the Serra da Lapa, falls into the
Paraúna, one league by water, or one and a half by land, above
this village. Eight leagues up-stream from the Ferry is the
Arraial de Paraúna, a place of no consequence. In 1801 Dr.
Couto declared that the Paraúna and its branches, as well as the
Pardo Major and Minor, in fact all the waters from the Great
Serra, would prove diamantine. This has lately been shown to
be the case, and there are now washings at the confluence of the
Cipó with the Rio das Pedras, near the south-west corner of the
Rotulo estate.

The ferry here belongs to Colonel Domingos, who lets it for
600$000 per annum and free passage for his tropas ; the toll was
not tollendus,· being only 0$500 for five mules and four men.
After the riverine valley on the right began the usual ascent,
winding round and up hills, whose tops and bottoms are earth,
whilst the sides are almost invariably ribbed with bare rock,
ledges of white grit, smooth as marble, and scatters of dark
blue sandstone.† These strike to the south-west, and are
raised at angles varying from 25° to 80°, giving a peculiar and
new appearance to the scene. The ascent of such places, often
made worse by tree-roots, is troublesome enough ; the descent
is still more disagreeable.

From the crest of this dividing ridge, the Black River, still
in its snowy bed, showed the Cachoeira do Paraúna, with three
distinct flashes down a rock wall, backed by the Nigger-head
Hill. The vegetation, like the pure white sandy soil, was a

---

* Burmeister is one mass of confusion.
M. Gerber makes the Cipó join the Paraúna close to the Rio das Velhas, and calls
the Junction "tres barras," the three
embouchures.

† To avoid this sandstone break, a road,
or rather a path, has been laid out to the
left, up a brown dusty hill, not yet worn
down to the stone, and at present offering a
little shade.

detritus of new "Itacolumite." For the first time in the Brazil,
I saw the Canelas de Ema, "Shank bones of Ostrich," the
Vellozias,* or tree-lilies, peculiar to these uplands.† They
take the place of the heaths so common in Europe and Africa,
and of which Gardner remarks, "not a single species has
hitherto been detected on the American Continent, either South
or North."‡ It is, like the tree fern, the bamboo, and the
Araucaria, an old world vegetation, suggesting the Triassic en-
crinitis, whilst the leafage was that of the Dragons'-blood
Dracæna. The field showed all sizes, from a few inches to ten
feet, the rough endogenous stems, mere bundles of fibres, were
quaintly bulged with abundant articulations, like those of a poly-
pus. This part of the plant contains resin, and the soft, high-
dried substance is prized for fuel where wood is scarce and
exceedingly dear. On the summit of each quaint stem was a
bunch of thin narrow leaves of aloetic appearance : as we brushed
through them, the mules snatched many a mouthful. In the
centre of the foliage was the lily-like flower, with viscid stalk,
quadrangular calyx, and blue and yellow stamens. There was
a smaller variety showing lavender-coloured blossoms, which the
people called Painera. This must not be confounded with the
Paina do Campo, or da Serra,§ from whose fibres are made
horses' saddle-cloths : it is probably the Composita named by
Gardner, Lychnophora Pinaster, a narrow-leaved, stiff shrub,
rarely exceeding six feet in height, but much resembling a very
young fir, and giving a decided feature to the peculiar vegetation
of Minas. It will be found taller in the upper levels. The
Carahyba do Campo, with tortuous branches easily formed into
yokes, lit up the scene, as if points of gamboge had been scattered over it : the naked form contrasted curiously with the well-
clothed Mimosa Dumetorum, one foot high, bearing a flower here
pink, there white, ten times larger than proportion requires, and

---

* So named from Dr. Joaquim Vellozo de
Miranda, Jesuit and botanist, born in
Minas Geraes.

† They flourish, I believe, on the Serra
de Ouro Branco. We shall find them again
on the middle course of the São Francisco
River, where they clothe the western
counterslopes of the Bahian "Chapada."

‡ I need hardly say that such is no
longer the belief of botanists. Australasia
alone has Epacrids instead of heaths.

§ St. Hil. (III. i. 247) mentions the
"Paineira" do Campo (Pachira marginata),
whose bark is scraped for bed stuffings. I
also heard the name Paina do Cerro (or
Serro) applied to a palm which extended
over the higher levels as far as the end of
this trip. The trunk is thicker above than
below, the general aspect is that of a huge
Sago, and the leafage, which resembles the
Indaiá, is useful for making hats.

with the pink, white, and scarlet tassels of the Cravinho do Campo, a shrublet whose root is a wild purge.* The people declare that Arnica is found in the uplands : † all know the medicine, none its plant.

Early after noon I descended the white hill into a red hollow, which grows a little coffee, sugar, and plantain fruit for the household. This is the place called Riacho do Vento—Windy Stream—a clean and well-wooded stream, flowing from the north. A certain João Alves Ribeiro was increasing his ranch, and the ground was strewed with timbers of the Aroeira, an Anacardium of several species : the heart was mahogany-coloured, and harder than any oak. The reception was not splendid, a tray turned up served for a table, a quarter-bushel measure for a chair, the food was as usual, and the dessert was snuff, either the coarse Rolão or the finer Pó de fumo. En revanche the bill, including breakfast and civility, was only 6$000.

I soon found out why my "Camarade" had dissuaded me from sleeping here. At sunset the east wind began to blow great guns, threatening to carry away the tiles—truly the place justifies its name. According to accounts the infliction is milder during the first and second quarters; it sets in violently with the full, and is most dreaded at new moon. It comes from the high and bleak meridional range to our right, and easily accounts for the regular morning gale on the Rio das Velhas. There was no "pasto fechado," and these "taboleiros" are proverbial for causing mules to stray : ours began locomotion at once, and were not found until sundown. They were necessarily tethered for the night in an empty ranch, and the tinkling of their bells proved that they were starved. Nor were the men better off.

We were glad to mount at 6 A.M., though the gale still howled overhead, and the stars were twinkling over hill tops, clearly cut and silver tipped. Crossing the Windy Rivulet, we struck up the Serra da Contagem,‡ or Range of the (diamond) tolls. This off-

---

* Probably a Myrtacea : of this genus several are called Craveiro da terra—native clove-tree.

† The Brazilians mostly mistake for Arnica a Composite known to us as Eupatorium Ayapana.

‡ These Contagens were established with the consent of the lieges in 1714, when, it will be remembered, the capitation-quints were raised. Dr. Couto tells us (1801) that the Villa do Principe was one of the four "Contagens dos Sertões," and says, "they call Sertoēs in this Captaincy the inner lands distant from mining villages,

set from the Espinhaço runs from east to west, and acts as
buttress to the Rio das Velhas. Our course was to the north-
east, and we wound from side to side with the blast catching our
ponches, and doing its best to blow down man and beast. Three
ascents, not precipitous, but rough with rolling stones, and mostly
using the rocky beds of streams, led to the summit: they were
divided by dwarf levels (Chapadinhas), scattered over with grass
and trees: in places water-sank, and during the rains transit must
be desperately bad. The soil was mostly red, set in patches of
glaring white sand, the detritus of the rock; in some places it
was blackened with vegetable humus, in others it sparkled with
pebbles and fragments of quartz. There were slabs and sheets of
the white gritty Itacolumite, yesterday so abundant: in places
long ridges crossed the path like the rock-walls that form a
Cachoeira, and nothing could be quainter than the shapes: here
they were gigantic frogs and " antediluvian," *i. e.* Tertiary beasts,
Megatheres and Colossocheles, seen in profile; there were magnified tombstones, erect or sloping, and there were fragments
pitched about as if in the play of giants.

After two slow miles up the south-western crest, we reached the
highest Chapada, and saw for the last time the plain behind us,
billowy with endless tossing of green-yellow waves. Here the
rocks and crags disappeared, and the compound slope was bisected
from north to south by As Lages, a tree-clad stream, running
over a bed of smooth slippery slab—an " ugly " spot; nor much
better were the ribs of fast or loose stone on the farther side
beyond a patch of rich ferruginous soil. On the right, a charming
Capão, which seemed to be traced by the hand, divided shade
from sunshine; whilst cattle, with clean hides, browzed the juicy

and where there is no mineration." Under it (Memoria, &c. p. 89) were,—

Caité (Caethé) Merim, with annual revenue of	766$400
Ribello	781$187
Inhacica (on Jequitinhonha River)	436$887
Pé do Morro	452$713
Contagem do Galheiro (of antlered stag) to south	1:146$437
Total	3:583$624

The profits of all four were but 5:446$562 (say = £544), without deducting the ex-

penses of barrack-repair, changing posts and so forth. The author justly ridicules a system which, for such paltry gain, did so much harm. Those who farmed the Contagens cared only for locating them where they paid best; when a new mine was discovered they surrounded it with a belt of obstacles, and thus they lost all, —like the husbandman who harvests before harvest-time. Of course the toll-gates should have been confined to the frontier, and collected from the imports; not inland where imports paid twice, or where dues were taken from those who had bought country-made goods.

pasture.* We then crossed a divide running east to west; the path was broken, and near it was a rib or dyke of dark stuff, which after rude testing appeared to be cobalt. The crest leads to the adjoining Limoeiro Basin, a formation similar to that just traversed, and cut by three waters flowing to the south-west.†

Two hours of dull riding placed us on the eastern edge of the Chapada, where the view suddenly changed. From our feet fell a long slope, or rather two slopes, a big one and a little one, of velvety surface, curiously contrasting with the hedgehog rocks around. At the base was a gleam of water flowing to the north-east; we are still in the valley of the São Francisco River. Below us, somewhat to the right, is a clump of oranges, spiky pitas and wind-wrung bananas, showing where stood the old Contagem das Abóboras, now desolate as the Inquisition of Goa. Further down is the Bocaina, or Gorge, seen from afar; on the right the Alto das Abóboras, and to the left an unnamed lump, form the huge portals of the lowland-gate. Masses of white sandstone, in places weathered to dingy blackness and queer shapes, and swept clean of everything by the wind, strike to the west, where they stand up in bluffs like river cliffs: the dip, from 70° to 90°, gives a quoin-like aspect, whilst the eastern backs are of gentle slope, frequently grass-grown. Scattered about are knobs, heads, walls, and saws, a peculiarly wild and hard aspect, and we look in vain for any correspondence of angles. Here Minas, always hilly, becomes extra-mountainous; and writers declare that the formation, generally arenaceous, turns to quartzose. In front are the distant lowlands, apparently plains dotted with dark hills, but really without half a mile of level, and the furthest distance is another line of fantastic rocks.

We now enter the true diamantine land, which older writers term the Cerro formation, thus distinguishing Diamantina of Minas from the diamond grounds of Bahia and from Diamantino of

---

* I saw no sign of the berne or worm. No one, however, breeds, and consequently the herds are small.

† The first is the Pindahyba, a muddy bed into which mules sink even in the "dries." An unpleasant path of white sandstone, with a pole serving as parapet to a precipice, leads to the Riacho da Vareda. The latter word here means a "Campina" or dwarf plain. The stream, coming from north to south, courses cold, dark, and clear over a rocky and slippery bed of sandstone, and on the left is a place where the tropeiros encamp. The third is the Limoeiro, dark and muddy, with a dense Capão a little beyond it. As a rule the water is of the best, a "pure vehicle for forming the finest crystallizations." In some places a white sand is spread over the black mud, reversing the usual process.

Matto Grosso. The view strikes at once. It is a complete change of scenery; everything is the image of bouleversement and aridity. The hills are no longer rounded heaps of clay, grown over with luxuriant vegetation. Here we have a dwarfed and pauper growth springing from the split rocks, a mean Campo flora, or yellow thickets based upon scanty humus, and even the hardy Coqueiro becomes degenerate.* It is a fracas of Nature, a land of crisp Serras stripped to the bones, prickly and bristling with peaky hills and fragments of pure rock separated by deep gashes and gorges; some rising overhead black and threatening, others distant with broken top lines, with torn blue sides, striped with darker or lighter lines. Here and there, between the stern peaks, lie patches of snow-white sand or a narrow bit of green plain, confused and orderless, a fibre in the core of rock-mountain. The land also is illiterate, and it is wild; fossils, those medals of the creation, do not belong to it.

After the first view of this country, and inspection of its material, I felt how erroneous was the limitation of the old men who confined the diamond to between 15° and 25° of north and south latitude, thus including Golconda, Visapur and Pegu, and making Borneo and Malacca the only Equatorial diggings. I at once recognized the formation of the São Paulo Province, in which many diamonds have been found.† My little trouvaille was that we may greatly extend the diamantine, as we have the carboniferous strata, and that the precious stone will be found in many parts of the world where its presence is least suspected, and even where the ignorant have worked the ground for gold.

But when, returning home, I looked at my newspapers, the trouvaille had been made for me. In one I read, " There are fifteen localities in California at which diamonds ‡ have been found in the course of washing for gold." The *Melbourne Argus* declared that " a small but very beautiful diamond had been found in a claim at Young's Creek, near Beechworth : the stone is perfectly white, and the crystallization well defined. It is the

---

* When clothed with sufficient humus, degraded Itacolumite is a very fertile soil.

† M. Barandier, a French artist, found a small diamond at Campinas in São Paulo. I have seen the "formação," or stones supposed to accompany the gem, in many parts of the Province, in the valley of the Southern Parahyba, and even near the city of São Paulo. A fine specimen of the black diamond, perfectly symmetrical, was taken from the Rio Verde, near the frontier of the São Paulo and Paraná Provinces ; moreover the Tibagy and other influents of the Paraná are known to be diamantiferous, and have supplied small specimens set by nature in the Cánga rock.

‡ The "California diamond" was formerly a bit of rock crystal.

second diamond found on that Creek. Again, the *Colesberg Advertiser* recorded the discovery of a diamond digging on the farm of Dr. Kalk, and asserted that some gems had been washed worth 500*l*.*

Old Ferreira, my comrade, used very hard words as he passed the ruins of the Contagem das Abóboras, which he called the Contagem do Galheiro.† The senior was a kind of Mr. Chocks, exceedingly grandiloquent till Nature expelled Art; he would call heat a "temerity of sun," rich ore a "barbarity of iron;" he told me to "charge to the right," meaning to take that direction; when uncertain he declared that "it did not constate," and when he ignored a thing, he was "not a great apologist of it." But, if tradition do not mightily exaggerate concerning the "days of despotism," as the colonial rule is popularly called, his bad language was justifiable. The soldiers and their commandant who occupied yon stone ranch, now ruined, held all the passes and watched the neighbouring Córregos, the only zigzags up which the Garimpeiro or smuggler could travel. Travellers were searched, and muleteers were compelled to take to pieces the pack-saddles where treasure might be concealed. Extreme cases are quoted. Men who bathed in the diamond rivers were flogged, and those found washing in them lost their hands. The tradition here is that the obnoxious system was abolished by D. Pedro I., that popular prince having accidentally, when disguised à la Harun El Rashíd, learned from a mule-trooper all its evils and injustice.

From the white soil we passed to a wave of reddish yellow ground, the "Mulatto" of the Southern States, and took the left of the huge portal on the right. The descent was gentle, but at the bottom came the usual troubles—tree stumps in the ground, holes whence roots had been drawn, banks up which the mules had to climb, a red soil forming puddle during the rains, and black earth even now a rivulet. We met a few mules about 9 A.M. Here the cold prevents an earlier start. Some carried for sale in the backwoods "Pedras de furno," round slabs of white Itacolumite, 2½ feet in diameter by 1 inch in thickness. For drying manioc

---

* When travelling in Virginia, I had heard of a true diamond picked up near Richmond; it weighed some twenty-four carats and cut to about half, and was sold for a small sum as it wanted "water."

† The Galheiro is to the north on the Rio Pardo Grande, six to seven leagues north of the Rio Paraúna. There is now a Fazenda do Galheiro, which belongs to many owners; it is drained by the Riacho do Vento.

they are preferred to metal pans or plates, because they cost 3$000 to 5$000. The manufacture is easy. They are prized up with levers, chipped into rounds or oblongs, and are ready for the oven. For convenience of carriage they are sometimes divided into semicircles. The quarry was shown—a mere dot on the hill side, a drop in the ocean that could supply all the Empire. Fine heavy soapstone is found in the torrent beds, and 1$000 procured for me a specimen in the shape of a candlestick.

Presently we reached a miserable hamlet of tattered wattle and dab huts, called Camillinho—little Camillus—after some "regulo da roça" who first settled there. An honest Rancheiro, Luis Monteiro, lodges man and beast. In his absence the wife gave us coffee and food, whilst the mules were sent to a good closed pasture hard by. Around the huts, which were jalousie-closed towards the road, and swarming with hens, pigeons, and black girls, grew a few coffee trees and wind-wrung bananas, whilst a single rose, which had learned to be a creeper, curled over a thatched roof.

From Camillinho we took a north-easterly course between two lines of rock. The soil appears to be always red clay upon the hill tops, with stony and ribbed sides, which sometimes throw lines across the road, and white or yellow tints in the lower parts. The huge Esbarrancados are here a mixture of water-breach and sun-crack; in places they cut up the country and cut off the roads. They are mostly elongated crevasses, whose projecting and re-entering angles correspond. Some form central islets, like St. Michael's Mount in miniature. The favourite site is the side of a hill, which will inevitably be eaten away, and often they moat the heights like the ditches of Titans. The old formations are known by their tarnish, and by the growth of trees in the lowest levels; the new are fresh, and generally bottomed with mud or flowing water. The whites and reds, yellows and purples, are lively as in other parts of the Province, and the feature is picturesque with light and shade, especially at times when the sun lies low. At first sight they suggest artificial models; the brilliantly coloured sections which are supposed to represent the earth's interior. We find even the "faults" and "dykes" which restrain percolation.

The line ran over sundry waves of ground, and wound round the hill sides, white with their small, loose, glaring stones. The descents and ascents were both bad, and led to and from waters either grey-coloured or crystal clear, flowing to the right, that is to swell the Paraúna River south-west. The huts appeared temporary, like mining villages, and here and there a manioc patch shows the capability of the soil. I presume that in many places the land would bear the short and strong-stemmed hill-wheat of Texas. The cool and shady wooded bottoms swarmed with the Carrapáto tick, and it was found advisable to send a man forward by way of "drawing them off." We are now approaching springtide, and the tints are prettily diversified. The pink Quaresma, dwarfed by cold, hugs the damp places near water; the golden Ipé, that local yew, also small, prefers the stony upland. In the hollows there is a flower that reminds me of the purple Aster. The stripped trees project their grey lean limbs against backgrounds of lightest-green, middle-green, and darkest green, and everywhere the bush is red, burnished with the new leaves of the Páu de Oleo,* a leguminous celebrity which prefers dry grounds and shuns stagnant waters.

The birds seem to be less bullied here than in most other parts of the Province. I saw for the first time a peculiar pigeon which extends down part of the Rio de São Francisco, and is found in the Highlands of Bahia. The people call it Pomba Verdadeira, or de Encontro branco, from the white marks on the wings. It is probably a variety of the Columba speciosa found on the seaboard, and its marbled neck and superior size suggest our blue rock. It looked like a giant by the side of the Pomba Torquaz,† the largest of the many doves (Jurity, Rôla, and others) which inhabit

---

* "Oil-wood," Copaifera officinalis, also written Copahyba, Cupaúba, and in other ways. The Caramurú (7, 51) describes it as,—

A Copaiba em curas applaudida—

"Capivi which oft works a certain cure."

The Indians, who knew the medicine well, collected it in sections of nuts, corked with wax, and during hot weather it used to sweat through the rude bottle, proving its excessive "tenuity." In 1787, according to Ferreira, a pot of nine Lisbon canadas (each two litres) cost 6$000 to 6$400, and "Capivi" was considered to be an important importation, having credit for many pseudo-virtues. Painters used it for linseed oil, but not in places exposed to weather, as it easily came off. Here it is sold in the shops, but it is held to be a

very violent remedy, and mostly confined to the treatment of cattle sores. The season for collecting the precious balsam opens with the new moon of August; the people say of the tree "Chora" (it weeps like Myrrha) "tudo o mez de Augusto," and a single trunk fills several bottles. The bark is cut, and pledgets of cotton are placed to drain the slit; the people have an idea that the greatest yield is when the moon is full, and that it gradually falls till the wane.

† The word is the Latin "Torquatus," and alludes to the ring round the neck; the vulgar corrupt it to Trocaes, and thus we find it written by Prince Max. (i. 396). Amongst the uneducated in the Brazil the unfortunate letter r is subject, amongst other injuries manifold, to excessive transposition.

these highlands. The Raptores are unusually numerous. There is the Caracará, which ranks with the eagles, and behaves, the degenerate aristocrat, vilely as a buzzard. A vulture (V. aura), probably the Acabiray first described by Azara, is here called Urubú Caçador, or the hunter. It resembles in form the vulgar bird, but it flies high. The head is red, and the wings are black with silver lining, like the noble Bateleur of Africa. Prince Max. (i. 75) makes the bird's head and neck to be gris cendre, which is not the case; he also guides its distant course by smell, which I vehemently doubt. Another hawk, known by the general name Gavião, poises itself in mid air, and is said to be a game bird, self-taught to follow and kill the Cadorna, or local partridge. If so, there would be no difficulty in training it. There is also a tiny raptor, hardly larger than a sandpiper. The first swallow seen during this year darted by in search of a warmer climate. The Scissar-tail (tesoura) turns sharply in the air, opening and shutting its forked tail; the pretty white and black Maria Preta, and the crimson Sangre de Boi or Pitangui, disported themselves amongst the stunted trees; while John Clay (João de Barros) hopped chattering before us as if he had some secret to tell, and the Tico-tico, tame as a robin, flirted with us like a little girl. At times the sharp stroke of a file upon a saw, sometimes singly and sometimes in quick succession, was heard. We recognised the voice of the bell-bird,* which has lately been introduced to England.

Ascending a slope after an hour's ride, we found a fresh change of scene. To the right, in a low, flat green bottom by the banks

---

* A drawing of a specimen which reached England lately appeared in the Illustrated News. It is the Campanero or bell-bird described in the last generation by Waterton, who makes its voice audible "at a distance of nearly three miles." The Chasmorhynchos nudicollis is popularly known as Araponga, a corruption of Guiraponga, from Guira a bird, pong onomatopoetic, and -a, what exists. St. Hilaire (III. i. 26) derives it from Ara, day, and pong, "son d'une chose creuse." He warns us not to confound it, like Mr. Walsh, with the "ferrador" or blacksmith frog, and, curious to say, for once Mr. Walsh is right. The T. Dict. explains Guiraponga by ferrador-ave. Castelnau mentions the ferrador bird (i. 274) and (in i. 169) the ferrador frog, which Prince Max. (i. 269) calls Ferreiro.

The Procnias (a genus formed by Illiger,) is called nudicollis from its thin green-patched throat, so conspicuous in the snow-white plume. It has no caruncle like the bird figured in the illustration to "Kidder and Fletcher," (edition of 1857) and called Uruponga; the bird with a tubercle is the white Cotinga, named Guiraponga or Ampelis Carunculata (Linn.). Prince Max. has described other species of this remarkable family, as, e.g., the Procnias melanocephalus (i. 260), and the Procnias Cyanotropes or ventralis, with blue green reflections (i. 291).

The peculiarity of this winged Stentor is the disproportion of the note to the size. We hear the blow of a hammer upon an anvil; we see a creature about the size of the smallest turtle dove.

of the Ribeirão do Tigre, another influent of the Paraúna, lay houses and dwarf fields; on the hill side was a tall black cross in a brand-new enclosure, a cemetery lately built, and already in active use. Around was a kind of prairie, high and subject to fierce winds, as the dwarfed Bromelias and the stunted Vellozias proved: the grass was thick but brown in the upper levels, and of metallic green below, suggesting fine pasture. The surface was pitted with termitaria, of which many had been mined by the Armadillo: mostly they showed annexes of a darker grey, clumsy projections like modern additions to some old country house. The prairie fires produced a dull glow in the sky, and the smoke folds crossing the sun had the effect of a cloud, and in places cast shadow upon the face of earth; we blessed the beneficent gloom. Far to the north-east lay our destination, Gouvêa—we are now about half-way—pointed out by its road, a red-brown ribbon spanning the sunburnt turf. To its left rose a massive, lumpy peak, streaked with horizontal wavy lines: on the right towered a cloud-kissing point, which some called Morro das Datas, and others Itambé.* The horizon in other places was bounded with bluff cliffs, which seemed to buttress an immense imaginary stream. Here and there was a "Pilot-knob," with strata regular as if built up, but defying human hands to build it.

The hill sides here showed traces of ancient leats, and heaps of clay stone grit which they had helped to wash. Within the Contagem all the soil is reputed to be diamantiferous, and the people delight to tell you that you may be treading upon precious stones. This, indeed, appears to be their thought by day and their dream at night. The surface was still disposed in waves, with abrupt inclines of red and yellow ground, deeply gashed, leading to three several waters,† which are struck perpendicularly. The watershed is from north-west to south-east, discharging to the Paraúna River. Mostly they are bright little streams, painted

---

* Ita-mbé, the big stone or rock. St. Hil. (I. i. 294) proposes as derivation, yta aymbe, pierre à aiguiser. There are two features of this name, as will presently appear.

† The first is the Agoa Limpa, on whose left bank rose a tall cliff, black as if volcanic—the effect of grass burning. Further on to the right is a silvery lakelet, containing a knobby islet. The Ribeirão das Areias spreads out wide, and has a rough bridge of eight trestles, some sixty-three yards long; at this season it is fordable. The Ribeirão das Almas showed a thread of pure water running along the main current, which had been made a dirty slate-coloured drain by washings in the upper bed. The soil is mostly red as if rusty with oxide of iron; it is fertile and produces oranges (remarkably good) and Jaboticabas, besides the normal coffee shrubs and bananas.

pink-red with iron, and set off by golden sands and avenues of leek-green trees. In the dwarf riverine valleys and the hill-sides were fields and huts, some of them tiled, and near the Areias a venda was being built.

We met on the way sundry parties of women coming from some local festival, a few whites, dressed in straw hats and rainbow-coloured cottons, with blacks carrying their children. They did not, as in many places, run away, and the tropeiros were unusually civil, seeing that I was still a recruiting-officer. The last divide led to the Córrego do Chiqueiro—of the Hogstye *— which is deep and dangerous during floods. We are now one league from our night's destination, and presently, after a long ascent and a leg to the east, we saw over a dwarf peak the conspicuous church of Gouvêa.

Women, all with the Caboclo look, carrying wood, entered with us as we passed the Cruz das Almas, which rose from a pile of stones. This cross, which recalls the souls in Purgatory, is here general. On the hill to the right was an unfinished building, Nª Sª das Dôres, undertaken by the vicar, Rev. Pe Francisco de Paula Moreira, and Sr. Roberto Alves, Jun., the son of a wealthy family. I thought that the grim, stone building, with what appeared to be a single chimney, was a fort raised for some inexplicable purpose; and it reminded me of the old Portuguese fane—

" Half church of God, half castle 'gainst the Moor."

We passed the Rosario, a detached chapel with a single palm tree, and rode northward, up a street of ground-floor houses and open Ranchos, each with its frontage of stakes towards the square, which apparently represented the town. After the sunny ride, and the high wind, which promised a cold night, I looked wistfully for a lodging, and saw none. Presently my guide remembered Dona Chiquinha, the wife of a Diamantina merchant, now at Rio de Janeiro: his name, Elizardo Emygdio de Aguiar, is written as pronounced by his friends, Elizaro Hemedio. Here began the civility of which I afterwards experienced so much in this part of the Province. The Dona at once admitted me, her

---

* A poetical name not rare. Near Ouro Preto is a place called Nª Sª da Conceição do Chiqueiro do Allamão (for Allemão). —Our Lady of the Conception of the Hogstye of the German.

married daughter brought oranges, her little granddaughter orange flowers, and her slaves coffee.

I presently walked out to view the place, and to escape being a menagerie. The people stared like the negroes of Ugogo: they could hardly gaze their full; they would, when tired, rest awhile, and presently take another "innings." The operations of shaving and of using a tooth-brush seemed to produce a peculiar edification. North of the town stands the chief church, Santo Antonio, occupying part of the square, which is rather a bulging in the street. It stands awry, having been built probably before Gouvêa was founded; it fronts south-west, unpolitely presenting to Jerusalem its dorsal region. On each side bits of Calçada line the red soil, and these incipient pavements lie here and there. About it are a few Casuarinas and Coqueiro palms, at this season, they say, always mangy; they feed a large caterpillar (lagarta) * which presently becomes a "borboleta"—moth or butterfly— after which they recover. The square shows one sobrado, belonging to João Alves, amongst the sixty-four houses east of the church: the fifty-eight to the west have sundry half-sobrados, and all the better sort are distinguished by shutters painted blue. The holy building is crooked from cross to door, apparently the people's eyes cannot see a straight line: it has four windows, and two weather-cocked towers, with roof covers upturned: there are two bells, and the eastern belfry has a bogus clock. Behind the temple is the God's acre, quaintly adorned with corner-posts of blue plaster, supporting rude and rusty armillary spheres.

The town is on a rough ridge, and water is scarce and distant. On the east, far below, lies the usual Lavapés: nearer is the Rua do Fogo,† a kind of chemin des affronteux, and in the distance is the Morro de Santo Antonio, a noble stone-knob based upon an earthen pedestal. No one has ascended it, yet it may be easily climbed on the south-east. Westward is the Rua do Socego or dos Coqueiros, with a few houses scattered and whitewashed, in compounds defended by dry stone walls. The growth is the Castor shrub, the Jaboticaba, the papaw, whose leaves are here

---

* The Curculio palmarum is relished in Africa, and greedily eaten by the S. American "Indians." I have never tasted it, but white travellers have informed me that it has a delicate and even a delicious flavour.

† The Street of Fire, not an uncommon village name in the Brazil, usually meaning that in it liquor and consequently quarrels abound.

used for soup, the plantain, a few good oranges, and the sweet lime with bitter placenta, called Lima da peça : the coffee looks thriftless and starving, as usual it is crowded and untrimmed. Provisions are excessively expensive, having to make the journey which we have made, and maize * costs 4$000 per alqueire.

On the next morning, when I called for the bill, the Dona refused everything, even a gift; such was her hospitable habit, and she declared that her sons also were wandering over the world abroad. We mounted at 7 A.M., a light east wind rising with the sun, whilst the sky was moutonné with clouds. Our course lay north-east towards the pyramids of dull grey stone, the smaller below the larger, and both sentinelling the richer diamond lands. A slippery hill, gashed with water-breaches, led to a wooded hollow, which sheltered a few thatched huts; to the right was a Sitio, belonging to Roberto Alves. It had outhouses, enclosures, and a coffee plantation, somewhat thin, but defended from the blasts and superior to all rivals.

Here began the Pé de Morro, or ascent, which will last till near Diamantina. The wheel-road winding round the western side is easy : the bridle-path to east seems made for goats, with its loose stones and its ruts petrified in hard pink clay. Presently the latter fell into the former line, and the slope improved. From the summit we had a good back view of Gouvêa, but soon the wind, chopping round to the north, drifted in our faces a thick Scotch mist. Old Ferreira complained that the Corrubiana † got into his bones and nearly made him lose the way.‡

The hill led to a plateau consisting of two plains divided by a water and a prism of rock. One of them was about two miles across; such an extent of level surface is here rarely seen. Cattle fine and plump, despite the Carrapatos, and probably strengthened by the highly ferruginous water, made it look like "a pastoral in a flat." The Capão, however, was not of the style "bonito," §

---

* In this country the alqueire of maize regulates prices like the quartern loaf in England. I have seen it at São Paulo, the city, fluctuate between 2$000 and 4$000 — more exactly between 1$940 and 4$160.

† This word is popular in Minas Geraes,

and also, I believe, in Rio Grande do Sul. Some Caipíras pronounce it "Cruviána."

‡ On the right hand a road sets off to Datas, the property of Colonel Alexandre de Almeida Silva Bitancourt ; it reaches the city, but after a very long round.

§ The "pretty tree motte" is often seen

it was coarse and ragged, whilst the land was much burnt. The road became excellent, broad, level, and fit for a carriage : unhappily, like that approaching Agbome, it is a mere patch.

At 9 A.M. we descended to Barro Preto, the first diamond digging which I had seen at work. The site is a stream bed, the head-waters (Cabeceiras) of the Córrego das Lages, which feeds successively the Corrégo das Datas (or the Cachoeira), the Córrego da Grupiára and the Paraúna River. The surface showed spoil-heaps of "saibro," clayey sand, varying in colour from dirty white to milky white, like the detritus of quartzum lacteum, turfy and vegetable matter, and pebbles mixed with fragments of rock crystal. A little thread of muddy water trickled down and served the "Serviçosinho."* We passed two huts and a half of thatch-wattle and dark-grey dab, whence the negroes stared, the dogs barked, the pigs grunted. The place, known for two to three years, has been worked during the last eight months by João and Manuel Alves, the sons of a centagenarian. It is said that they have several diamonds exceeding two oitavas (say each =280l.), and there are vague rumours of a large stone which is kept a profound secret. In these diggings all is mystery, and not without reason ; an exceptional diamond generally counts in the wild parts at least one murder.

Pushing across the sterile diamantine land, where the windwrung trees acted as anemometers, I again remarked the fantastic forms of the sandstone, especially on the north-east, whence the weather comes. Here were watch-towers and pyramids, there were walls which no Cyclops could have raised ; now we passed peeled skulls, then mouldering bones. Between them the surface was mottled, sand-patches white as kaolin, or stained with humus and soil, yellow, purple, and dull crimson with ochre and hæmatite, dotted the expanse of warm-red brown land ; the latter was comparatively fertile, and clothed with black ashes, from which sprouted grass of metallic green, spiky as a stiff beard. The expected eclipse came on, the sun diminished to a crescent, but the mist was so thick that the effect passed away almost

---

in the Province of São Paulo, where the grass, like the nap of yellow or green velvet, sweeps up to the clump, which is of tall and regular growth.

* A small Serviço. The latter is an old name still applied in Minas Geraes and Bahia to diamond washings worked by a tropa or slave-gang under free-men.

imperceptibly.   No one paid any attention to it, nor would
they

Si fractus illabatur orbis ;

not because over-just or tenacious of things proposed, but from
mere incuriousness.   Old Ferreira, it is true, remarked that it
might be the cause of the " confounded Corrubiana," * but then,
he could think of nothing else.

Still ascending, we crossed three waters flowing to the westward, † and divided by bulges of ground.   Near the first was a
clump of huts and signs of industry.   A rough " Báco," ‡ or
three-sided trough of planks and sandstone-slabs, awaited the
rains to wash the heaps lying near it.   After four miles of barren
soil we made " Bandeirinha," § a whitewashed house, surrounded
by a few trees, and a close pasture fronted by an open ranch.
Maria Augusta de Andrade, in the absence of her husband, José
da Rocha, miner, "merchant," Rancheiro, and so forth, rose up
shivering and prepared breakfast for us : the south-east wind had
blown for five days, and on my return, five days afterwards, I
found it blowing still.

Now remained only ten miles.   In half an hour we ascended a
stony hill of red and white soil.   This is the great dividing line
between the Rivers São Francisco and Jequitinhonha ; from this
point it trends in a northerly direction, bending to the west.   On
the left was a cross-road leading through the little villages O
Guíndá, the Brumadinho, and the Rio das Pedras to the Mine of
São João. ‖   In front lay a huge brown slope, patched with
snowy, glittering, dazzling sand, and here and there growing grass
of a lively green : in places there was an abundance of the
ground-palm, here called Coqueirinho do Campo, dwarfed by the

---

* Perhaps this was the case. On my
return the mist tried to gather thick, but
was soon dispersed by the sun.

† The first is the Córrego de João Vaz,
so called from an old settler whose descendants still gamble in diamonds ; they have
seven huts, one neatly whitewashed. It
flows to the Córrego do Capão, and thence
to the Rio Pardo Pequeno ; during the
rains it is dangerous. The second is known
as the Braúna (Melanoxylon Grauna), a
rocky bed with the bulges called Caldeiroës,
and at this season a trickle of water, which

also feeds the Córrego do Capão. A single
house is built near its bank.

‡ This trough corresponds with the ca-
noa used in gold-washing.

§ Dr. Couto, in 1801, mentions the
Sitio da Bandeirinha, the little Bandeira,
or Commando.   Burmeister erroneously
writes " Bandeirinho." This and Bandeira
are common names in the Province of
Minas, dating from the days of the slaving
expeditions.

‖ See Chapter 9.

gales. Near the horizon, scatters of tall stone, heads, shoulders, knobs, piles and lumps broke the outline, and far to the right rose the long blue wall which bears the majestic pyramid Itambé.

Presently we passed, on the left, O Guindá, so called from a broad, shallow, and sandy stream, once very rich, and still worked : it feeds the northern Rio das Pedras, the Rio do Caldeirão, the Biribiri, the Pinheiro, and the Jequitinhonha Rivers. It is a miner-town, surrounded by red excavations, and looks from afar like an ant-hill; has a single small square and large black cross, sheds for tropeiros, and decent houses, hugging the left bank of the water. Beyond it is the Brumadinho, a similar settlement, but smaller. Presently we sighted, far ahead, a grim rocky wall, with a white path winding up its darkness ; this is the good new road leading to Medanha on the Jequitinhonha River, and thence to São Salvador da Bahia. Crossing the northern Rio das Pedras, a crystal water-babe in a sandstone cradle, I crested a hill, and saw to the east a big white house, garnished with a few brown huts, and standing apparently on the edge of a precipice—the Episcopal Seminary.

Diamantina was within musket-shot, but a long northerly detour was necessary in order to gain the main road. I forded the Riacho das Bicas, so called from an old and rich gold mine on the hill behind the Seminary : this Lavapés flows to the east, and falls into a little Rio de São Francisco, south of the city. The hollows were rich in the large and deeply digitated Aroid with an edible fruit, known as Imbé, or Guaimbé, and in Tupy, Tracuans (Philodendron grandifolium). It loves damp places, and has an extensive range between sea-level and 3000 feet of altitude. A stiff ascent—the last—and a line of stunted Araucarias, led to a hill-crest and the usual Cruz das Almas. Here the traveller first sights the city, falling in perspective below his feet. It is a Brazilian " Pangani "—a settlement " in a hole." The first glimpse suggests—

> Dirarum nidis domus opportuna volucrum.

Yet sings of it its local poet, the late Aureliano J. Lessa—

> Vês lá na encosta do monte
> Mil casas em gruposinhos

Alvas como cordeirinhos
Que se lavaram na fonte ?
Qual dragão petrificado
Aquella serra curvado
Que mura a cidadesinha ?
Pois essa cidade é minha
É meu berço idolatrado.*

---

* See'st thou upon yon slope of hill
     A thousand houses grouped together,
       White as the yeanling of the wether,
All freshly bathed in summer rill ?
And see'st not in far background
     Like to a serpent turned to stone,
     The range in regular curving thrown,
That walls the little city round ?
     Behold my own dear walls arise,
     The cradle which I idolize.

CHAPTER VII.

AT DIAMANTINA.

CITY DESCRIBED.—SOCIETY.—POPULARITY OF THE ENGLISH IN THE BRAZIL.—
THE DIAMOND IN THE BRAZIL, ITS DISCOVERY, &C.—VALUE OF EXPORTED
DIAMONDS.

"The temperate climate enjoyed by the inhabitants of this part of the country
renders them more healthy than those who dwell in the Sertão (Far West); the
women are the most beautiful I met with in Brazil."—*Gardner*, chap. xii.

THE site of Diamantina is peculiar: it is almost precipitous to
the east and south-west, whilst the northern part is a continuation of the broken prairie-land. This incipient Haute Ville is the
best and healthiest locality, and here the settlement will spread.
The " Cidadesinha " runs down the western face of a strongly
inclined hill to meet on the sole of the deep valley the Rio de São
Francisco, or Rio Grande; its water, draining the lowlands, feeds
the main artery of this basin, the Rio Jequitinhonha, distant three
leagues in a straight line, and five to six indirect.* The breadth
of the torrent-bed, here running from north to south, is patched
with red-brown soil and brilliantly green herbage: the middle is
white with cascalho heaps thrown up by the old diggers: a mere
thread of water now trickles down it, but after rain it becomes
dangerous: a dwarf bridge has been put up to save servile life
from the frequent inundations. The further side of the ravine is
a grim broken wall of grey rock, white under the hammer; the
rampart springs steeply from a base encumbered with spoil-
banks, washed many a year ago, and is raggedly clothed with
grass now brown.†

Viewed from the " Alto da Cruz," the city has a well-to-do

---

* The course is southerly to the Southern
Rio das Pedras; it then turns by east to
north-east, and joins, or according to some,
forms the head waters of the great Jequitinhonha.

† It is advisable to walk up the new
Bahia road, which commands an excellent
prospect of the city.

and important look. It is much changed since 1801, when as the "Arraial do Tejuco"—the village of the mud-hole,[*] it had nothing but wooden tenements; nor can it be recognised in the pages of Gardner and M. Barbot,[†] who described it as it was during the last generation. Below us lies a sheet of houses dressed in many colours, pink, white, and yellow, with large green gardens facing broad streets and wide squares, whilst public buildings of superior size, and a confusion of single and double church-steeples, testify to the piety of the place.

From the Alto da Cruz we make the Largo do Curral, the best building-site in, or rather out of, the city. Formerly cattle were here stabled and slaughtered; now a tall black cross has converted it into a respectable square. Descending the good new Calçada of the Rua da Gloria, formerly "do Intendente," we passed on the left the Sobrado da Gloria, which began life as the Intendency of Diamonds, then became the provisional Episcopal Palace, and now lodges those Sisters of St. Vincent de Paul whom we met upon the road near the Caráça. Inside the carpenters are at work pulling to pieces timber still sound after a century of use: an old-fashioned wooden verandah looks upon a large back-garden of the richest soil, supplied with the purest water. Opposite is the tall sobrado belonging to the Lieut.-Col. Rodrigo de Souza Reis, whose mine we shall presently visit.

The Gloria strikes at right angles a street called, no one knows why, the Macáo do Meio. It must not be confounded with the Largo do Macáo, where stands the Caridade Infirmary, a long, broad, white building belonging to a "brotherhood." The roughly paved Middle Macáo contains good shops, the "Hotel Cula,"[‡] and the Church of São Francisco, whose doors and windows are set in a framework of very unpretty streaky red—here a fashionable tint, supposed to resemble marble. A six-faced and two-spouted fountain of Egyptian grotesqueness, set in the wall and dated 1861, begins the normal Rua Direita. "Straight Street" is exceedingly crooked, steep, and badly paved. Most of the houses are new and boast of windows: some preserve the shutter, and one retains the hanging gallery and Rotula or

---

[*] The word is explained at length in Vol. I. Chap. 10.

[†] Traité Complet, etc., p. 218.

[‡] In full Sr. Herculano Carlos de Magal-haens Castro, a delegate of police. Breakfast at 9·30 A.M., a table d'hôte (mesa redonda), at 4 P.M., and 0$800 per meal.

lattice-work of dingy, chocolate-coloured wood. It will soon be removed: these antiquities are very properly despised in the Brazil: here Temple Bar would be photographed, and no longer allowed to cumber the ground. The sooner the old Pillory is demolished, the better for progressive Diamantina—let me suggest.

In the Largo da Rua Direita or de Santo Antonio is the Town-hall (Casa da Camara), a humble building, displaying the Imperial Arms.* It has latterly been used as a Masonic Lodge. This was forbidden, justly enough, because a Portuguese priest, Padre Luis, became a brother. Opposite the Camara, and facing with the Course of Empire, is the Matriz, whose "Orago" is Santo Antonio. It is an "insula," with a raised platform towards the northern slope of the hill. A stone wall shows the cemetery, to be banished quam primum. The two-windowed front, with two rose-lights pierced in the rude Taipa-conglomerate, is bound in neutral-tinted sky-blue french-grey, whilst the doors and shutters are daubed chocolate. All above the cornice is of board work, even to the belfry, the first instance of the kind which I have seen in the Brazil. The single window of the steeple shows a gilt bell. There is a clock which, wondrous to relate, goes, but goes wrong, and the finial is the usual armillary sphere with the normal extensive weather-cock, more often a dragon than a cock. There is nothing to be described in the interior of this or of any other Diamantine Church, and the "lumber" work gives them generally a look of instability.

We are in the heart of the city, the centre of business-circulation. On the left of the Square is the Intendencia de Sousa Reis.† "Intendency" here means a substantial market shed, the embryo of the Pisau Sotto borgo. Sousa Reis is private property, and under the deep dark verandah are shops which sell everything, from flour to snuff, required by the wild country. Below and to the east is a large open square, the "Cavalhada Nova," as distinguished from the "Velha," further down and almost outside the city. These clear spaces were so called from the Portuguese carrousels, which, like bull-fights, once accompanied every festivity. They are obsolete in the Brazil, though they preserve vitality in

---

* The lower story is not the normal prison, which has been removed to a building near the theatre.

† There are two other Intendencias, de Sebastião Picada, and the Lages; the latter has five stores.

Italy, in Portugal, and even in Anglicized Madeira. The last "tournament" I saw was at the Island of Fogo, in the Cape Verde group.

Crossing and leaving on the right the Rua da Quitanda, I found the house of my host, Sr. João Ribeiro (de Carvalho Amarante), on the northern side of the Praça do Bomfim. The ground floor is laid out in a dry-goods store and an inner writing apartment, where the diamonds are kept. The dining room and kitchen affect the back part of the tenement, and above are the apartments of the family. The hospitable Lisbonese freely confesses that he began life with driving a few mules; he is now the wealthiest merchant where all are merchants, and he supplies goods even to Guaicuhy and Januaria.* At the Pé de Morro, near the Curumatahy influent of the Jequitinhonha, he owns a large fazenda, where he breeds cattle, grows provisions, and manufactures sugar and rum. He is in trouble about his 50 slaves, and nowhere, as far as I know the Brazil, are negroes so troublesome as those in and around Diamantina. Many of them take to the bush and become "Quilombeiros," black banditti, ready for any atrocity which their cowardice judges safe. Here no one travels even by day without having his weapons handy and without looking round the corners. They are skilful as Canidia or Locusta, and much addicted to the use of Stramonium.† A common symptom is an intense pain in the legs, a medical man assured me, causing a drawn and anxious countenance. Many a slave-owner has suspected malingering, till undeceived by the sufferer's speedy death. A case has lately occurred at Pé de Morro; the owner will presently visit it and make a terrible example of the poisoner. Thus a threatened servile mutiny was summarily crushed in 1865 by flogging and the galleys;‡ nor did anybody meet with the fate of Governor Eyre.

Sr. João Ribeiro consigned me to his bachelor guest-house in the Rua do Bomfim, so called from a Church dedicated to Our Lady of Good End. The street is a kind of ragged irregular

---

* See Chapters 13 and 17.

† The System says that its alkaloid principle is well known to the negroes, who prepare from the plant their "philters," that is to say, charms and poisons, love-draughts and other devilries. May not the seeds of the Stramonium have been brought from India viâ Africa? St. Hil. (I. ii. 97) determines

that the plant has here followed the footsteps of man from N. America.

‡ The "Quilombeiros" of Medanha had a Maroon settlement within a league of the village, and threatened the suburbs of Diamantina. When their stronghold was attacked and taken, whites as well as blacks were found in it.

square; it boasts of a good barber, a watchmaker, and an apothecary. Of course all imported articles are sold at an extravagant
price, and considering the transport, this is not astonishing.* From
the Bomfim the Rua do Amparo, tolerably paved, runs to the east,
and strikes the Valley of the Rio de São Francisco. It passes by
the Church of Nª Sª do Amparo—Our Lady of the Refuge. The
front was adorned with coloured glass lamps, and the Sunday
morning squibs told us that a Novena was in progress there. The
best drinking water is brought from the bottom of the ravine,
where a few houses and huts, plantations and fields, are scattered
about, leaving abundant building room. If not afraid of snakes,
ticks, and thorns, you may fight your way far down the Rivulet
banks.

My three days spent at Diamantina left upon me the most
agreeable impressions of its society. The men are the " frankest,"
the women are the prettiest and the most amiable that it has yet
been my fortune to meet in the Brazil. Strangers everywhere in
these regions receive cordial hospitality, but here the welcome is
peculiarly warm. Perhaps the wealth of the place has something
to do with it. Where lodged I was at once called upon by some
young men from Rio de Janeiro, here popularly called Cometas.
Sensible, obliging, and well-informed, they had none of that offen-
siveness of the European Commis-voyageur, or travelling bagman.
The calling is honourable as any other. It may be said with
truth, and greatly to the credit of the Brazil, that no man feels
degraded by honest industry, however humble. Consequently
society ignores the mauvaise honte about professions which distinguishes the old world, where I have seen a man blush to own
that his father was a "doctor," and where Faraday was lauded
because he dared to confess in public that his brother was a gas-
fitter.

My first evening was spent at the house of John Rose, a Cornishman, originally a miner at Morro Velho, afterwards a diamond-
digger, carpenter, mason, architect; his last job was at the Bishop's

---

* My test bottles having been broken, I bought—

3 oz. muriatic acid	.	.	.	1$040
3 oz. nitric acid	.	.	.	1$040
2 oz. tannin, in alcohol	.	.	.	6$500
Total	.	.	.	10$580

At that time about one guinea.

Palace. By sobriety and good conduct he has cleared some 5000l., and now he can amply enjoy his propensity for independence in word and deed. Not so pleasant was another stranger, who at once showed the cloven foot by loudly abusing the Brazilians, and by declaring that they allowed none but themselves to thrive. I will not mention his name, for, although he must have turned the half-century, he may still find out that it is never too late to mend. He is a well-educated man, knowing German and English perfectly, Portuguese well, French tolerably ; he can teach languages ; he can keep books ; of course he has a gold mine ; he has been a doctor— still a popular character ;* and he still practises homœopathy. But he prefers to "loaf about," borrowing 100$000 from this and 160$000 from that acquaintance, whose charity he expends, not on raiment but upon drink. When in liquor he is addicted to the free use of knife and pistol. He attributes his habits of sleeping in the streets to the infidelity of his spouse. He had left her at Rio totally unprovided for, and she was persuaded to accept the protection of a Portuguese, who offered to, and who did, maintain, educate, and settle her children. The latest little game of my unpleasant acquaintance has been Freemasonry, to which he has, for a consideration, admitted the least worthy aspirants. He proposed, moyennant the payment of 5l., to make me a P.M., and he had the impudence to deliver a message from me to a certain ecclesiastic, begging that Freemasonry might not be preached against ; it was necessary to call, and to explain the affair.

This man was a Hanoverian, consequently a Prussian, but he called himself an Englishman. Britons in the Brazil are wont to complain that they and the Portuguese are exceedingly unpopular. The fact is that we frequently suffer not only for our own sins, which are manifold, but for those of our European neighbours, which are not few. Foreigners also exaggerate our unpopularity. " Les Anglais sont détestés au Brésil ; on regarde comme appartenans à cette nation tous les étrangers chez lesquels des cheveux blonds et une peau blanche indiquent qu'ils sont origaines du

---

* The Diamantists did not seem to me satisfied with the gifts of their Esculapiuses, as everywhere in the outer Brazil a stranger is expected to be a medicine-man. I was at once consulted for a simple hepatitis, which the leech, after the normal treatment of cupping and blistering, was attacking with anti-spasmodics. In vain I assured the patient that my favourite profession was rather to kill than to cure ; he seemed satisfied that he had already run the very greatest risk of killing without murder.

Nord," says, in 1815—1817, Prince Max. (i. 119). M. Dulot (p. 62) speaks of " la brutalité traditionelle envers les faibles qui fait détester partout l'Angleterre ;" and here he would be justified if he alluded to the " Aberdeen Bill." St. Hilaire (III. i. 219) remarks that "grace à leurs compatriotes, Mawe, Luccock et Walsh," the English became unpopular in the land. And it is almost a truism to say that if perhaps we hear too little good of ourselves from others, we, like other nations, hear far too much good of ourselves from ourselves. This puffery and clap-trap about our own perfections is still held to be patriotism, and at last the " genial, broad-shouldered Englishman" has learned to bear without a murmur gigantic weights of " Buncombe." *

The Brazil, also, like other people, has met with a small amount of merited praise, and a large amount of unmerited abuse. But the travellers of one nation have hardly been more polite to her than those of the others.† The result of my experience at present is that, despite the Aberdeen Bill and the silly Abrantes-Christie affair, the Empire respects us, and even likes us as much as, if not more than, her other visitors. It is not pretended that strangers are favourites anywhere in the Brazil ; the country expected from them far too much, and they justified considerably less than the most moderate expectations. In our case they complain of the "insular manner," now happily waxing obsolete, as the Frenchman of Goldsmith and Sterne, the coarse roughness of the uneducated, ‡ and the shy pride and haughty reticence of their " betters," are ever gall and wormwood to the Brazilian spirit. And we have lost esteem by the

---

* It has lately been judged advisable in British India to consult high officials concerning the appreciation of our rule by the natives, not by ourselves. Many men, myself included, have since 1850, written and repeated in the plainest English, what now comes before the public in a decorous foolscap form. The only result was that we were pronounced by the few who took the trouble of reading us, to be either ignorant or impertinent, and ignorance and impertinence in such matters can expect very little mercy.

† Nor have the French tended to improve the entente cordiale. The Comte de Suzannet (Souvenirs, 1842), M. de Chavaignes (Souvenirs, p. 160), the unjustly treated M. Jacquemont, and MM. Biard,

Expilly, and D'Abbadie, may be quoted versus MM. Reybaud, Ferdinand Denis, and Liais. I cannot explain, except by the influence of an outrageous nationality, how St. Hilaire (III. i. 263), defends and applies the terms " homme de beaucoup d'esprit," to M. Jacques Arago, author of the "Voyage autour du Monde," and one of the most disgraceful charlatans that ever appeared in the Brazil.

‡ " This is a free country, and any man therefore may take any freedom he likes with any other man, and protest is simply Quixotic. But we are a coarse people." Thus writes a popular author, who has never yet been called a "degenerate Englishman."

great country's little wars, which began the dotage of a liberal policy, and which led it to shirk the duties of its position, and to retire from the business of the world. An Abyssinian Expedition benefits England as much in the Brazil as in Hindostan, and may be pronounced to be worth the two-pence.

I paid a visit to the Rev. Michel Sipolis, at the Episcopal Seminary, the staring white building with unfinished outhouses, before mentioned. The Government assists the establishment by paying salaries for the several chairs, and the three French priests receive, per annum, only 400$000 for clothing and all wants ; this salary of £40 must raise them above all suspicion of interestedness. At 1 P.M. the bell rang and we went to the Refectory ; there were twelve pupils, a considerable number during "long vacation," and these young men spoke French during the meal, and ended it with a long prayer. M. Sipolis then led me to the Episcopal Palace, which is opposite the Carmo Church, a white building picked out with blue, plastered concrete below and boarding above. The diocese of Marianna formerly extended here : Pius IX. created the bishopric by the Bull "Gravissimum Sollicitudinis," June 6, 1864. The Exmo and Rev^{mo} D. João Antonio dos Santos,[*] of the Council of H. I. M., is an old elève of the Caráça Seminary ; he naturally patronises, in preference to the Propaganda of Lyons and the Capuchins of Rome,[†] St. Vincent of Paul, who must find it hard work to answer all the calls upon him. The Bishop was a man about forty, with a gentle, feminine voice and manners : I found him diligently engaged with M. Mirville on Magnetism (not Faraday's), and he did not take part with M. Sipolis when the latter proved to me that table-turning and "rapping" are the works of evil spirits.[‡]

From the Palace we passed over to the house of a fazendeiro, at whose door an Agent de Police sat comfortably in the shade. He had had with a neighbour some trifling dispute about a water-

---

[*] In the Brazil it is often impossible to tell the family names of ecclesiastics, who mostly adopt some technical or theological cognomen, somewhat after the fashion, though not quite in the style, of "Praise-God-Barebones."

[†] Here the Capuchins have assumed as instructors the place held by the Jesuits. I need hardly say that they have never done so in Europe.

[‡] Nec deus intersit, etc. We may add nec diabolus. As regards the spirit theory I may again remark that, if after this life my psyche or pneuma, or whatever it may be, is to find itself at the mercy of every booby who pays half-a-crown to his or her medium, evidently the future state of this person will be much worse than the present.

course, which ended in a " shyuting," and he was expected to purge himself before a jury. The antagonist having fired into his side and mangled his thumb, which required amputation, the wounded man cried out to his son, who discharged a barrel or two into the hostile face, and then sensibly took to the bush. Of course there was another and a contradictory account, which declared that the fazendeiro had snatched the gun from his antagonist, and that it had exploded, hurting his hand. I could not but think of the true or apocryphal story touching Sir Walter Raleigh and the " History of the World:" he would have found it impossible to settle the rights of this little affair at Diamantina.

Meanwhile the hurt man was in great pain, restless, and fearing tetanus. Yet the room was darkened, the windows were shut, the air was oppressive, five silent ladies sat pensively looking on, and just outside the doors were half a dozen muttering male friends. When a patient is held to be sick unto death, the popular Brazilian idea—of course the rare sensible scout it—is to visit and console and condole with him. Such an apparatus would injure the most robust; surely it would be humane to publish a Portuguese version of " Notes on Nursing." The vile Caldo de Gallinha, or hen-broth, which it is indispensable to swallow every two hours, is an infliction to be compared only with the "beef-tea" of the old-fashioned priestess of Libitina in Great Britain.

My last appearance in " Society" was at a ball given by a wealthy widow, the Sra. Dª Maria de Nazareth Netto Leme, in honour of the baptism of a grandson, the second child of a very charming young person, wife of Sr. Joaquim Manoel de Vasconcellos Lessa. When this pretty lady was married, she was attended by twenty-four bridesmaids in dresses from Paris; the merry-making was kept up for a fortnight, and it is said that 750 bottles of Bass disappeared every night. This rain of meat and drink at the City of Diamonds is a great contrast to the ascetic " tea and turn out " of Southern Europe.

The whole of the City of Diamonds was in accurate black raiment before 3 P.M., the hour for the religious ceremony. As evening approached, I accompanied Sr. João Ribeiro with the most amiable Dª Maria and his daughter up the Rua das Mercês, so called from its church, to the Alto da Gupiara.* The rooms were crowded, and many had sat down to a preliminary supper. The toilettes were remarkably good, a contrast to the times described by 'Gardner, when ladies went abroad in men's hats, and " black seemed the most fashionable." Every neck sparkled with diamonds : the other ornaments were the solid and honest, if not tasteful, jewellery of Diamantina. The ball seemed to be a family party, infinite in merriment: here, as amongst the Catholics of England, all are related or connected, more or less, and those who are not, intend to be, or are " gossips." The dancing was chiefly quadrilles. I excused myself on the plea that my last performance had been with Gelele, King of Dahome : thus the proprietress of No. 14, St. James's Square wore for life a glove upon the hand saluted by a former Prince of Wales.

Supper seemed never to end, and a stiff shower of rain only added to the mirth within. The life of the party was " O Dia-mantino," curtly for Sr. José Diamantino de Menezes, son of the late Barão de Arassuahy.† I stole away at 2 P.M., leaving all " merry and wise." This is specified, because the country mice around give the city mice a bad character, and declare that every morning the ladies and their slaves sally forth to pick up their husbands from the pavé, where " tangle-leg" had put them to bed. Of this I saw nothing.

Of course in a place where money is abundantly ‡ spent, and where visitors flock in for pleasure, after the toils and the dulness of the out-station, there must be some debauchery. The many smiling faces, protruding from small casements, cheeks blooming with the juice of a certain Hibiscus and a squeeze of lime, tell their own tale. But such things have nothing to do with society. The " hell," moreover, that usually accompanies the modern growth of mining cities, does not exist in these

---

* I have already explained Gupiara (corrupted Grupiara), to mean the slope of a tilted shed ; hence in gold and diamond diggings it is applied to a ledge projecting caves-like over a stream. The Alto, seen from the entrance of the city, is a conspicuous hill, crowned by a building that resembles a fortress or redoubt. This property originally belonged to Sr. Luis Antonio, and then passed to Sr. José Joaquim Netto Leme, whilome husband of the present proprietrix. It is still rich in gold,

which no one takes the trouble to disturb.

† The river rising about twelve leagues east from Diamantina, passing by Minas Novas (do Arassuahy), and forming the eastern gate of the Jequitinhonha. The word is Araçu, a kind of bird, and -hy, water. There was also a Baron of Diamantina, of the Lessa family.

‡ Here, as in Australia and California, the miner is mostly poor, whilst the merchant or storekeeper is rich.

regions, except when a stray Frenchman starts a roulette table, and makes his fortune after a few months.

An Englishman, who had spent thirty years in and about Diamantina, told me that of late years its prosperity had diminished.* Formerly diamonds were easily washed from the surface diggings : now the works are confined to capitalists. In early days the stones were sold in the city, at present they are sent to Rio de Janeiro,† and to Europe. The slaves have been traded off to the coffee-growing Provinces, and the free man, white or black, will not, or cannot work. Hence fortunes now average 4000l., whilst the highest may amount to 10,000l. ; these figures, however, represent very different values in Minas Geraes and in England.

But so far from the diamonds being exhausted, I believe that the true exploitation of precious lithology has still to begin, and that it will extend 800 miles along the Serra do Espinhaço.‡ There are also rich gold-diggings, which men hardly take the trouble to work; with gold they justly say you may be poor, with diamonds never.§ When the rail shall have reached Sabará, and the paddle-wheel shall connect the Rio das Velhas with the great São Francisco, the immigrant may be expected, and the Diamantine country will attain its full development. "The Lord bring them!" say the mine-proprietors, alluding to the Southerners of the Union, "and they will soon use up our useless slaves!"‖ And whilst Golconda and Visapur have failed, and the Cape of Good Hope, Australia, and California are but beginning, and whilst men sink capital in the trash manufactured in

* The population in 1800 was about 5000 ; in 1840, it was 6000 ; and now it is not increased.

† Diamond cutting was attempted without success by a Sr. Carvalho, at Bahia. There are three or four lapidaries at Rio de Janeiro ; the best is, I believe, Sr. Domingos Moitinho (at the corner of the Rua d'Ouvidor and the Rua dos Ourives). Some of his workmen are descendants of the artists brought from Portugal by D. João VI. The machinery is driven by an engine of five-horse power. The diamond is here cut exactly as in Europe, and the Brazilians ignore the flat slab-like shapes of Hindostan. Of late years Boston has attempted the industry, but it cannot, I am told, compete with Amsterdam.

‡ The portion which has been explored

begins at the Rio do Peixe, nine leagues south of Diamantina, and extends to the celebrated Serra de Santo Antonio, forty to fifty leagues to the north, or between N. lat. 16° to 19°. All was found to be diamantine, but not continuously so, as in the Demarcation Proper.

§ According to Dr. Couto (p. 112), who settled and died at "Tejuco," the city is built upon slabs of red copper, and the metal is found in the pavement and the garden walls.

‖ "The pride of man makes him love to domineer. Wherever the law allows it, therefore, he will generally prefer the service of slaves to that of freemen." (Wealth of Nations, iii. 2.) My experience is diametrically opposed to this dogma of Adam Smith.

Paris and Birmingham, the Brazil may still hope to do great things in the " diamond-line."

The sounds of every city leave upon the traveller's sensorium their own impression. At Diamantina my brain connects the church-bell and the Araponga, or blacksmith bird. The sharp, sudden cry, which to the stranger seems artificial, charms in the dead silence of the forest alcove, tempered by the distance of the tallest tree-top, and when the little white form is not visible in the verdant gloom. Caged, and in a street, the thing is quite out of place. The situation of Diamantina, as has been seen, renders the rumbling of the cart and the rolling of the carriage impossible : here, as at São João d'El-Rei, the hammock is the only conveyance, and it is seen in the hall of every rich house. As usual in the Brazilian interior, the city is guiltless of club, café, Mechanics' Institutes, Christian Young Men's Association and Mutual Improvement Societies, except for musical purposes ; the bands, however, are, all things considered, good. There is neither library, literary cabinet, nor bookseller, but of course there is a photographer. About three years ago, the only news-paper " O Jequitinhonha," which was devoted solely to politics, expired, and now the city does not contain a printer. Yet the citizens—the Brazilian is a citizen, not a subject—are wild for education, even for church education. The "Sisters" have already had offers of 100, and have accepted 30 pupils.

The site of the city is one of the highest in the Empire,* and to reach it we have ascended seven distinct gradients. The coldest months are June, July, and August, when frosts are common in the lower levels ; they do not, however, prevent the maturing of the Pitanga berry.† The wet season opens in October or November, with thunder storms from the north ; the heaviest downfalls came from the west, but sometimes the warm south-west winds bring rain and hail. The fertilizing showers of the dries, which abound in other parts of the Brazil, are here

---

* The altitude ranges, according to travellers, between 4000 feet and 1730 metres (5702 feet) above sea level. The steps of ascent from the Rio das Velhas are seven, viz., first, to the Paraúna stream ; second, to the Riacho do Vento ; third, to the Chapada ; fourth, to the Contagem summit ; fifth, to Gouvêa ; sixth, to Band-cirinha ; and seventh, to Diamantina.

† The well-known Eugenia pedunculata (E. Michelii, Linn.), whose quadrangular red fruit ripens well at Madeira, and makes good jellies. When raw it has a drug-like flavour, which is disliked by strangers. In this part of Minas Geraes it is rare, but it flourishes at S. Paulo, 2200 feet above sea level, though not so kindly as on the coast.

rare. The east wind is the mildest and the most agreeable; the north is cold and raw, causing sickness like our east. From November to February is the hot season, and the annual range of the thermometer is from 64° to 88°. Water of the best quality is supplied by almost every hollow. In the clear, bracing air European fruits and vegetables thrive; the soil is sometimes rich and deep, and the abnormal expense of provisions would make the neighbourhood an excellent market for an agricultural colony.

"'Tejuco," the village in the Comarca do Cerro, became a Freguezia September 6, 1819, a Villa Oct. 13, 1831, and the Cidade Diamantina by the Provincial Law, No. 93 of 1838. It owes its prosperity solely to the diamond. This valuable stone was used, it is said, by the Indians as playthings for their children.* The first man who sent it to Portugal was one Sebastião Leme do Prado, in 1725; he had washed certain brilliant octahedrons in the Rio Manso, an influent of the Jequitinhonha. They found no sale, and the same happened to Bernardo (or Bernardino) da Fonseca Lobo, who hit upon a large specimen amongst others in the Cerro do Frio. There is a local tradition that the latter was a friar who had been in India, and that about 1727, seeing the curious, brilliant little stones used as counters at backgammon by the gold miners of the Jequitinhonha, he made a collection of them and went to Portugal. Others attribute the discovery to an Ouvidor or Auditor Judge, fresh from service at Goa; the specimens were sent to the Netherlands, then the great jewel-market of Europe.

The official account of the exploitation is that D. Lourenço de Almeida, the first Governor of Minas Geraes (August 18, 1721—Sept. 1, 1732), reported the new source of wealth to the Home Government. Portugal at once declared the diamond to be Crown property (Carta Regia, Feb. 18, 1730), and established the celebrated Diamantine Demarcation, forty-two leagues in circumference, with a diameter of fourteen to fifteen leagues.† Gold

* It is generally supposed that in Europe Louis Van Berghem, popularly written Berquen (1456—1475), invented the practice of making diamond cut diamond, and established a guild in Bruges. But the Hindus must have been long beforehand, and the working of diamonds in Europe is mentioned in 1360. It is possible that the industry had a little before the fourteenth century drifted, like the cholera of modern days, westward.

† John Mawe's Map gives a sketch of the "Diamantine Demarcation." It is an oval of eight by sixteen leagues, and "Tejuco" was nearly in the centre.

digging was forbidden within the limits, and a tax of 20$000—subsequently raised to 40$000 and 50$000—was placed upon every head of negro. To arrest the many and repeated disorders, an Order, dated Sept. 30, 1733, created the "Intendencia Diamantina;" the washing-grounds were marked out, and no one might enter without a licence. In 1740 (Henderson says 1741), the lands were farmed out, with great restrictions, for 138:000$000, but this first contract was much abused. In 1771 (1772, John Mawe), the great Pombal reformed, with characteristic thoroughness, the diamond mines, by taking the management into his own hands. He abolished ruinous leases, and governed by an Intendant-General, under whom worked a board of three Directors in Lisbon, and three Governors in the Brazil. The scheme failed, and so energetic was action against the "extrariadores," that the place became almost a desert. In 1800 to 1801 the gold supply began to fail, and the lands about the Villa do Principe, where diamantine was mixed with auriferous matter, yielded only 2½ instead of 25 arrobas. Thus the Government lost by reducing all industry to the diamond, and the people fled because they could not afford to buy iron, steel and gunpowder.

I have not been able to find out exactly at what period of Tejucan history occurred the event alluded to by Sr. Joaquim Norberto de Souza Silva: *

> E o filho de Erin, que em duros ferros
> Pagou seu pasmo por um novo imperio.

The name given in the foot note is "Nicolas George." He was, we are told, of Irish extraction, and employed in the Junta of the Arraial do Tejuco. Admiring the fertility, the wealth and the vastness of the Brazil, he declared that her shores contained everything necessary for a mighty Empire, and that she might become free and independent as the United States. The sentiment made him share the pains and penalties of the "Conspirators of Minas."

According to John Mawe, from 1801 to 1806, both years included, the expenses incurred by the Government in exploiting

---

* In the Cantos Epicos—a Cabeça do Martyr—

> "And Erin's son who in the eating irons,
> Atoned the purpose of a free-born realm."

the district were 204,000*l.*, and the diamonds sent to the Treasury amounted to 115,675 carats. During the same period gold was washed and valued at 17,300*l.* Thus, he says, the carat cost 23*s.* 9*d.* At length the Decree of Oct. 25 (1832) abolished the monopoly with its Junta Administrativa dos Diamantes, and the industry assumed its present form.

If the Portuguese doubted the existence of the diamond in the Brazil, the English did the same. There is a difference in specific gravity between the noble Vieille Roche of India and the produce of the New World.* In the last century, Jeffries and other lapidaries contended that the Brazilian were unformed gems exported from Hindostan. The miners cleverly turned the tables upon their scientific antagonists by sending their stones to Goa, whence they were forwarded as true East Indian to Europe.

According to John Mawe, during the first twenty years some 1000 oz. of diamonds were annually extracted from these diggings. Castelnau (ii. 338), in 1849, estimates the total value of the Minas Geraes exportation at 300,000,000 francs. The subject is also treated by José de Rezende Costa, in the Memoria Historica sobre os Diamantes (Rio, 1836). I will not trouble the reader with details, as all such estimates are the merest guess-work, and even the modern appliances of Custom-house collection and statistics are powerless against the general rule of contrabandism. The following table, however, taken from Mr. Nathan's annual report (Rio de Janeiro), will show the

EXPORTS OF DIAMONDS AND ESTIMATED VALUE IN YEARS 1861 TO 1867.

Years.			Oitavas.			Price.			Total Value.
1861	.	.	4,696	.	.	500$000†	.	.	2,348,000$
1862	.	.	5,019	.	.	,,	.	.	2,509,500
1863	.	.	5,824	.	.	,,	.	.	2,912,000
1864	.	.	4,861	.	.	,,	.	.	2,430,500
1865	.	.	4,962	.	.	,,	.	.	2,481,000
1866	.	.	5,695	.	.	,,	.	.	2,847,500
1867	.	.	5,704	.	.	,,	.	.	2,852,000
Total	.	.	36,761						18,880,500$‡

* The difference of weight is attributed to the mineral oxides that colour the stone. The following are the popular figures :

	Golconda (Indian).		Brazilian.
White, spec. grav.	3·524		3·442 (M. Barbot, 3·444).
Yellow,	,,	3·556	3·520 (,, 3·519).

Lapidaries generally agree that the old or E. Indian diamond has more lustre and brilliancy than the new or Brazilian.

† This is too low.   ‡ £1,888,000.

CHAPTER VIII.

TO THE DIAMOND DIGGINGS OF THE SOUTHERN RIO DAS PEDRAS, ALIAS THE JEQUITINHONHA.

THE RIDE.—QUAINT STONES.—SÃO GONÇALO OF THE GOOD GIRLS.—THE SERVIÇO MINE DESCRIBED.—EXPENSES.—WANT OF MACHINERY.—PLUNDER.—DR. DAYRELL.—THE "LOMBA" MINE.—THE MARAVILHA MOUNTAIN.—RETURN TO DIAMANTINA.

Οὔ χειμὼν λυπεῖ σ᾽ οὐ καῦμ᾽ οὐ νοῦσος ἐνοχλεῖ,
Οὐ πείνη σ᾽ οὐ δίψος ἔχει σ᾽.

SHORTLY after my arrival I was introduced to a Brazilian gentleman, Sr. Francisco Leite Vidigal, who lost no time in inviting me to visit his "Serviço," known as the Canteiro or "pot-stand." This season, the height of the dries, is the best for exploring the diggings, which are now all activity.

We breakfasted perforce and set out late, although the sun is hot, and we had four to five leagues of total work before us. We rode down the Rua do Bomfim to the southern suburb, past a very small post-office in the Largo do Rosario, and a fountain with cocks sticking out of steatite faces. Here is a negro church, as usual mean and gaudy, and a large unfinished theatre, a carcase of timber and brown clay. A splendid Gamelleira fig, whose natural grandeur did not set off the dwarfishness of the Art around it, led us to a Calçada winding down a stiff descent. Here the site of the city falls into the riverine valley, and the slope of fine soil is rich in oranges, plantains, myrtles, and trees that give more shade than fruit.

Beyond the bank the place is called La Palha; here are the large ranch, the venda, and the camping ground belonging to a Frenchman, M. Antoine Richier. I failed to find him at home, but the thumbing of his photographic manuals showed an interest in something civilised. We then crossed a confluence where the

Pororuca or Pururuca,* translated "Stream of the sand and gravel," flows from the west into the little Rio de São Francisco. The banks were a mass of loose amygdaloid, pebbles of water-rolled quartz; and they "paint gold," which no one cares to work. In the evening my host showed me many oitavas lying in the corner of his hut; they had not even been washed for market.

We then ascended to Campo ground, and struck the highway which leads to the Provincial Capital, viâ the city of Cerro, now Cidade do Principe, distant ten leagues.† Before us rose the grand Peak Itambé, said to be 6000 feet above sea level. Its head was in a cap of clouds, ever similar, never the same, and the shoulders were clad in ruddy grass and gloomy forest. On the eastern horizon rose the hilly mass called the Curralinho, and held to be very rich in diamonds. Around us were outcrops of the usual granular quartzose Itacolumite, hard and soft, finely laminated or coarsely agglutinated, greyish outside, and overgrown with lichens; the inside is snow-coloured or slightly yellow. In places the masses are horizontal, forming regular walls; in others they become ridges of slabs disposed at every possible angle. During the day we saw a man in a liberty cap, a sphinx, a frog-like labyrinthodon, an old mutilated lion, gravestones with inscriptions, stones with hands, gaps, arches, circular holes, and every variety of outlandish shape. The degradation of this grit forms the frequent patches of snowy sand, which are of course sterile, whilst here again the red-brown soils which separate them are often exceptionally fertile.

The road proved to be especially vile, and at the most precipitous narrows we were certain to meet strings of horses or unruly mules laden with large square boxes, generally labelled "Louça," equivalent to "Glass, with care." How anything ever reaches Diamantina unbroken is beyond my comprehension.

---

* The word is here applied to a large sand and pebbles, either water-rolled or not; the formation is not agglutinated by paste or cement (gomma), and has no body (corpo). In the diamond mine it is more watery than the "desmonte," which will presently be explained.

† St. Hil. (I. i. 330) says that Cerro is more than ten leagues from Diamantina. Dr. Couto (p. 1) makes it ten leagues to the south-south-west. The people say it is less, but their leagues are of the longest.

Cerro (or Serro, perhaps a more modern form) is a rare word applied to particular places where there are lines of hills or mountains. Originally it signifies a hillock or rising ground; Constancio explains it "Monte Alto;" and Moraes "Outeiro," as well as "Monte Alto." The Cerro do Frio, which is more usual than Cerro Frio, is supposed to be a translation of the Tupy "Yviturui," from "Yvitu," wind, and "tuy" cold.

After fording sundry streams, we crossed by a neat bridge the Ribeirão, called by the early travellers do Inferno on account of the difficulties which it offered. Its source to the west is known as "As Porteiras," and the yellow rocks and blue skies make it a "Rio Verde." Above the bridge were the "casas palhoças," the poor thatches of sapé and walls of stick and clay that tell the presence of miners.

Beyond the stream we found a few men tinkering up a very bad ascent, and we remarked with indignation a mile-post which told us that we had finished one league—such here are leagues—after two hours of sharp riding. We then pricked across a taboleiro coberto,* or wave of ground, beautified only by the view. In addition to the fronting Itambé, we had now to the left or west the Maravilha, or Marvel, a local Sugarloaf, just the place where a Maharatha Rajah or an Abyssinian Dejaj would build his Durg or Amba. The Ribeirão do Palmital, bridgeless, and rolling its pellucid waters over a dwarf cliff of sandstone, veined, dyked, and ribboned with lustrous-white quartz, dashed to meet the "Rivulet of Hell." Of course a house was near the ford; linen hung in the yard to dry, but no amount of shouting would open the door. It was the same at the next bridge, although near it was a large ranch and a staked camping ground.

The hills resembled those about the Paraúna River, rough above, whilst the lower folds were of earth, here light, there stiff. On the flanks about half-way up were zones of stone piercing the soil, weathered and trodden into ledges, gutters, and deep hollows, whilst here and there lay loose rounded boulders. The head was generally spread into a dwarf plateau of thin soil, with more or less of vegetation. On cresting a summit we suddenly saw across a long green valley traversed by the long red line of highway, the church and village of the "Marriage-maker of Old Women."† The place is remarkable for its order and industry; not a "lost girl," I was told, can be found in it, and the inhabitants have many small industries. They do not care to work, where diamonds are, a hill of rock crystals which lies near their doors. When these six-sided prisms of pure silicic acid,

---

* Not Taboleira coberta as Gardner wrote. This "covered plateau," a modification of the Campo, is thinly clad with gnarled trees; the term is opposed to the Taboleiro descoberto, a formation of greater altitude, growing only the hardier shrubs and grasses.

† Casamenteiro das Velhas, the title which S. Gonçalo bears in the Brazil. John Mawe, with his usual inaccuracy about names, calls the village "San Gonzales."

terminating in hexagonal points, have unbroken pyramids, which is rare after travelling, and when the interior contains the water of crystallisation or heterogeneous bodies, the larger blocks are valuable as museum specimens.

This wave of ground ended at the Córrego do Jacá (of the Pannier),* which boasts a small bridge. Another ridge brought us to the Descida do Córrego do Mel (the Descent of the Rivulet of Honey). On the further slope the sandstone slabs were so steep and slippery that my companion, a very light man, dismounted from his good new mule. When a Brazilian does this it is generally wise to follow his example. All the ground which we have traversed is rich in diamonds, but it cannot be worked for want of water; near the Córrego which feeds the Rio das Pedras many white heaps were waiting to be washed during the rains. The Gurgulho † or breccia, here sometimes so sharp (gurgulho bravo) that it cuts the hands, is peculiarly rich in stones, and about the bridge the torrent banks produce gold.

We then turned to the left, and made two miles of "picada" or bridle-path. The country was as before rocky on both sides, and poorly clad. The greenest and shadiest tree was the Canella (Laurinea). I remarked also an abundance of the large-leaved Congonha do Campo (Ilicinea), and a tree with green berries, called by my friend "Mata Cavallo," a general term for all things that bear "wild," that is to say poisonous, fruit. The herb called Arruda do Campo, because supposed to resemble the European rue, scented or tainted the air.

The last descent led us to the Southern Rio das Pedras, here running from the south. It is one of the head waters of the great Jequitinhonha ‡ River, a lesser rival of the Rio de São Francisco

---

* The Tupy Dict. explains Jacá by Cesto (basket) de Cipós. It is more usually made, I believe, of woven bamboo-bark.

† The word is pronounced like, but not written, "Gorgulho," which means a weevil (Curculio). It is described as a loose or compact pudding of angular stones mostly found in Campo ground, and thus distinguished from the water-rolled Cascalho. Some apply the term to a collection of Cascalho, others to a larger formation than Cascalho. An English writer on precious lithology has followed John Mawe's misprint, which corrupted gurgulho to "burgalhao."

‡ The name is written in many ways;

the old style is Gectinhonha. Then came Giquitignogna, Gigtinhonha, Gequitinhonha, Jigitonhonha, and so forth. The trivial and popular explanation of the word is "Jequi tem nhonha," the fishing crate has caught a nhonha fish. Jequi is a Tupy word meaning a fish trap (armadilha). Nhonha, according to some, in the local dialect meant any fish; in the Lingoa Geral the word is Pyra or Pira. St. Hil. (I. ii. 142) says it was explained to him by une nasse (creel) pleine; "Juquiá" being the nasse. This reminds us of such derivations as Capivarhy from Capivara ahi, Arassuahy from Ouro só ahi (gold only here), and so forth.

(Maïor). It rises a mere torrent in the mountains to the north of the Cidade do Principe. It is joined by many streams, amongst which is the Lomba or Jequitinhonha do Mato ; about two leagues below the Canteiro it becomes the Jequitinhonha do Campo, and finally the true Jequitinhonha. According to others, the Southern Rio das Pedras is the Upper Jequitinhonha do Mato, which, after receiving the Ribeirão do Inferno, is *the* Jequitinhonha, and absorbs the Jequitinhonha do Campo. The course of this river, which upon maps looks so well, is said to be much obstructed by rapids. I have not visited it. At last it takes the name of Rio Grande, divides into several arms, unites with the Rio Pardo, forms a delta, and buries itself in the Atlantic about forty-five miles north of Porto Seguro in the Province of Bahia.

After six hours' work we entered the little mining station of a dozen huts, built upon a rough stubby slope that lines the left side of the Rio das Pedras. Under the circumstances, a " Roxo forte," or cup of café noir " laced " with rum, was excusable ; this taken, we went off without further delay to inspect.

We began with the beginning, a proceeding which, say the Germans, we English rarely adopt. The descent to the mine is a narrow unrailed path, winding down the precipitous left bank of the Rio das Pedras. It was crowded with double meeting lines of black and whitey-brown labourers, free as well as ser- vile, whom the presence of the master had galvanized into a momentary " spurt." Those ascending carried on their heads Carumbés, or cedar-wood platters, about twice the size of soup-plates, containing " desmonte,"* or the useless sand and gravel which is washed down by the greater inundations of the year, and which underlies and overlies the strata of true diamantiferous Cascalho. Planks, rough ladders, and inclined planes, led to the bottom of the long pit, whose southern extremity was 80 feet deep by 19 to 20 broad. It was evidently the river bed in bygone ages before the channel was filled up to its present height. Each talhião, or rock-wall of the underground channel, was wonderfully worked into pit holes and convex curves, regular as though the latter had been used, by the grinding action of gravelly water. †

---

* Desmonte is sand and gravel, with more or less consistency (liga). In gold mining " desmontar "—literally to un- mount—is to remove the vegetation and the humus from over the auriferous cas-

calho. In Portugal it is synonymous with " roçar " or " desmoutar," to clear the land for cultivation.

† We shall find many of these " pit-holes " in the bed of the São Francisco River.

These are the richest pockets, and each *may* yield a hundred contos of reis. The hanging wall, and the loosened blocks on the sides, were carefully timbered wherever a joint was inclined to open.

The negroes, watched by overseers stationed at every angle, were removing, with the usual merry song, the valueless stratum under which they expected to find the gem-bearing yellow Cascalho. Some bored, others broke away the interfering rock with huge pyramidal-headed crow-bars (alavancas). These loosened the gravel with the almocafre,* an oval-shaped, blunt-headed iron, whose handle was about two feet long; those scraped out of the fendas or fissures the likely sand, with an "almocafre de frincha," a bent blade one inch broad by four to six in length. I was shown in situ the curious formation called "Cánga preta," which is found in hundreds of pounds' weight, though rarely of large size. At first it was mistaken for coal, but it became red-hot in the fire without being consumed. It looks fibrous, like asbestos, and in appearance much resembles graphite. Here also are found loose fragments of polished sand-stone, turned by the water into curious shapes. I saw a child's foot perfectly imitated, and many leg bones and shoulder blades were of monstrous size.

All this work is going on far below the water level. A strong dyke of ashlar and earth has been run out from the right bank to the mid-stream of the Rio das Pedras, which here runs from south-east to north-west, bending north. Above the pit the waters are all collected into solid wooden launders, some 400 feet in length. The trough bifurcates below the mine; one fork discharges its load of foaming yellow water into the lower channel; the other turns a wheel which works the syphons and drawing pump, †—a " sack " or wooden tube, with leather joints, which should be replaced by caoutchouc. ‡ The mine, though somewhat wet, is thus kept in order.

The name is "Caldeiroẽs," not "Caldrones," as John Mawe writes ; he justly, however, describes them as "les creux, qui étaient auparavant des remous"(ii. chap. 2).

* Not Amocafra as written by Castelnau. Tavernier mentions "little iron rods bent at the end," and used to "draw diamond sand and earth from the veins."

† The usual pump is called Bomba, the one above mentioned is known as Buxa de Saco.

‡ In this part of the Brazil several trees are supposed to be capable of supplying caoutchouc. In 1785 — 1787 Ferreira noted the "India rubber" of the Hancornia speciosa. " Resina elastica e concreto succo lacteo arbor vulgo Mangabeiras—in hac observantur proprietates ususque gummi elastici." The people seem to think highly of this source of caoutchouc. I do not.

These works must be renewed every year. At the end of the dries the moveable plant is taken down for use during the next season. In November, when the rains set in, the dam is swept away; the height of the inundation here averages twenty-five to thirty feet, and has risen to forty. The uncertainty of the seasons renders diamond mining far more precarious than any other industry which depends upon the weather. Of course, the longer the dries last the better; and miners gratefully remember 1833-4, whose prolonged drought followed closely the Anno do Rato, or Rats' year, when those rodents appeared in swarms. * Usually the wet season ends in April; in 1867, however, showers fell even in July. This incertitude, combined with many other hazards, serves to explain the gambling nature of the pursuit. "If I hit upon a pocket of diamonds," said an Englishman to me, "I will go home next year." But the "if" points to a contingency far less to be expected than breaking the bank at Baden-Baden.

In former days, the diamond diggers, like the gold diggers, contented themselves with washing the rich superficial Cascalho; after which they removed to another place. It is but a short time since "deeper winnings" have been commenced, and the originators had to endure the usual amount of ridicule, in addition to the great expense. They have now silenced the laugh by winning the day: the "old school" revenges itself by predicting that the "luck" cannot last. This Canteiro mine was held to be exhausted, valueless, when Sr. Vidigal, who deserves to become a Podre de Rico,† took it in hand. A most energetic and progressive man, he ventured £6000, here a fortune, before getting the mine into proper working order. Some 6400 pounds of gunpowder are annually expended in blasting. The outlay during the last year was 25:000$000, and the income was 80:000$000; this year it may rise to 100:000$000.

My host employs during the digging season 300 slaves, worth £120 to £150 per head. The hire of each hand, food included, is about 1$200 per day, and the monthly expense is £750. As is general amongst Brazilians engaged in any pursuit that requires head-work, Sr. Vidigal complained bitterly of the servile

* In parts of the Brazil rats are supposed to swarm every seventh year, when the bamboo flowers.

† "Rotten with riches," an expressive conversationalism.

labour-market; he wishes to dig by night as well as by day, but the smallness of his gang compels him to begin at six A.M., and to end at six P.M. Another especial grievance is the prevalence of theft. Some mine owners go so far as to declare that almost all the finest stones disappear. A receiver of stolen goods settles near every new digging, as surely as a public-house follows the Hydropathic Establishment; and here, as elsewhere, the broker is generally richer than the diamond proprietor. President Jefferson, of Virginia, desired that a sea of fire might roll between Europe and the United States. Sr. Vidigal would prefer, and justly, to see a tunnel or a bridge.

The desmonte which we have just seen carried up in platters is disposed of in the readiest and most suitable way. When the rich Cascalho,* or Cánga,† is struck, the labourers transport it up the left bank, and dispose it in heaps (amontoadas) near the Lavadeiro, or washing place. In this shed I at once recognised the drawing familiar to my childhood, and copied from John Mawe into every popular book of travels. I remembered the long thatched roof of the Mandanga mine, with a stream of water passing through a succession of lengthy boxes; the four inspectors in straw hats perched upon the tallest of stools, and armed with the terriblest of whips; whilst the white-kilted sable washers, in a vanishing line, bent painfully to their tasks, and one of them, in an unpleasantly light toilette, was throwing up his arms, to signify " Eureka." It was written that " when a diamond is found weighing seventeen and a half carats (my innocence did not remark that "half"), the negro is entitled to his liberty—is crowned with flowers, and is entitled through life to look for diamonds on his own account." How I used to sympathise with that happy black person, little thinking in my simplicity, as does many a philanthropist, that he was likely to die an early death from a disease which may be described as consisting mainly of want, drink, and debauchery!

---

* Generally called "Cascalho corrido" (water-washed), opposed to Cascalho virgem, the pudding stone. Its substance is quartz of many varieties and colours, clear as crystal, yellow-white, slightly transparent, opaque and dark.

† The Cánga of Diamantina is a conglomerate of quartz, mica, and other components pasted together with red-yellow iron clay, and covered with the dark,

ferruginous, shining, metallic coat which gives to it a name. It is eminently diamantiferous as well as auriferous. M. Sipolis showed me a fine stone embedded in it, of course the result of water washing. This amygdaloid has always consistency or body (corpo). When broken up it becomes "gurgulho de Cánga." For other particulars, see Vol. I. chap. 21.

"Cánga," in its agglutinated form, is

The reality of the Lavadeiro is an open thatched ranch, built "convenient" for the master's eye, and one end, which is slightly depressed, is set off for the use of the panner. The total length may be 35 to 40 feet by one-third of that breadth; but the size is of course proportioned to the number of washers at the Canteiro. One of the long sides is occupied by a line of nine "bacos,"* three-sided troughs of rough wood; the poorer owners make them of flat stones, clay slates, or slabs of the granular, quartzose and laminated Itacolumite. The troughs are each four feet long, three feet broad, and one deep; they open with a little slope towards the inside of the shed, where the water is, and there is a cross piece to arrest the heavier material.

As the Brazil borrowed her gold mining through Portugal from the Romans, so she has taken her system of diamond washing from Hindostan.† There the season was in January when the rains had ceased, and the rivers ran clear. The diamantine earth was carried into an enclosure, surrounded by a wall from two spans to two feet high, with little drains at the foot; this served as a "baco" or "batedor." Water was added, and the mixture was left for a day or two till it became mud. The mass was again watered, and loaded with soil to press down the mud, after which the drains were opened, and the earthy matter flowed off. The residuum of gravel was again covered with water if not clean; when dry it was sifted in baskets like grain for the sand to drop through. It was returned to the enclosure, spread out with a rake, and beaten with long staves or wooden pestles; pebbles had been used, but they flawed the stones. After this it was resifted, spread out again, and collected in one spot, when the diamonds were picked from it. ‡

The washing here begins with the rains about November. The upper parts of the troughs are charged with Cascalho, and a hand standing before the open end or at the side of each "baco" dashes water from a shovel, often a bit of wooden platter, upon the contents; he then stirs with the fingers the mass to relieve it of the worthless earth, dust and clay, till the water runs

---

often applied, says Dr. Couto, to ochres of copper. When Mr. Emmanuel writes "Takoa Carza," I presume that he means "Tauá," felspathic clay, and "Cánga."

* These in older books are called Cuyacas; they seem to have been then made

larger, often three yards long by two broad.

† The stone there occurred in soil, gravel, and silicious grit (Itacolumite?)

‡ Dr. John Francis Gemelli Careri's Voyage round the World. 1683.

clear, and this washing may be repeated. Thus a pocket of diamonds is sometimes, but very rarely, hit upon. The fortunate slave no longer claps his hands in the old style of signal. He may receive his freedom after finding a stone weighing more than an oitava and a half; not by law, however, but in order to encourage the other labourers.

This preliminary ended, the Cascalho, now technically called "arêas" or sands, is made over to the panner. His implements are two wooden basins like those used in gold-washing. The peneira or sieve-pan is fitted at the bottom with a bit of tin pierced with holes, averaging six to the inch, and arresting stones of one vintem (half a carat); the sizes, however, vary as required. The other is the common batêa with the central depression (pião) into which the diamond, like gold dust, sinks by its superior specific gravity.

The washing (lavagem) begins in the batêa. It is charged with the rich Cascalho, mixed with sand and water to form a paste in which the gem will sink; the usual rotatory motion is given to the pan, the surface water is poured off and the upper useless matter is removed with the hand, more water is added, and the operation continues. The next process is sifting (peneirar), the pierced pan being held over the other batêa. After this the finer sand which falls into the under pan is washed and becomes "côrte," from "córtar," to cut or stop. When washed once more it is "recórte." The gravel may be thus treated a dozen times or more, and precious stones, of course very diminutive, will still be found in it. A good washer takes from half to three-quarters of an hour in order to exhaust a single pan-full. After sifting the sand is called no longer arêas, but canjica grossa, and the pieces are smaller in the latter than in the former.

Magnifying glasses are not yet in use, yet they would save much trouble and prevent loss. The present rude system is very severe upon the sight, which soon fails; past twenty-five few eyes can be trusted, and children are always the best washers.[*] It is during this treatment that robberies are mostly effected. Few swallow the diamond, not because it is considered poisonous, as by the Hindu,[†] but on account of the difficulty of doing so

[*] Thus in Hindostan Tavernier tells us that children were the best judges of the water, weight, and clearness of the diamond; he gives a pleasant description of the boy purchasers and their boy principal.

[†] The Hindus, it is well known, consider powdered diamond to be a deadly poison, and all old Indians remember the

unobserved. In India the miner jerked the stone into his mouth, or stuck it in the corner of his eye; twelve to fifteen overseers were required per gang of fifty light-fingered men. The civilized thief pretends to be short-sighted, and picks up the plunder with his tongue-tip. A favourite way is to start as if frightened by a snake, and thus to distract the attention of the superintendent, who, if " clever," is wide-awake to the trick. Most of the stones disappear by being tilted or thrown over the lip of the pan during the washing, and are picked up at leisure.* They are easily sold to the huckster, the pedlar, or the keeper of the nearest groggery. Thus may be explained the number of slaves who have purchased their liberty and taken to the bush. Even the white man has owned that his first impulse is always to secrete the diamond.

In the evening I met Mr. Thomas Piddington, a Cornishman, who, thirty-two years ago, came out as a miner, and who during upwards of a generation has not seen his wife or children. Yet, to do him justice, he always talks of returning " home," and perhaps he might do so, but for an unhappy habit of being generous to the extent of double his means. He has turned his hand to and from everything between a pump and a bridge, and he is generally consulted in their difficulties by the mine-owners of all the country side. A fine-looking man, with straight features and jovial countenance, he is still the model of a Britisher, and he would hardly be persuaded that I was not an American; in fact he probably still preserves his opinion. He urged me to visit one of his chums, a Mr. Aaron, who is diamond washing at Quebra Lenha near the Santa Cruz village, on the Jequitinhonha River, twenty-three leagues from Diamantina. Time, not inclination, was wanting to me.

The night was cold, the stream was dark and sullen, and heavy clouds gathered in the north, making my host look glum; a few showers at this season are sadly damaging to the owners of diamond mines. On the next morning we arose early, for we had hard work " cut out " for us. After coffee we rode down the very rugged and troublesome left bank of the Rio das Pedras; a shorter and better path runs along the right. Close to the Can-

---

case of the great Commissariat Agent who came into court with a small packet under his waist-shawl, determined to swallow it if cast in his suit. It can only act mechanically like coarse powdered glass, formerly given to dogs as an anthelmintic, by abrading the surface which it touches. I have known cases in which the latter has been tried in the Brazil.

* Many a wager has shown that the black can rob his master under the latter's eyes.

teiro is a smaller "Service," also belonging to my friend Vidigal;
at this season it employs about a score of slaves. Above it is a
good site for a house, with the essentially useful capability of
overlooking the work; but my host is a philosopher, satisfied
with his hut as long as it brings money; he will never have a
better building until it is built for him. The country here is
pretty, and the contrast of blue sky, white sands, and a profusion
of the purple Quaresma, which grows about in clumps, makes it a
Wady in the waste. The land, where not stony, is productive, as
was proved by the fields around the Fubá Mill. My guide
pointed out to me certain red cuts and spoil banks at the bottom
of a small Gupiára on the further side of the stream. Here, some
years ago, one José Joaquim da Souza saw the true diamond
formation thrown to the surface outside the nest of the large
plantation ant (Atta cephalotis, the Taó of the Tupys and the
Formiga da Roça of the Brazil). Before purchasing the ground
he cleared 150 oitavas (nearly four lbs.) of diamonds, and at his
death he left £6000.

After half an hour we forded the Rio das Pedras, a notoriously
dangerous stream: but lately it had drowned two boys. I readily
recognised from afar our destination. The house looked neat,
and the orchard-garden, rich in oranges and other fruits, was
prettily laid out; in fact there was some flavour of the old country,
pleasurable—when not too strong—in a new land. The most
curious growth is the Cipó Jiboia,* the "boa," or "snake"
creeper, so called from its form; the juice they say forms excellent cement, and cracked china mended with it will, when thrown
on the ground, fracture in another place. This would be a boon
to many a notable house-wife.

Dr. Dayrell, my countryman, of Barbadoes family, originally
from Bucks, can correct Rokeby in the matter of his ancestor
"Wild Darrell" of Littlecot Hall, who burned the baby. After
taking a London degree and marrying, he came out in 1830 to
the Cocaes Company, and he can tell many a curious tale touching
the early mines. For the last thirty years he has been settled at
Diamantina, where a large family of sons and daughters has grown
up around him, and where, much to the detriment of his professional
prospects, everybody is now his "gossip." He has a house in

---

* Or Giboia, the boa constrictor, from "ji" or "gi," an axe, and "boia" or
"boya," a serpent, because it is supposed to strike like a hatchet.

the city, and a fazenda of some 1200 acres; all his sons have found employment, and he looks with indifference even at the prospect of becoming lord of the old manor-house.

Dr. Dayrell kindly consented to accompany us, threw his holsters across the mule saddle, and whistled his dog, a half-bred English mastiff of the Morro Velho breed, now unfortunately becoming extinct. He had learned to be cautious, having been twice shot at in the Serra de Grão Mogor, once by mistake and once with malice prepense. We rode down the right bank of the Rio das Pedras to a little Lavra where one of the doctor's sons, Mr. Felisberto Dayrell, was working with a score of hands. The property is hired and has produced daily 2$000 per head; with industry and economy it may turn out well. The "Corrida" is a miniature of the Canteiro mine; there is the dam, but of trifling size, and the pit is still very shallow.

Beyond this point we found the road rough, and the river valley much turned up. After about a league we reached the Ponte de Santo Antonio, named after a rich Córrego, which has caused the growth of an Arraial. The troughs worked last year by Sr. Antonio Baptista still lay on the ground. The Córrego do Mel joins the Rio das Pedras above this Devil's Bridge, and the joint channel is hideous with jagged cruel rocks extending almost across. The blocks are of the hardest crystalline Itacolumite, showing a distinct cleavage: one kind is the green (Cabo verde), whilst the other has a ruddy, purplish blush, the effect of iron. Both glitter and sparkle with mica.

Accompanied by Mr. Carlos Dayrell, another of the scions, we reached the Barra da Lomba Mine. This Serviço, worked by the concessionists, José Bento de Mello, José Julião Dias Camargos and others, deservedly enjoys a high reputation. During the last year a single share yielded forty-one oitavas, or above five ounces, worth £4000. The system was that of the Canteiro, but the works are larger, the pit is deeper, and the labour is more dangerous. The dam extended half across the Rio das Pedras, here a much more important stream, and cut off the water from the excavation on the left. I descended about 180 feet along a slope of 45°—50°, and found the subterraneous part very narrow and close, as the workmen were obliged to use lights, and those lights were torches.

The Lomba was unwatered by a pump which John Mawe

sketched in 1801, and which Caldcleugh compared with the irri-
gators of China. This Caixão de Rosario, or Macácu,* borrowed
from the Hunde, or Hundslauf of Freyberg, is on the principle
of elevating-buckets : squares of wood disposed at intervals in
endless string, passing up a long narrow trough, which they fit
tightly, and working over the axle of a water-driven wheel, raise
the drainage. As I have before remarked, the only labour-saving
machine bequeathed by Portugal to the Brazil is the wretched
old Monjolo-mill, rudest of Oriental contrivances. The art of
mechanics is at as low an ebb as on the southern shores of the
Mediterranean, and we still recognise the appliances described by
Piso and Marcgraf in 1658. I found in the most civilized
diamond-diggings of Minas Geraes no trace of kibble, crane and
pulley, or rail, no knowledge of that simplest contrivance a
tackle ; the negro was the only implement, and he carried as
much as a schoolboy would stuff into his pockets—a pair of
buckets would have done the work of a hundred such men.
Even the Hindus used great wooden wheels turned by hand
labour to work the steel plates upon which the diamond was
cut. Important improvements, however, can come only from the
example of a more constructive race. I was asked my opinion
about the system, and suggested a few of the simplest modifications ; they were found to be unpractical, and did not meet with
favour. In this point many Brazilians resemble the phreno-
logical patient, who will swallow unmoved the largest draughts
of "soft sawder," but who makes wry faces when it is suggested that a single organ may be "somewhat deficient in
development."

We breakfasted at the Lomba with new appetite. The meal
is usually eaten at a late hour by mine-owners and diamond-
diggers, who give the greater part of the forenoon to their work.
The style is very patriarchal. The head man sits at the top of
the table and drinks from a silver cup, whilst all his overseers are
ranged along the sides, and disappear immediately after coffee.
Despite the "difference" about machinery there was no want of
cordiality on the part of my hosts.

From this Serviço we made for Diamantina by a vile line some

* Former travellers describe the Macácu as a "series of wooden cogs passing
up a square trough." Mawe, vol. i.,
French Edition, has given a sketch of the
machine.

twenty miles long, leaving the highway on the west. Happily for me I was mounted upon a mule as good for bad as it was bad for good, roads—not an unfrequent case. The only bridge was broken, and the muds were deep; the bridle-path was all up and down, and the banks were unpleasantly steep. The vegetation, Peroba and Copahyba, Monjolo and Braúna,* seemed to be as hard and stony as the soil, here justifying the popular belief in the concomitance—or perhaps I should say, the consequence. We passed to the left of the Maravilha, or Wonder-Mountain, which here appeared to be divided into two lumps. That to the north-west had a sheer fall of immense height, a grim, dark wall, up which only an insect could creep; from the south-east the ascent is probably easy. At the base were white holes and heaps awaiting the rains, and the summit was feathered with vertical slabs of stone emerging from the thin scrub.

Under a broiling sun we pursued our way over the barren hills that bear the diamond. We passed sundry forlorn-looking thatched hovels, at this season all deserted. The first stood near the Ribeirão do Inferno, where certain wet-weather diggings called Mata-Mata,† belong to Sr. José Juliano and Company. The next were the washings on the tributary Ribeirão do Palmital ; they are the property of the Collector Sr. Venancio Morão. Shortly afterwards we struck the southern highway by which we had left Diamantina, and between the gloaming and the mirk we found ourselves once more under the hospitable roof of Sr. João Ribeiro.

After this experience of two days we may venture to set right Mr. Harry Emanuel, who, in his carefully written book, ‡ almost ignores the Diamantine formations of Minas Geraes in favour of Bahia. Thus for the last three years the cotton of São Paulo has, much to the disgust of the Paulistas, appeared in the London market misnamed " Rio Cotton." § Minas began her labours

---

* Often written Graúna. The latter is also the name of a bird with shining black plume, from Guira (avis) and una contracted from pixuna (nigra).

† " Lorsque l'on découvrit des diamans dans cet endroit, le peuple s'y précipita en foule ; des rixes s'engagèrent, et de là vient, dit-on, le nom de Matamata (Tue-tue)." St. Hil. (II. i. 64), from Spix and Mart. Reise i. 452.

‡ Diamonds and Precious Stones, by Harry Emmanuel, F.R.G.S. London : Hotten, 1865.

§ "Provinces like São Paulo, where a foot of ground had never before been planted with cotton," says Prof. Agassiz (A Journey in Brazil, p. 508). But the Province of São Paulo has ever been celebrated for her cotton cultivation.

with the seventeenth century, and in 1732 the Lisbon fleet carried to Europe 1146 ounces of precious stones. We read (p. 59)
" In 1754 a slave who had been working at (?) the Minas Geraes
was transferred to the district (?) of Bahia," and that thus emi-
gration set in and exploitation began. But the great Province
of Bahia commenced to work her Chapada or diamantine plateau
only in 1845—1846. In the same page we find "the most
productive district is at the present time the Province of Mato
Grosso, in the vicinity of the town of Diamantina." This must
refer to the city which we have just visited in Minas Geraes;
the Mato Grosso diggings are called (Rio, Arraial or Sertão)
" Diamantino." *

* Memorias Historicas (Pizarro, ix. 19, 20, 21, &c.).

CHAPTER IX.

THE DIAMOND MINE AT SÃO JOÃO.

" C'est dans ces lieux sauvages que la Nature s'était plu à cacher la précieuse
pierre qui est devenue pour le Portugal la source de tant de richesses."—*St. Hil.*
II. i. 2.

MR. GORDON had supplied me with " recommendations " to the
brothers Lieutenant-Colonel Felisberto Ferreira Brant, and Major
José Ferreira Brant. The family is descended from an ancient
governor-at-arms of Bahia, and, as may be seen in Southey and
St. Hilaire,* has taken a prominent part in the exploitation of
diamonds. The Major has a store at Diamantina, and the
Lieutenant-Colonel, during the temporary absence of his son-in-
law, superintends the important digging of São João. It lies
north-north-west of the city. I was threatened with the worst of
journeys, but the reply was, " There is no good pasture or bad
road in the dries ; there is no bad pasture or good road in the
rains."

About noon I set out, " convoyed " for a short distance by
Major Brant; M. Sipolis had half agreed to join me, when the
theft and flight of the negro slave-cook who fed the Episcopal
Seminary required his presence at home. Passing through the
Curral and by the Alto da Cruz, where the prospect was the
more enjoyable because now I understood its details, we struck
the high road to the west of the city. A party of young chas-

---

* Joaquim and Felisberto Caldeira
Brant, says Southey (iii. 624), were rich
miners of Paracatu. Under the Count de
Bobadella, the second became the Third
Administrator of Diamonds in Tejuco of
Minas Geraes, and both were bound to
organise a " Serviço " of 200 negroes to
work the two Diamantine Rivers of Goyaz.
Felisberto, accused of malversation, died
in prison at Bahia.

seurs, with guns thrown across their shoulders, was leisurely
sloping along. An over-love for "sport" has done as much
harm in the Brazil as the ridiculous "sparrow clubs" of a former
day threatened to do to England. I have mentioned the prevalence of the ant plague since the ant-eater has been killed out,
and the destruction of birds has increased the host of Carrapátos.
The scenery, too, has lost in artistic beauty; the brilliant birds,
as the Arára (Macaw), have disappeared from the coast, and taken
refuge in the Far West. It is to be desired that amateurs would
give ear to the sensible advice of Padre Corrêa, and attack vipers
and jaguars, instead of slaughtering the Tanager and the Orpheus-
thrush.

The cantonnier is not abroad in this part of the Brazil. The
ascents and descents over the normal waves of ground, subtended
by streams in sandy or rocky beds, with pure water or current
dyed slate-colour by the washer,* were of the worst. The land
was by no means deserted; many little mining stations were
scattered about, and frequent snowy heaps denoted "Serviços."
At 2.15 P.M. old Ferreira and I crossed the Córrego dos Morrinhos, and halted for coffee at the nearest ranch. The mistress of
the house sat coiled up on her bed like a Hindostani woman, but
her extreme communicativeness, and an approach to what we call
"chaff," made up for want of graceful posture. The semi-Oriental
and old Portuguese reserve begins to vanish as we enter the
interior, and to a Northron the effect is decidedly pleasant. I
did not ask the names of host or hostess, as they openly told me
that I was the Chief of Police from Ouro Preto, and they were
most anxious to know my business. They laughed to scorn the
idea of my being an Englishman. "If this be true," they asked,
"how is it that you do not know 'Nicholas,' † your countryman,

---

* The drainage is to the Rio Penheiro,
which falls into the Jequitinhonha, six
leagues below Diamantina. On the left
bank of the Ribeiraõ dos Caldeiroẽs is the
Serviço known as the Retiro de João Vieira.
The next important stream is the Córrego
da Prainha; then comes the Córrego da Se-
pultura, an ill-omened name, common here.

† Amongst the Southern Latin races
generally, and especially the Hispanian,
the individual is known by his Christian
name only; and as this must be taken
from some saint, and as saints are few,
nicknames are common. The family name,
which we use, is mostly neglected, espe-

cially in the case of northern strangers,
whose cognomens are so often unpro-
nounceable by southern organs; and thus
the foreigner is perpetually in a fix.
Even neighbours who have known one
another for years often ignore all but the
prenomens. The practice is of old date.
"Quinti," puta, aut "Publi," gaudent
        prænomine molles
    Auriculæ.
The surname also was rarely used
amongst us in the days of the Plantagenets,
and until the last fifty years the Christian
name was that of the people in certain of
our rural districts.

who is living within musket shot of us?" He was, they insinuated to me, one of the "perdidos," the lost ones, a poor wretch who spent his life in squalor and in liquor, when obtained by some precarious job. However, they gave me a good brew of coffee, and sent us on our way rejoicing.

We then crossed a long plain, a most likely place for game : only one Campeira, or prairie deer (Cervus campestris), showed at a considerable distance; giving good venison, it is much hunted. Castelnau mentions the Campeiro, and Prince Max. (iii. 109) suggests that it may be the Mataconi of Humboldt, the Cerf du Mexique (C. mexicanus) of naturalists, and the Guazati of Azara, who speaks of a white variety (albino ?). It prefers plains to forests, and runs with frequent bounds. The size is about that of the roebuck ; the tail is short, and the coat is a reddish brown. Here the people declare that it is the female of the Galheiro, whose large antlers prevent it from entering the bush, and whose flesh is fetid. It is the Çuçuapara* of the Tupys, and the Guazupucu of Azara; according to the older writers, it attacks man at certain seasons. This deer haunts the prairie and the marsh. It is short tailed, and about the size of a yearling calf. Its flesh is eaten in January, February, and March, after which it is said to be offensive. The favourite form is "Moquendo,"† roasted on the embers. The Mateiro, or forest deer, the Guazupita of Azara, called by the Tupys "Cuaçu rete," or "true deer," is of all the most common species ; it is white tailed, and stands about the height of a sheep; the dry, hard, lean flesh much resembles that of the cow (Carne de Vaca), especially the old cow. The Catingueiro, literally the Stinker‡ (C. simplicicornis), the Guazubira of Azara, lives, like the preceding, in woods and well-clothed valleys. It is supposed to shed

---

* More correctly Çuaçú-apára, a word applied to both sexes. The Tupy Dict. declares that it has large horns, and feeds in the Campos.

† Amongst the Botocudos, "bacan," pronounced "bacoun," meant flesh,; and the Tupys had "mocaém," to toast in the flame. In Tupy also, according to Sr. J. de Alencar, Bucan was the implement with which meat was roasted, and the origin of the French boucaner. The indigenes smoke-dried their meat-provision for journeys or campaigns by hanging it upon a little gallows over a wood fire, or by suspending it to the fuliginous thatches of their huts. Hence is derived the Brazilian " moquem " and the verb Moquiar (St. Hil. III. i. 269), synonymous with the boucan of the buccaneers. Moquem has become the name of many country places in the Empire.

‡ So the word was explained to me by Dr. Alexandre. The Tupy Dict. writes Çuaçú-caatinga, the deer of the second growth (Mato rasteiro). St. Hil. (I. i. 337) makes the fetor proceed from "une matière d'un vert noirâtre que remplit une cavité profonde que l'on trouve entre les deux sabots des pieds du derrière."

its very short, straight, branchless horns; it is dock-tailed, and the brown-coated body is apparently too heavy for the slight legs, which are disposed at an angle fitting the animal for long high buck-jumps. In shape it resembles the Pallah, or hog deer of Sindh, and even the Brazilian rodent " Paca " (Cœlogenys Paca). Besides this, I heard of a marsh deer (C. paludosus), the Çuaçú-pucu), sometimes erroneously written Guaçu pucu, and the rare Bíra, a small red deer which is said, when pursued, to leap upon a tree branch. But the fallow deer mentioned by Mawe have not yet been discovered, nor have the antelopes which Koster has placed in the New World.*

Creeping up a bad hill, pitted with the deep gutters, and dotted with the loose stones of the normal Itacolumite, we saw, far to the left or west, amongst the peaks of the Cerro Frio group, the curious formation known as the Tromba d'Anta, the Tapir's trunk.† From this point it much resembles the Itacolumi of Ouro Preto, a huge monolith raised at an angle of 50°. Another hill, and below us on the left was the large mining establishment known as the Chapada. Yet another long slope and we struck a high grassy plain, where nothing taller than a foot could face the fierce north wind, which caused the leaves to droop in the lower levels, whilst the fiery sun made the wild flowers shrink and wither. Here we sighted the Arraial of São João do Descoberto, considered to occupy the highest site in the Municipality.

The village lies in a shallow hollow close to the mines which made it. To the west is the " Morro Redondo," a dwarf quoin crested with a tall cross; eastward is the cemetery, also with its cross. The single street boasts of a humble wooden chapel in a dwarf square. The " Almanak " (1864) gives it 2000 souls and 300 houses, a figure which I should divide by two. The tene-

---

* I have seen a large red skin brought from Rio Grande do Sul ; the people had no name for the beast but Cervo. The Tupy Dict. gives as the native names of the Cervidæ : 1. Çuaçú-tinga (white), the smallest. 2. Çuaçú Cariacu, so called from its sleeping in the thickets, and showing only its back. Ferreira explains the word as " Caa," foliage, " ri," many or much, and " acú," that exposes itself. 3. Çuaçú Anhanga, the devil-deer, so called because its flesh is held to be injurious to those suffering from fever or

syphilis.

† The Portuguese, who ignored the Tapir, called it Anta, or buffalo (F. Denis, Anta or Danta " buffle ") : thus their ancestors had named elephants Lucanian bullocks. On the other hand, the Tupys, never having seen black, called the bull Tapy'ra oçu (big Tapir), and the calf Tapy'ra Curumim Oçu (Pappoose of the big Tapir). We have corrupted the word Tapy'ra to Tapir ; Brazilian purists prefer Tapyr.

ments are the usual taipá, mostly whitewashed, of the door and window order, very narrow and somewhat deep, roofed with thatch or tile. Each has a large "compound" to defend the vegetation from the rudest Boreas; the material is puddle or dry stone, here and there eked out with stakes and other contrivances.

Turning to the right we made a crest our "espigão mestre," whose watershed is north to the Jequitinhonha, and south to the Rio das Velhas. On clear days it commands a view of about eighty miles in diameter. To the west is a bald Campo, eastward lie piles of jagged rock; in front, placed for shelter a little below the hill, stands a long, low, single-storied house, with a small chapel at one end, and looking upon a tall black cross, a pit full of muddy water, and a vegetable plot enclosed to keep off animals.

According to custom my Camarada had ridden forward with my letters. The mistress of the house met me at the door, and hospitably asked me to dismount. I found the host dining with sundry men and youths, relatives and employés. The work of refreshment soon over,* we repaired to the digging. It is known as the Duro Mine, because when the diamond was first "won" the sinker had met hard ground—presently to become soft and soppy as that of the neighbouring pit.

We found a large hollow, which at first glance suggested the Esbarrancados, or water-breaches, so numerous in Minas Geraes. The shape was an elongated horse-shoe, with the major axis disposed from south-west to north-east, and the heel draining towards the Jequitinhonha River. The maximum depth may be ninety feet, the breadth 300 yards, and the length about double. The material is a hardened paste of clay, whose regular and level stratification argues it to have been deposited in shallow water. The eastern side of the gap is the more ferruginous formation

---

* Brazilians eat nearly as fast as the citizens of the United States. I have met only one who "took time over his meals;" and indeed this is the rule of the world. In the nearer East a man sits down with a pious ejaculation, swallows his quantum, ends with drinking water, rises with another pious ejaculation, washes his hands, and with frequent eructations, applies to his pipe. Those who amongst us write "Manuals of Health" never forget to dwell pointedly upon the necessity of food being thoroughly well insalivated before it is swallowed, and they allow at least half an hour to each meal. I presume that the necessity, if it exists, arises from the artificial habits engendered by civilization, and the practice of eating frequently and at regular hours when the stomach does not call aloud for another supply.

(terra vermelha); on the west it is mixed with beds of white sand. Below one foot of brown soil the argillaceous matter has the usual staining and marbling, glaring white like fullers' earth with felspar and kaolin, chocolate-brown or rapé-coloured with organic matter, blue-green with traces of copper, pink and rose-purple and dark yellow with various oxides of iron, especially hæmatite, and dark steel colour with oxide of manganese. Thus old travellers describe the diamantiferous pits of the "Mustapha nagar circar" as a peculiar fat white clay associated with iron-stone.

We zigzagged down the easy slope of the eastern wall, which everywhere bore marks of the pick. Here the "hydraulicking" of California, where a fall of water hollows out chasms 250 to 300 feet deep, might be applied with great advantage. The richest lode (corpo) is No. 3, or the highest. The strike of the ribboned clays is north and south, bending eastward. The lode inclines towards the higher grounds, and thus the owner hopes to find the gem-bearing strata spreading over the crest or watershed ridge which forms his property. Through the ferruginous sand-stone (borra) and the white felspathic matter run dykes and lines of fragmentary rock crystal, sometimes fibrous like arragonite, and often finely comminuted. Large pieces of imperfect specular iron and thin strata of quartz, yellow and brown at the junction, thread the argile, and I was shown a specimen of fine sandy conglomerate, blackened and scorified by the injection of melted matter. The characteristics of this upper lode are a drier clay, silica, a trace of copper, of iron-cement, and of Cánga in small pieces; when the specular iron is in large pieces and abundant the rock is rich in gems. Its "agulhas" are iron-like bundles of needles welded together by intense heat: some are double, the fibres coming at obtuse angles. The "Agulhas Côr de Ouro" have a burnished coppery surface, whence the name.* Through-out all these corpos the diamonds are small, averaging perhaps a little under one grain, or 64—72 per oitava; they are mostly crusted superficially with a light green tinge.

Lower down we came to the middle or second body. Here the "tauá" (felspathic clay) was stiff and sandy, marbled with a fat, blue, muddy marl, which leaves upon the fingers a greasy steely

* The owner informed me that he had sent specimens of all his minerals to the Institute of Civil Engineers, London.

streak. It also yields a dark olive-green argile harder than the rest; like all the others it has consistency in situ, but when removed it crumbles to pieces after drying. Lieutenant-Colonel Brant gave me from this corpo a fragment of hard large-grained clay, reddish coloured with oxide, and showing a small brilliant imbedded in it.

We then descended to the lowest formation. Here the clay contains very little sand, and much stained; the colours are white and blue, red and yellow, rosy, spotty, and in places dyed as with blood. Here also are found the "Agulhas" in streaky bundles of iron like asbestos. The sole of the pit is uneven with working, and in places "horses," "old men," and long walls of stiff clay have been left standing amongst the holes and gashes. From this point the several lodes were distinctly traceable in the walls of the basin. A deep draining trench divided the length, and at the north-eastern end was a washing place, a shallow, muddy pool, faced by two concentric circles of staked fascines, to prevent the slime from falling in.

We then walked to the north-eastern end, and found traces of Messrs. Rose and Piddington. Rails, 600 fathoms long, had been laid down, and a white-washed towerlet denoted the engine-house, where a raising pump of three-horse power enables the mine to work throughout the year. The washing apparatus under the neighbouring shed consists of a "batedor," or stone-faced pit, eighteen feet long, nine broad, and eight deep; the clay tilted in it by the "trolleys" is here first puddled. Thence a stream of running water washes it down a succession of bolinetes or bulinetes,* coffin-shaped troughs like Canôas, but much larger. They are revetted with masonry, and each is provided at the lower end, where the slope is, with a batten or cross piece of wood to prevent the heavier substances from being carried down stream. Very few hands were at work. Formerly the Duro employed upwards of one hundred negroes, a number now reduced to half, and looking very "small" amid the vast area.

In the evening the host discussed the celebrated Rabicho of the Jequitinhonha River, seventeen leagues from Diamantina City. The "crupper" takes its name from a saco or bend, across which a cutting of one mile would expose five miles of highly

---

* St. Hil. (I. i. 255) makes the diamond "bolineté, un canal de bois beaucoup plus court et plus étroit que ceux dans lesquels on lave le cascalho,"

adamantine bed. A plan of this place has been made by Mr.
Charles Baines, C.E., and also a concession to exploit it has been
granted to the Commendador Paula Santos. Unhappily the law
in its unwisdom requires that companies for working diamond-
diggings must be composed of at least an equal proportion of
Brazilians to strangers. This is verily a relic of the old narrow-
minded colonial exclusiveness—it is not easy to see why the
diamond-coin should require an especial regulation.

Early on the next morning Lieutenant-Colonel Brant took me
to visit the Mina do Barro, belonging to Lieutenant-Colonel
Rodrigo de Sousa Reis, a wealthy mine owner, who is part concessionist of the Caéthé Mirim. We gained the Espigão mestre,
the great "Wasser-schied," and found lying *dos à dos* with the
Duro, another similar quarry, but somewhat larger and deeper.
A narrow slip of land was preserved for a path between the two,
but this will probably soon disappear, as Lieutenant-Colonel
Brant's prospects are best in this direction. It was a strange
view to one standing on the crest, with the two painted pits yawn-
ing on either side, and stretching away into the distance. On the
further bank of the artificial ravine lay the owner's house; the
large, pale clay square of buildings, with courts and outhouses
enclosed, as if for defence, reminded me of a fortified village in
Ugogo. We found nothing new in the "Barro;" like the Duro
it was drained by a trench; the washing pit was prevented from
caving in by stakes and fascines. A few negroes were removing,
under an overseer, the clays, coloured and white (Jiz), which
serve as guide to the diamond formation; and there was a
steam pump of four-horse power, with a tall useless engine
turret.

This diamond digging was discovered at a time and place when
and where no one dreamed of looking for the gem. An old
woman, who was in the habit of panning Cascalho gravel in a
little trickle of water from the gap, found that the precious stones
extended into the blue argile (barro azul). About thirty-three
years ago the digging was begun with a will, and presently it
passed into the hands of the actual owner, who has employed as
many as two hundred head of slaves. Other similar diggings
came to light, and the wealth was such that sometimes an owner
would exclaim, "O my God, are you doing this to cause my loss?"
The Duro is the legitimate offspring of the Barro, begotten, seven

to eight years ago, by Lieutenant-Colonel Brant, who judged, naturally enough, that if one side of a clay slope be productive, so might be the other. As has been seen, the progeniture has thriven.

       *   *   *   *   *

I left the Diamantine region, including the Duro mine, with regret. Socially speaking, it is the most "sympathetic" spot in the Brazil, according to the light of my experience. With an "enemy in the fortress" traitorously urging delay, it was not easy to escape from its hospitalities. My plea was the absolute necessity of an Englishman being punctual; I had promised to be at Bom Successo before the eleventh day, and the promise must be kept. This requirement is universally recognized throughout the Empire. Lieutenant-Colonel Brant accorded to me a reluctant dismissal, and the amiable Senhora charged me to return, and loaded me with kind messages to an unknown, and what might have been a theoretical, or even a hypothetical wife.

Old Francisco Ferreira was in no hurry to take the road once more. He was paid by the day—1$000—and thus interest combined with inclination to urge a little laziness. But neither cough nor groan, nor euphuistic phrase of the old eloquent, nor muttered anticipations of "Corrubiana in the bones," was of the least avail. I struck the direct road viâ Guindá to Bandeirinha, and on Thursday, September 5, 1867, after a day's ride of forty miles upon jaded beasts, that now fell twice every twenty-four hours, I found myself within the pleasant walls of Bom Successo.

As my Jaguára pilots did not profess to know much of the stream below this point, I engaged, with the assistance of Dr. Alexandre, a third paddle. He answered to the name of Antonio Marques, but was better known as "O Menino," the "Little 'un," because he was peculiarly tall, broad, and raw-boned, "a long, hard-weather, Tom Coffin-looking fellow;" moreover, he was grim and angry-looking as a Kurdish "irregular cavalryman." He had begun life in English employment at the Vâo Mine, near Diamantina, and he had mastered more than one northern habit, such as drinking and brawling. He had learned the world, he had travelled half-way down the São Francisco, and had struck overland to Piauhy; he had run up north as far as Maranham, and he had even seen a steamer. His price was somewhat exorbi-

tant—2$000 per diem, and he vainly attempted to instal himself
as pilot by ousting the good old "Chiko Diniz," who was worth
a dozen of him.   He greatly preferred conversationizing to rowing, and drink to both.   My temper was sorely tried by him, but
I kept it till we reached Varzéa Redonda.

CHAPTER X.

NOTES ON THE DIAMOND.

"The substance that possesses the greater value, not only among the precious stones, but of all human possessions, is adamas, a mineral which for a long time was known to kings only, and to very few of them."—*Pliny*, xxxvii., Chap. 15.

DR. COUTO (p. 127) described the diamond diggings of Bagagem which he visited, and named Nova Lorena, after D. Bernardo José de Lorena, Count of Sarzedas, and eleventh Governor or Administrator of the Minas Geraes captaincy. These lands, he shows, are of greater antiquity than the countries near the coast, as is proved by their degraded and water-washed forms. They are also the easier to work, having more of plain ground and larger rivers. The crystallisations of the Cerro or Diamantina diggings have smoother facets and sharper angles, whilst the yield is more regular and constant. On the other hand, the stones are small; 1000 oitavas hardly produce a single gem of one oitava. From Bagagem many stones, varying between three and six oitavas, have been taken, but by jumps, as it were. The water is fine and brilliant, but the shapes are more rounded and more deeply flawed, the effect of longer weathering and more water-rolling. Castelnau (ii. 231) describes, in 1844, the diamond diggings of Goyaz, on the Araguaya or Rio Grande. We lack, however, a modern description of the Diamantino diggings near Cuyabá, in Mato Grosso, and of the Bahian Chapada. The latter Province extends its wealth almost to the seaboard; gems have been found within one or two leagues of São Salvador, at the Engenho do Cabrito, and at other places near the railway. The Caldeirinos of Parahy, thirty leagues from the São Francisco River, and the lands between Crato and Icó, in Ceará, require inspection. I shall presently allude to the formation on the lower waters of the great artery. In the Provinces of São Paulo and Paraná, the rivers Parahyba do Sul, Verde, and Tibagy, have produced diamonds, whilst the best indications are found near the coast about Ubatúba.

Evidently the Brazil has a vast extent of diamantine ground reserved for future generations to work with intelligence, and especially by means of machinery.

Prospecting for diamonds is done as follows: The vegetable humus, the underlying clay, and the desmonte, or inundation sand, are removed with the almocafre, till the labourers reach the gem-bearing "cascalho," or "gurgulho." This first work is usually an open cut of a few feet square. The larger fragments of quartz are then removed by the hand, the gravel is washed in a " báco," " canôa," or " cuyaca," and, finally, the batêa is used.

After the prospecting (provas) a concession to work diamantine ground is directed to, and is easily obtained in these days from Government. The applicant specifies the limits of the extent which he proposes to exploit. The land is put up at public auction, any one may bid, and it is knocked down to the highest offer. The owner of the soil has the right of pre-emption, and if only 0$200 per braça (Brazilian fathom) be called, the proprietor can take it. After the death of the concessionee, the digging is inherited by his wife, his children, or, in default of other heirs, by his brother. For the use of the reach * in the Rio das Pedras, 13,000 braças long, Sr. Vidigal pays a tax of 1$000 per thousand, and Dr. Dayrell, within whose limits the Canteiro is, might for that sum have exploited it had he so pleased.

The diamond,† say old writers, unites all perfections: sparkling limpidity, lustrous brilliancy—the effect of its hardness—

---

* " Tiro do rio."

† M. Caire (La Science des Pierres precieuses, Paris, 1826,) observes that the word is derived from αδαμας (in Arabic and Persian, almas), " indomptable"—nullâ vi domabilis, because not to be conquered by fire. This is true only when the oxygen of the atmosphere is excluded from it, and when the heat is under 14° Wedgewood. He also notes that our modern word " diamond," " diamant," &c., by rejecting the " alpha privative," etymologically signifies the reverse.

the accidental colours of the rainbow, reflections that come and go with the vivacity of lightning; and, finally, it has "as many fires as facets." The structure is of thin shining plates closely joined, and thus it is easily split along the line of cleavage, which is parallel with the planes of the octahedron or dodecahedron.* The substance has been proved to be crystallised carbon,† but the origin is still debated. Some believe that the vapours of carbon, so rich during the sandstone period, may have been condensed and crystallized into the diamond. Newton, it is well known, argued from its great refractive power that it is "probably an unctuous substance coagulated." For reasons which will presently appear, it is evidently younger at times than the formation of gold, and it is possibly still forming, and with capacity for growth. Others have conjectured that the Itacolumite matrix may have been saturated with petroleum which has gradually disappeared from oxidation or otherwise, except where the carbon has collected into nodules, and has formed the gem by gradual crystallization.‡

As has been shown, the specific gravity of the diamond varies from 3·442 to 3·556, quartz being 2·600, and water 1·000; hence it is easily washed, and a practised hand distinguishes it by the weight. The index of refraction or quotient, resulting from the division of the sine of the angle of incidence in the vacuum by the sine of the angle of refraction in the vacuum, is equivalent to 5·0,§ water and plate-glass being 1·50, sulphur 16·0, and bi-sulphide of carbon, the most refractive liquid at present known, 37·0. According to Sir D. Brewster it slightly changes the light passing through it : older authorities remarked that it decomposes light into its prismatic colours, and shows a distinct phosphorescence after being exposed for some time to the sun, imbibing luminosity even through leather. Rough or polished it acquires by friction positive electricity, other precious stones

---

* Thus the test of striking with a hammer, often applied by those who have heard that the diamond is of extreme hardness, has destroyed many valuable gems. They were split with the grain or in the plane of the crystals. That "shocking the diamonds" (with iron levers) "causes them to be flawed" was taught by the Hindús to Tavernier. The file roughly applied to the girdle or edge is likely to chip it.

† It was, and perhaps still is, believed that a dissolvent of carbon is alone wanted to make the artificial diamond.

‡ I have seen it popularly stated that flexible Itacolumite is the matrix of the diamond, which is undoubtedly incorrect. Nor I believe do any of the Itacolumites contain petroleum.

§ It has been stated to be as low as 2·439 (Brewster).

being negative in the rough, and positive only in the polished state.* Old authors remarked that the gem when placed in the magnetic line of the loadstone neutralizes the attraction to a considerable degree. Most precious stones will scratch glass; the diamond cuts it with a peculiar creaking sound, hence this is a favourite test. † Another is the peculiar shock of diamonds rubbed together, which is more or less sonorous according to the hardness of the stone : ‡ this, however, requires long practical acquaintance. It gives to the hand a sensation of cold, a property shared with it by many other stones, and notably by rock crystal. Finally it is said that the diamond is the only stone which can scratch the sapphire.

As regards the matrix of the diamond, many popular errors are still afloat. It has been washed mostly in the " Cascalho " gravel brought down by streams and deposited either on the banks or in the beds. Hence books have determined that " the diamond is always found imbedded in gravel and transported materials whose history cannot be traced." Others are of opinion that the diamond was formed in the alluvial and arenaceous matters that accompany the Tertiary and Quaternary epochs. The accurate M. Damour, who wrote two conscientious papers § upon the diamantine sands of Bahia, tells us (p. 11) " Ces roches crystallines, servant autrefois de gangue au diamant, ayant été brisées et en partie détruites par l'effet des commotions qui ont remué et sillonné la surface du globe, à certaines périodes géologiques, *ne se montrent plus qu'à l'état de débris et de matières arénacées.*" Professor Agassiz (A Journey in Brazil, 501), " is prepared to find that the whole diamond-bearing formation is glacial drift." This, however, is qualified by—" I do not mean

<hr/>

* The electro-magnetic current strongly affects the diamond. I spoiled a fine rose-cut stone by allowing the ring to remain upon my finger when using a Meinig's chain. My attention was aroused by a peculiar rasping sound, and I found the corners of the diamond chipped and ground off as if a rough file had been applied to a bit of glass. Perhaps this may prove a labour-saving method of treating stones which require to be much cut. The " Odylic Sensitives " of Reichenbach see when "magnetized" a brilliant white light proceeding from the diamond ; and hence probably the idea that precious stones had specific virtues.

† Diamonds, especially those with acute angles, have been injured by violent rubbing upon hard substances. Pliny's process of testing them by anvil and hammer may easily split them.

‡ I have heard this asserted by some diamond merchants and denied by others.

§ Bulletin de la Société Philomatique, 5 Février, 1853, and Bulletin de la Société Géologique de Paris. 2ᵉ Série. Séance du 7 Avril, 1856. It is regretable that sands from other parts of the Brazil, from the Ural, from Hindostan, and from the Borneo have not been sent to this savant.

the rocks in which the diamonds occur in their primary position, but the secondary agglomerations of loose materials from which they are washed."

Many authors have mistaken the secondary for the primary formation of the diamond. The gangue, about Diamantina at least, is the white and red, granular and quartzose Itacolumite, which has been weathered and worn down by geological commotions.* This was suspected by Dr. Gardner, who observed that the matrix of the stone is not the " diluvial " gravelly soil, but the metamorphic quartzo-schist rock. It is not unknown to the people : the general idea is that the hard sandstone " pissarra " or psammitic grit bears diamonds when old, but not when new. The fact is easily proved. All the diggings which are not near or in rivers, lie at the base of some stony mass.† Diamonds have been found in the Itacolumite by several hands, and finally I have sent to England a specimen embedded in Itacolumite. Perhaps the day will come when the rock will be spalled, stamped, and washed for diamond-dust as if for gold.

According to miners in this part of the Brazil the best diamantation (to borrow the native term) is found in the gurgulho, breccia, or loose pudding of angular stones.‡ Wonderful tales are told of its wealth, how the discovery of five or six gems was made by pulling up a handful of grass—the picturesque detail has, since the days of Potosi, become a favourite legend, and has ever been carefully collected by the popular writer. The choicest specimen of a digging of this kind is said to be " O Pagão," at the head waters of the Caéthé-Mirim near São João. The next best supply (Mancha de diamantes) comes from the " Cascalho," which has been compared with boiled beans : of this the Rio das Pedras is an instance. The third habitat which we have visited at São João is the " barro " formation, which seems to contain all the others, mixed and degraded. It must, however, be borne in mind that the diamond grounds greatly vary in a country so immense as the Brazil. §

* In the crystalline Itacolumite I have not seen the diamond, but I can hardly doubt that it exists there.

† So Tavernier, speaking of the Gani or Coulour Mine, under the King of Golconda, where 60,000 souls were employed, remarks, "The place where the diamonds are found is a plain situated between the town and the mountains, and the nearer they approach the latter the larger stones they find."

‡ Castelnau (ii. 323) declares of the diamond-diggings of Diamantino (Mato Grosso), "Il n'y a jamais de diamant dans le gorgulho " (gurgulho).

§ Dr. Dayrell described it to me in the

As various are the indices of diamonds (pinta em diamantes), and almost every digging yields some novelty.* The chief signs of many are here given in order of importance, and their name united is the Formação Diamantina, Diamantine formation.

Cattivo (the Slave), of old called "escravo do diamante," and supposed to accompany it, as the pilot-fish does the shark. This includes at Diamantina bits of transparent, semi-transparent, or rusty quartz, silex, rock crystal, and especially spinelle.† The latter is transparent or semi-transparent, octahedrous (Cattivo oitavado), and with tolerably regular facets (facetas); it is distinguished from the diamond by its want of fire and inferior hardness. The "Cattivo Preto," or black slave, is probably Titaniferous iron, and the miners believe that when occurring in quantities it betrays the presence of black diamonds. These Cattivos in places are found strewed over the ground; they show that the diamond may be there, not that it is there. The same has been said of quartz, the "flower of gold." The word is applied to very different formations. Dr. Pohl translates it "thonseisenstein," oxidised hydrate of iron or the limonite of Bendant (St. Hil. III. ii. 144). A practical miner assured me that at the Chapada of Bahia "Cattivo" includes zoned

Serra de Graõ Mogor of Minas Geraes, a lode of soft sandstone, one foot broad, in containing walls of hard Itacolumite. He gave me a specimen of sand from Brocotú or Brucutú, near Cocaes, where spongy nuggets of Jacutinga gold abound ; it contains a small diamond, a ruby, a sapphire, and iron pyritiferous as well as specular. The curious formation called "Boart," and of which I shall have more to say, is also local. At Diamantina of Minas jt is unknown, and Bagagem produces small quantities. It is found at Sincorá, the Diamantine Chain of Western Bahia, and the largest supply is from the Chapada of the latter Province. I have remarked that in many places gold accompanies the diamond. Plato believed that the diamond is the kernel of auriferous matter, its purest and noblest pith, condensed into a transparent mass. Thus also we may explain Pliny's statement that "adamas" is a "nodosity of gold." Itacolumite is also the matrix of the topaz and the ruby. A specimen of the latter was shown to me : it was a small square stone of tolerable water, but too light in colour, not the real "pigeon's blood" of Asia. Garnets are found in handfuls, but they are valueless.

* John Mawe (ii. chap. 2) describes the diamond-accompanying substances as " Un mineral de fer brilliant et pisiforme (ferragem), un mineral schisteux silicieux ressemblant à la pierre indiqué 'Kiesel-Schiffer' de Werner (?), de l'oxide de fer noir en grande quantité, des morceaux roulés de quartz bleu, du cristal de roche jaunâtre, et toutes sortes de matières entièrement différentes de celles que l'on sait être contenues dans les montagnes voisines." Castelnau limits the "formação" to three kinds—Cattivo do diamante, Pedra de Osso, and Pedra Rosea, a violet-coloured grit. According to Tavernier the Hindús judged the land diamantine when they "saw amongst it small stones which very much resemble what we call "thunder stones."

† The Brazilian name of this crystal is, I believe, "Sаruá." Under this word, however, are probably included the hexahedrous fluor spar, corundum, and perhaps also certain titanates. The chrysolite suggests Pliny's description, " never larger than a cucumber-seed, or differing at all from it in colour."

quartz, chrysolite, bits of magnetic iron ore, iron pyrites, and so forth.

With Cattivo we must associate "Siricória," elongated prisms of chrysolite (Chrysoberil, Werner, and Cymophane, Haüy), of a faint yellow-green, sometimes almost white. Amongst the Cattivos on the São Francisco River I found a large proportion of straw-coloured topazes,* with sharp angles, and readily leading to error.

Pinga d'agua (St. Hil. I. ii. 6, "Pingo de agua") "drop of water." It is applied to rounded and cylindrical pieces of every size from a pea to a pigeon's egg; some are white, others rusty; the drops are transparent, semi-transparent, opaque, or zoned. They include cornelian, white topaz, and more especially quartzum nobile. The small diamond-shaped stones are the most prized. With the Pinga d'Agua we must associate the balls of quartz, called from their shape Ovos de Pomba, or "doves' eggs," and the pedras de leite, "milk stones," rounded and water-washed bits of silex calcedonius and agates. Both are clear and diaphanous, dull and opaque, or zoned and prettily marked with concentric undulations.†

Fava, a stone shaped somewhat like a broad bean, and varying in size from a pea to two inches in diameter. As a rule it is jasper, blood-stone, or one of the many varieties of white, brown, and yellow quartz. Many "favas," however, are clay revetted with iron, one-half to two lines deep.‡ The fava branca and the fava roxa are sometimes of pure silex or of crystallised quartz. Several appear likely to supply good blood-stone for seal rings.

Feijão, a haricot-shaped stone, rounded and rolled. It is also of different sizes, and is mostly of tourmaline (Schorl) or hyalo-tourmaline, like that which accompanies the tin-mines of Cornwall. The colour ranges between dark green and black, and the people believe it to have been glazed by great heat.§

* The Cattivos may be compared with the Bristol or Irish diamonds so often associated with bog-oak. They have been frequently taken to Europe, but with little profit. It is said that they break when being cut.

† Mr. Emmanuel (p. 126) says, "These topazes (i. e., of Minas Geraes), found in rounded pebbles, are perfectly pure and colourless, and are termed 'pingas d'agoa' or 'gouttes d'eau;' they are also termed Nova Minas (?). The Portuguese call them 'slave diamonds.'" Here there is evidently a confusion between the quart-zose "pinga d'agua" and the crystal "Cattivo." The term "Minas Novas" is taken from John Mawe (ii. chap. 3.)

‡ Marumbé, or Pedra de Capote.

§ I believe that the feijaõ is sometimes of jade, axe-stone, nephritis or nephrite, because used by Hindûs against "the pain of the kidneys." The aborigines of the Brazil employed it as labrets and other ornaments, and made their hatchets of this fine apple-green mineral, which is known to be soft when first taken from the quarry, and to become tough and compact by exposure to the atmosphere.

Caboclo, mentioned by Dr. Couto (p. 64) as Pedras Cabocólas, and explained to be Ferrum Smiris and rubrum, red with dark stains (mesclas). This jasper or petrosilex takes its name from the dull yellow tinge caused by oxide of iron. It is compact, and feebly scratches glass. The surface is polished and lustrous, as if it had been in contact with excess of caloric; the usual colour is of dark or light yellow, opaque, and verging on brown; and there is no peculiarity of shape except that the fragments are mostly flat. There are many varieties of the Caboclo. The C. Oitavado is that which has angles. The C. bronzeado, common in the Barra da Lomba, is dark yellow. The C. Comprido is an elongated bit of jasper. The C. Roxo is a compact red sandstone, possibly altered by heat. The C. Vermelho, common in the Caéthé-Mirim, is apparently cinnabar.

Esmeril,* in shape resembling the feijão, is mostly oxydulated iron. According to the miners, some stones contain eighty to ninety per cent. of metal. Of this stone, also, there are many varieties. The Esmeril Caboclo has a dull yellow tinge. The E. preto, in Gardner's opinion, is a kind of tourmaline. The E. lustroso is almost pure iron, often welded by heat to a fine breccia; it sometimes resembles a black diamond, but it is amorphous. The E. de agulha is a long, thin strip of iron-stone.

Ferragem, or Pedra de Ferragem, is either flat, bean-shaped, nodular, or rounded like a bullet. It is mostly of oligistic or specular iron, of dark purple or lustrous black. I have seen some specimens which are iron pyrites, and others are bullets of silex, making good touchstones of velvet-black colour.

Pedra de Santa Anna, squares and cubes of magnetic iron that affects the needle. The name is also applied to copper pyrites, and this is often found degraded to a mere sand.

Osso de Cavallo,† "horse's bone," which it resembles in appearance and consistence. The shape is long or round like an osseous fragment, and it appears to be pure sandstone (granular Itacolumite?) which has long been buried.

Palha de arroz, "rice straw," a fragment of light yellow sublustrous chlorite, slate or hardened clay-slate, resembling a cucumber-seed.

---

* Not Ismirim, as Castelnau writes (iii. 178). "L'oxide noir de fer, appelé ici émeri," says John Mawe (i. chap. 12). Spix and Martius explain the word by "Eisenglanz."

† Pedra de Osso (Castelnau, ii. 323). This "horse-bone" must not be confounded with the "Pé de Cavallo" or "horse-hoof," a yellow jasper, which merits its name.

Agulha, or Agulha de Cascalho, Titanic iron, in bundles or in single needles.

Casco de telha, cinnabar or reddish clay, yellow inside, and showing mica and talc.

Pissarra folhada, schists of different colours,[1] varying from a dull yellow white to black.

Pedra Pururucu, a light-coloured friable grit.*

---

* The following note is taken from the valuable paper of M. Damour (Soc. Geol. p. 542, April 7, 1856), describing the diamantiferous sands sent to him from Bahia. The numbers show the formations which occur most frequently.

1. Hyalin Quartz (the yellow is the occidental topaz, the blue is the occidental sapphire).

Jasper and Silex.

Itacolumite.

Disthene or Cyanite. This substance is easily distinguished; it is infusible by the blow-pipe, consists of little needles or thin-bladed crystals, the edges are rounded by rubbing, and the colours are pearl-grey, light blue and pale-green.

Zircon or Hyacinth, also found in the auriferous soil of California. This silicate shows well-preserved crystals more than a millimetre in diameter: it occurs in squares and prisms ending in four-sided pyramids, with the angles and crests sometimes modified. Some are colourless, others are brown, yellow, violet, or clear red.

Felspar, in rare water-rolled fragments of reddish matter, cleavable in two directions, which meet at right angles. It is not affected by acids, but is fusible before the blow-pipe. Melted with carbonate of soda it proves to be composed of silica, alumina, and a little oxide of iron, with probably some alkaline earth.

2. Red Garnet (almandine or precious garnet).

Manganesian Garnet (spessartine or deep red garnet). Density, 4·16. In dodecahedral rhomboids, very small bright crystals of a topaz yellow. The blow-pipe fuses it to a glass which becomes black and opaque in the oxidizing flame. The glass made with salt of phosphorus (microcosmic salt), and heated to redness with a little nitre, shows manganese by assuming a dark violet tinge.

Mica.

Tourmaline (green and black.)

3. Hyalo-tourmaline (feijão). Density, 3·082, scratches glass feebly. Under the microscope it looks like a number of small needles crossing one another: the fracture is fibrous. The dust is of greenish grey. Heated in a glass tube it disengages a little water: melted with borax, it gives a reaction of iron, and before the blow-pipe it swells and fuses to a brownish black or dark green scoria, which, after being subjected to burning charcoal, becomes slightly magnetic. The scoria can be decomposed by boiling in sulphuric acid; and burnt in alcohol it gives a green flame, showing boracic acid. Analysis also yields silica, titanic acid, alumina, magnesia, a trace of lime, soda, water and volatile matter. It differs from black tourmaline only by the presence of water and titanic acid.

Talc.

4. Hydrous phosphate of alumina, or Wavellite (Caboclo). Density, 3·14 in Diamantina and Abaété, and colour a coffee brown. Density, 3·19 in Bahia; tint rosy or brick-red, and shape rounded galets. Composition, phosphoric acid, alumina, a little lime, barytes, oxide of iron, and 12 to 14 per cent. of water.

Phosphate of white yttria, which M. Damour previously called Hydro-phosphate. Before the blow-pipe it becomes white without fusing; the lustre is the fat adamantine, and the colour white or pale yellow: it scratches fluorine and is scratched by a steel point. The irregular and rounded fragments have a double cleavage leading to a rectangular or slightly oblique prism. One incomplete crystal showed a pyramid with four faces, two large

As regards shape the rule is that the smaller stones are the most regular. The larger specimens seem to have no constant form or crystallisation; they are round, flat, or elongated, and generally truncated abruptly at one end, as if a piece were wanting. The facets, which when cut appear flat and even, are, in the natural stone, concave, convex, or rounded : hence the Abbé Haüy observed that the component molecules may be regular tetrahedra. Wallerius (quoted by M. Caire) assigns to the diamond three shapes, the octahedron, the plane, and the cube.* The normal form of the diamond, here as elsewhere, is the regular

and clean with an angle of incidence at the summit, amounting to 96° 35′ ; the two others, narrow and mirrory (miroitantes), had the angle of 98° 20′, whilst that of the neighbouring facets was 124° 23′ 30″.

Phosphate of titaniferous yttria, previously termed silicate of yttria, silica having been confounded with zirconium. Density, 4·39 : it feebly scratches glass ; it is opaque and of cinnamon brown. The rounded grains are pierced with surface holes ; it is also in square-based octahedrons, with facets like those of zircon. Boiling sulphuric acid decomposes it, leaving a white residuum. This substance is found in the auriferous sands of Georgia and North Carolina.

Diaspore, or hydrate of alumina. Density, 3·464 ; composed of bright crystalline blades of greyish white, resembling certain felspars. The composition is alumina, ferric acid, and water ; when this is disengaged by the blow-pipe, it becomes opaque and milky white.

5. Rutile, in small rolled grains or quadrangular prisms, with striæ along the major axis, ending in a four-sided pyramid with modifications.

Brookite, differing from rutile in having the crystal type. It is entirely composed of titanic iron. The only specimen examined was a flat prism striated along the major axis and ending in the dihedron, like the formations found in Wales.

6. Anatase (titane). Density, 4·06 ; bright, octahedrous, transparent or semi-transparent, and distinguished

from the diamond by inferior hardness and reactions before the blow-pipe. It becomes opaque, brown and reddish after an epigene, which converts it wholly or partially to rutile. These transformed crystals are hollow, and composed of a multitude of needles which cross in all directions.

Hydrated titanic acid ; of this substance no quantitative analysis was made. The whitish yellow concretional matter crepitates strongly, and disengages water in a glass tube ; and with salt of sulphur it gives reactions of titanic acid.

Tantalate. Density, 7·88 ; it is a black amorphous substance, which scratches glass.

Baierine, or Columbite (Niobate of iron) ; in flat striated and often regular crystals ; the dust is reddish brown.

7. Iron, titaniferous. Density, 4·82. Formula, 3 Fe O + 8 (Ti O³, Ta O³). It scratches glass ; the fracture has a semi-metallic lustre, and the dust dark olive-green. The black grains are almost all water-rolled ; a few crystals show rhomboidal oblique prisms of 123°.

8. Iron, oxydulated (Esmeril.)

9. Iron, oligist (rhombohedral, six-faced prisms).

10. Iron, hydroxydated.
  Iron, yellow with sulphur.
  Tin, oxide of.
  Mercury, with sulphur ; heated in a glass tube it gives a black sublimate.

11. Gold, free.
  * Mr. Emmanuel (p. 49) says, "The Indian diamond is generally found in octahedral, the Brazilian in dodecahedral crystals."

octahedron (Adamas octahedrus turbinatus of Wallerius), composed of two four-sided and equilateral pyramids, springing from a common base.   This is called the Diamante de pião, and it loses much in cutting.   With this primary are found the modified forms, the hexahedron or cube, the dodecahedron (twelve rhombic faces), the pyramidal hexagon (tetrakis-hexahedron of twenty-four faces), and others.   When the table and the culet of the fundamental system are worn down, the octahedron becomes a decahedron; the abrasion of two other points or angles (quinas) makes it a dodecahedron, a geometrically allied form, but approaching the spheroidal, and when two other edges at the girdle or base of the double pyramid disappear, it will number fourteen facets.   These rounded stones (tesselladas or boleadas, Adamas hexahedrus tabellatus of Wallerius) are locally known as the primeira formula, and they are preferred by the trade, as they lose least by lapidation.   There are all manner of derivations from the normal octahedron and dodecahedron, as the flat and triangular hemi-hedral, or half-sided diamans hemiodres maclés, the effect of secondary cleavage, called diamantes em forma de chapéo (hat-shaped); these find no favour.   The tetrahedrons (four-sided) are pyramidal, little valued when the vertices are acute.   There are also diamantes rolados (water-rolled stones, reboludos, M. Jay), which lose all their "pointes naives;" these are held, when round and oval, to be a good form. They may, when elongated, explain Pliny's "two cones united at the base;" they are often covered with opaque crust, and rugged like ground glass; in this state they are not to be distinguished, except by their power of scratching softer substances, from the Pinga d'agua.   Some of the latter, on the other hand, especially when of pure opaque quartzum nobile, so much resemble the gem in its "brut" or rough state, that many an inexperienced man has lost his time and his money.

The form of the diamond greatly influences the price, and thus it is that the merchant makes his profit.   He pays for size, weight, and water; he gains by the shape.   Purchasers on a large scale have boxes of metal plates pierced with holes, and acting as sieves (crivos).   Those shown to me were in sets of nineteen, and bore upon them the mark of Linderman and Co., Amsterdam.

The diamond greatly varies in colour.   Those mostly prized

are nitid as silver plates, clear as dew-drops, lively and showing
the true diamantine lustre.   All that are deeply tinted with
oxide are called " fancy " or coloured stones.   A light yellow is
very common, and detracts from the value ; the decidedly yellow,
the amber-coloured, and the brown are worse.   The rose-tinted
are rare and much admired, the red are seldom seen.   At
Diamantina I was shown a fine green specimen, but the price
was enormous.*   The black or rather steel-coloured diamond
being very rare, and rather curious than beautiful, is valued by
museums ; as the shape is often a good double pyramid, it should
be mounted uncut.†   The dead-white is not prized, and the same
may be said of all " false colours," especially the milky and
undetermined tints.   The violet is still, I believe, unknown.   I
heard of blue diamonds, and many of those brought from Caéthé-
Mirim are coloured superficially with a greenish-blue coating.
This and the various oxides of iron must be removed by burning
at a loss of about one per cent.‡   The " Duro " stones are
distinguished by a light green colour, crusting sometimes thickly
outside, but they cut white.   Tavernier learned in India that the
colour of the diamond follows that of the soil in which it is dug ;
red if it be ruddy, dark when the ground is damp and marshy,
and so forth.   This has been copied into our popular books.

To discover the flaws so frequent in diamonds, the purchaser
has several simple contrivances, such as to breathe upon the

---

* Mr. Emmanuel relates a case of £300
having lately been paid for a diamond of
vivid green colour, weighing 4¾ grains ;
had it been of the normal colour the value
would have been £22.   " Until lately,"
says Tavernier, " the people of Golconda
made no difficulty in buying diamonds,
externally of green colour, because when
cut they appear white and of a very fine
water."

† " One (diamond) was jet black, a
colour that not unfrequently occurs."
Thus says Mr. Gardner (chapt. 13), speaking of the " Serro " formation.   I have
only seen one in the Brazil, and that was
brought from Rio Verde of São Paulo by
my friend Dr. Augusto Tiexeira Coimbra.
It came to a bad end : he dropped it from
his waistcoat pocket, and it was swallowed
by a fowl.   In rich and new districts the
crops of all poultry when killed are carefully
examined, and are often found to contain
diamonds—another proof, if wanted, that
the gem is not poisonous.   Possibly this

may explain the fable believed by Marco
Polo in the middle of the thirteenth
century—" Such as search for diamonds
watch the eagles' nests, and when they
leave them, pick up such little stones, and
search likewise for diamonds among the
eagles' dung."   Hence too " El Sindibad of
the Sea " (Sindbad the Sailor), whose adventures are a curious mixture of fact
distorted to fable.

‡ At the Chapada of Bahia the gems
are placed with saltpetre in a crucible
which is closed and kept over the fire,
usually for about a quarter of an hour :
this, however, is a " kittle " point.   When
sufficiently roasted to have lost the oxide
of iron or the earth colour, the stones are
thrown into cold water, and of course they
are found to have lost a little weight.
Heating the diamond and then throwing it
into cold water was a Hindu test of sound-
ness and freedom from flaws.   These
crusted stones, according to John Mawe,
generally cut well.

stone, when defects and deficiencies of colour appear; or to place it in the palm of the hand, and to look through it towards the light, turning it in all directions.* The Jaça (in French Givre, or Gerçure) is a shallow line or speck, often of a dark colour, such as is seen in crystallized quartz; it is also a semi-opaque imperfection, which we call "milk," or "salt." The Natura (glace) is a want of continuity, or a void where the planes meet; the Racha is a fissure, or vein; and the Falha is a serious fracture, where two flaws join as if cemented together. In cutting these flaws they open out, and the diamond is split (estalado). The "ponto" is a strange body which has entered into the crystallization. Grains of sand have been observed in the diamond by many writers. I heard of a stone which contained a spangle of gold, and the same peculiarity has before been noticed.†

This formation shows the comparative date of the stone, whose crystallizations of carbon, or protoxide of carbon, must have arranged themselves round the metal; and favours their opinion who believe with Brewster, that the diamond, like coal, is origi-nally vegetable matter which has passed through Nature's crucible. A stone was lately found at Bagagem, with a loose piece nailed (cravado) as it were into the body of the gem; a similar "implantation of crystal" was suspected in that cele-brated stone the "Estrella do Sul." The flawed diamond generally is called "fundo." Possibly many of these defects may be removed, and tradition dimly records that the Comte de Saint Germain, and others who have displayed immense wealth, had mastered the art.

The diamond-merchant in the Brazil still cleaves to the old system of money-weights, introduced by the Portuguese in the

* The Hindus tried the goodness of the diamond by cutting one with another, and if the powder was grey or ash-coloured, it was held sufficient test, "for all other pre-cious stones, except the diamond, afford a white powder."—(A Description of the Coasts of Malabar and Coromandel, by Philip Baldæus, 1670.) They also exa-mined them by night, and judged of the water and clearness by holding them be-tween the fingers and looking through them at a large-wicked lamp placed in a wall-niche.

† "Nous y avons constaté des paillettes d'or," says M. Charles Barbot (Traité Complet des Pierres Précieuses). He calls the flaws caused by metallic molecules, "crapauds." M. Damour, speaking of "boart," remarks, "Des paillettes d'or sont quelquefois implantées dans les cavités de certains morceaux de ces diamants." Sir J. Herschel (Phys. Geog. 291) quotes M. Harting, who in 1854 "describes a diamond from Bahia, including in its sub-stance differently formed crystalline fila-ments of iron pyrites—a fact unique in its kind, and, taken in conjunction with the affinities of iron and carbon at high tem-peratures, likely to throw some light on the very obscure subject of the ultimate origin of this gem."

days of colonial ignorance. The Brazil has, like ourselves, an especial diamond weight; * but practically, and amongst miners, one hears of nothing but "grain" and "oitava." Quilate, or carat,† is not popular. Thus, in selling "fancy" or coloured stones, such as the blue, green, rose, or yellow-coloured, the old French lapidaries said, for instance, "eighty grains," not "twenty carats."

The following is a complete list of weights :—

Dezreis	= 1 grain (0·892 gr. Troy). This is the lowest of all weights: below this all becomes "fazenda fina," or diamond dust.
Vintem	= 2 grains (2·25 Portuguese) = 20 reis = ½ a carat. The Vintem (plural Vintens, not Vinteis as St. Hilaire writes) is the unity of measure.
Meia-pataca	= 16 grains = 160 reis = 8 vintens.
Meia oitava	= 32 grains = 320 reis = 16 vintens.
Cruzado	= 45 grains = 400 reis (an old weight).
Sello	= 480 reis (quite obsolete).
Oitava	= 64 grains‡ (72 grs. Portuguese) = 640 reis = 17·44 carats = 32 vintens = 16 carats.

Above four vintens, the diamond is considered large. Many miners have dug all their lives without finding a stone that exceeds twenty vintens. The most useful size is probably six vintens or three carats. The smaller stones are known in the trade as "pedra de dedo," stone of the finger, because they can be raised by pressing the tip upon them. The "cuberta" is when the lot consists of the larger gems; e.g., "Partida (parcel) de diamantes que tem cuberta."

---

* The Brazilian measures (found in books) are—

	Lisbon lb. 233·81 grammes.	Brazilian Custom-house lb. = 458·92 grammes.
4 grains = 1 quilate (carat) = 0·203		= 0·199
6 quilates = 1 escrupulo (scruple) = 1·218		= 1·195

Our diamond scale is—
16 parts   = 1 grain = 0·8 grains Troy.
4 grains   = 1 carat = 3·2 ,,
151·50 carats = 1 ounce Troy (8 oitavas, or 256 vintens).
16 ounces = 1 pound.

† The word carat is derived from the Arabic قيراط (Kírát), through the Greek κερίτιον. It is the small, red, black-tipped bean of the Abrus precatorius, a tree probably indigenous to Hindostan, but which has migrated to Eastern Africa, where it grows wild. Mr. Emmanuel (p. 55) says, "The origin of the carat weight is from the Arabic word 'Kuara,' the name of the seed of a pod-bearing plant (?) growing on the Gold Coast of Africa" (?). The "Kuara" of Bruce grew upon a region adjoining the Red Sea. The Hindu equivalent is the Rati (Ruttee), which Tavernier makes = ⅞ths of the carat = 3½ grains.

‡ Some make the oitava = 60 grains English.

Of late years, the price of diamonds all the world over has prodigiously increased. In 1750-4, when David Jeffries wrote, a perfectly white and spread brilliant of one carat was worth £8; it now fetches from £17 to £18.* The reason is easily found. The influx of gold has raised the price of stones. The market has greatly extended; † in the United States, for instance, these gems are eagerly sought by those who have made money. And lastly, in unsettled countries, as the Orient has long proved, and wherever political troubles threaten, the diamond is used " en cas," or " en tout cas ; " its extreme portability—the fact that its currency is nearly at par all the world over—and the difficulty of destroying it, raise it to the category of a coin of the highest value. ‡ In the Brazil, as in the Atlantic cities of the United States, where every one that can afford them, even hotel waiters and nigger minstrels, wear diamonds in rings and shirt fronts, demand has produced the same result, which is, moreover, exaggerated by the want of slave hands, and by the exhaustion of the superficial deposits. Thirteen years ago the oitava sold for 320$000 ; now it fetches from 800$000 to 1:000$000, nearly three times its former value. § In 1848, during the European convulsion, the price of brilliants at Bahia was reduced to fifty per cent.; but the market lost no time in recovering itself. || Castelnau (ii. 345) predicts that at the end of the present century the diamond will be worth only twenty per

---

* A "specimen stone" will rise to £20 or £21.

† "Amid the sumptuous articles which distinguish the Russian nobility, none, perhaps, is more calculated to strike a foreigner than the profusion of diamonds," says Coxe, writing in 1802. California, after 1848, developed the demand for diamonds in the United States. During the ten years following 1849 the various custom-houses registered a rise from an annual average of $100,000 to about $1,000,000. The duty was kept as low as 4 per cent. to discourage smuggling ; but it was paid, they calculated, by something less than one-sixth of the importation. The stones are mostly small, weighing under the half carat, and jewellers ask 25 per cent. more than in Paris. A good article on "Diamonds and other Gems" (Harper's New Monthly, February, 1866) declares "it is doubtful whether there is any diamond in the United States of over twelve carats in weight." It states that a marked advance in price took place between 1863 and 1864, when gold

rose above 200. Good diamonds of three to four carats then sold for $3500 to $4000. Finally, it assures us that "ninety-nine out of every hundred diamonds sold in the United States are what are called brilliants," as opposed to the rose, the table, and the brilliolette.

‡ Thus only can we explain the fact that many noble but reduced families have sent their diamonds from Hindostan, the very home of the diamond, to Europe, and have brought them back because they could find a better market in the older country. On the other hand, the general style of East Indian cutting, making the gem lustreless and glassy from want of depth, injures it in public esteem. I have seen a fine stone placed like a bit of crystal over a portrait, and even thus it was valued at £1000.

§ In 1867-8 the fall of the milreis has produced other complications in the diamond trade of the Brazil. At the present moment (July 23, 1868) the oitava may average 1:000$000 at Rio de Janeiro.

|| During the first French Revolution,

cent. of its value in 1800. I venture to say that, unless the stone can be manufactured, the reverse will approach nearer to the truth.

In producing the diamond, Nature preserves her regular proportions; the small are comparatively numerous, and the larger stones are progressively rarer. In rough diamonds, the ratio of value more than doubles with the weight. Thus, supposing a stone of one vintem to be worth 18$000 to 20$000; and one of 16 vintens will fetch 400$000 to 500$000 when the oitava is at 1:000$000. At Bahia the price is thus ascertained. Assuming, for instance, the unworked stone to be worth £2 per carat, the worth of a heavier diamond is known by doubling the square of the weight (e. g., 2 carats × 2 = 4 × 2 = £8.) For worked stones, double the weight, square it, and multiply by 2; for instance, 2 carats × 2 = 4 × 4 = 16 × 2 = £32.

Lieut.-Colonel Brant gave me the following list of prices in brute stones, showing that the value at Diamantina differs little from that of England. Diamonds, I should remark, are divided for facility of pricing into first, second, and third waters.

Grain diamonds* 12 to 18 per carat = 75 shillings.				
„ 6 — 9 „ = 77 shillings.			1st water.	
For single Stones.		Paris, 1863.	Paris, 1866.	
1 to 5 grains	= 83 shillings.	96 francs.	110 francs.	
6— 7 „	=107 „	125 „	140 „	
8— 9 „	=120 „	145 „	160 „	
10—11 „	=148 „			
12—13 „	=160 „	156 „	180 „	
14—15 „	=185 „	175 „	200 „	
16—17 „	=195 „	190 „	220 „	
18—19 „	=210 „	205 „	235 „	
20 grains	=220 „	250 „	290 „	
24 grains	=280 „	285 „	325 „	
8 carats†		2500 „	2750 „	
10 „		4650 „	5100 „	
12 „		5650 „	6200 „	
16 „		7800 „	8000 „	
20 „		12,500 „	„ „	

panic and a want of demand sunk the value of the gem 25 per cent. in the shortest time, but the assignats assisted it to recover. In 1848, "portable property" was in requisition all over Continental Europe, and the price of the diamond rose greatly.

* The Parisian table, March, 1853, gives—

First water, 25 to 30 to the carat, per carat, 72 francs.
Do.   18   „   „   78 „
First water (defective) and 2nd water ,,   60 „
Third do.   „   „   45 „
Eight stones, per carat   .   .   „   90 „
The "Mélés" in Paris are stones that weigh less than half a carat.

† Above five carats the price can hardly be fixed; it depends upon the demand, the

The curious substance called by the English "boart"* and "graphite,"† by the French "boort" and "diamant concretionné," that is to say having no cleavage, and by the Brazilians "carbonato," was formerly valueless. In 1849 it became worth from one to two francs per carat, and now it fetches 56$000 per oitava. It is supposed to be the connecting link between carbon and diamond; its hardness is that of the true gem, and its specific gravity ranges from 3·012 to 3·600. The granular amorphous mass appears under the microscope distinctly crystalline, in fact an aggregate of granules or lamellas of diamond analogous to a grit of quartzose sand. In some specimens are cellular cavities like pumice, empty or full of sand, and geodes lined with small regular crystals of colourless diamond. It is black and lustreless, and when burnt it leaves a residue of clay and other substances. This "diamond-carbon" accompanies the diamond in sandstone and in cascalho; it appears in angular and rounded galets; the irregular lumps being often as large as a walnut. Castelnau speaks of a piece weighing more than a pound. I have heard of 2:500$000 (£250) being paid for a single fragment. When "boart" is of large size it is generally broken to find if it be full or hollow. It is known by the great weight, by its diamond-like coldness in the hand, by the sharp peculiar sound when bits are scratched and rubbed together. The miners sometimes steep it in vinegar, as we do lard in water, to augment the weight, and it so resembles a piece of common magnetic or pyritic iron ore that without great care the best judges are

circumstances of buyer and seller, and so forth. The larger stones often remain on hand many years before they find a purchaser. I have heard of a Brazilian gentleman who expended nearly all his property in buying a "great bargain," in the shape of a diamond, of which he has never been able to dispose. The larger stones are always sold singly. Tavernier gives the following rule for estimating their value :—

15 carats (perfect stone)
15
———
225
150 (value of a single carat)
—————
33,750 livres.

15 carats (imperfect stone)
15
———
225
80 (value of the single carat)
—————
18,000 livres.

* Wonderful to relate, the diamond merchants of Bahia could not agree upon the meaning of "boart," which books apply as in the text. One of the oldest and most experienced insisted that it was the cheapest and worst kind of perfectly crystallised diamond, worn by attrition into spherical globules, like shot grains. This kind is mostly unfit to be cut, and when crushed the dust is used for polishing gems and for engraving on hard stones.

† Graphite is usually applied to the pure debitumenised carbon found in the Laurentian, and associated with anthracite in the Cambrian systems. Its vegetable origin is not thoroughly established.

deceived.* It is pounded and used principally in diamond cutting. Drills pointed with this mineral have, I am told, been employed with great success in driving tunnels through hard rock.

Of this little known substance three kinds are distinguished by the trade. The worst is the "Carbonato;" a finer kind with better formed crystals is the "Torre," which fetches 60$000 per oitava; the best occurs in small rounded balls of shining metallic appearance, and is therefore called "Balas," this may rise to 80$000 per oitava.† Some Chapadista miners have not yet learned to sort the varieties.

The Brazilian diggings have produced some large and valuable gems, which have all been sent out of the country.

The Braganza diamond was worn by D. João VI., who had a passion for precious stones, and possessed about £3,000,000 in value. Now amongst the crown jewels of Portugal, it was extracted in 1741 from the mine of Caéthé Mirim.‡ Authors differ touching its weight,§ and no drawing of it has, I believe, been published; it is supposed to be larger than a hen's egg, and it has long laboured under the suspicion of being a fine white topaz, a stone which in the Brazil, as elsewhere,∥ often counterfeits the diamond.

---

* The boart or carbonato, however, has no attractive power. It is tried by striking it between two copper coins, and if it breaks or does not dint the metal, it is held valueless.

† Dr. Dayrell gave me a specimen of "boart" from Sincorá. It much resembled pyritiferous iron-sand. The substance is found in pieces varying from one grain to half an oitava. I have heard it called "bolo redondo," and was told that the colour is sometimes of an opaque white.

‡ M. Barbot specifies the place as the little river "Malho Verde," in the vicinity of "Cay-de-Mérin."

§ John Mawe and the Abbé Reynal make the weight 1680 carats (12¼ French ounces). Romé de l'Isle, who estimated its value at 7 milliards 500 million francs, gives 11 ozs. 3 gros. and 24 grains of gold weight. M. Ferry says 1730 carats, estimating the Brazilian carat at 0·006 less than the European. Mr. Emmanuel gives it 1880 carats in p. 78, and 1680 in p. 128, the former being probably a misprint.

∥ Mr. St. John (Forests of the Far East, vol. i. 48) mentions a nóble in Brunei who for £1000 offered a diamond about the size of a pullet's egg, which proved to be a pinkish topaz.

In reading these two pleasant and instructive volumes I could not but regret that the author had not given us an account of the celebrated diamantation of Borneo. In old authors we find that the sands of the "Succadan" River produced fine stones of white and lively water, but that the Queens of Borneo would not allow strangers to export them. We remember, too, that in Borneo was found, in 1760, the largest diamond known. The weight was 367 carats = 1130 grains. It caused a war of nearly thirty years' duration, and it remained with the original possessor, the Rajah of Mattam. The island, with its core of granite and syenite which protrude in the vast mountain mass known as Kina Balu, the "Chinese Widow," through the secondary limestones and sandstones, much resembles the Brazil. We read also of the pot holes washed by sand-water, the gravels, and the rocky streams which characterise a diamantine country. There are curious resemblances in minor points. For instance, the people of the Sulus Islands keep their small stores of seed-pearls in hollow bamboos. These are the "Pequás," so well known to the Brazilian mine-owner.

The Abaété* brilliant was found in 1791, and the circumstances of the discovery are related by John Mawe, M. F. Denis and others. Three men convicted of capital offences, Antonio da Sousa, José Felis Gomes, and Thomas da Sousa, when exiled to the far west of Minas, and forbidden under pain of death to enter a city, wandered about for some six years, braving cannibals and wild beasts, in search of treasure. Whilst washing for gold in the Abaété River, which was then exceptionally dry, they hit upon this diamond, weighing nearly an ounce (576 grains = 144 carats).† They trusted to a priest, who, despite the severe laws against diamond washers, led them to Villa Rica and submitted the stone to the Governor of Minas, whose doubts were dissipated by a special commission. The priest obtained several privileges and the malefactors their pardon, no other reward being mentioned. A detachment was at once sent to the Abaété River, which proved itself rich, but did not offer a second similar prize.‡ D. João VI. used to wear this stone on great occasions attached to a collar.

The "Estrella do Sul" brilliant was found in July, 1853, at Bagagem of Minas Geraes by a negress.§ In the rough state it weighed 254½ carats. The owner parted with it for 30 contos (£3,000); at the Bank of Rio de Janeiro it was presently deposited for 300 to 305 contos, when it was worth £2,000,000 to £3,000,000. After being cut by the proprietors, Messrs. Coster of Amsterdam, it was reduced to 125 carats, and now it belongs, I believe, to the Pacha of Egypt. Though not perfectly pure and white, its "fire" renders it one of the finest gems extant.‖

The Chapada of Bahia also produced a stone weighing 76½

---

* M. Buril (427) calls the Abaété diamond "O Regente."

† In some books the weight is given at 138½ carats; in others it is made 213.

‡ This stream has already been mentioned. The diamond was described by John Mawe as octahedral in shape, weighing seven-eighths of an ounce Troy, and perhaps the largest in the world. It passed through the hands of the Viceroy, and was sent in a frigate to the Prince Regent.

§ A story far too long to tell here belongs to the Estrella do Sul, which appeared at our Great Exhibition in 1851. Exceptionally, for few diamonds with names can make such boast, it has caused no blood-shed; even the finder was not murdered—only ruined, and died broken-hearted. Of the score or two of persons who made fortunes by the discovery, Casimiro (de Tal), whose negress (not a negro, as the writer in "Harper's" says) brought it to him in order to obtain her freedom, was the only one disappointed.

‖ M. S. Dulot (France et Brésil, Paris, 1857), p. 20, seems to confound the "Star of the South," which was found in 1853, with the "Braganza," dating from 1741. Mr. Emmanuel (p. 61) rightly makes the Estrella do Sul the largest found in "the Brazils."

carats, and when cut into a drop-shaped brilliant it proved to possess extraordinary play and lustre. It was bought by Mr. Arthur Lyon, of Bahia, for 30 contos, and it is now, I am told, in the possession of Mr. E. T. Dresden.

Briefly to conclude. As yet the Diamantine formations of the Brazil have been barely scratched, and the works have been compared with those of beavers. The rivers have not been turned, the deep pools (poços or poçoẽs) above and below the rapids, where the great deposits must collect, have not been explored, even with the diving helmet; the dry method of extraction, long ago known in Hindostan, is still here unknown. All is conducted in the venerable old style of the last century, and the fiend Routine is here more deadly than Red Tape in England. The next generation will work with thousands of arms directed by men whose experience in mechanics and hydraulics will enable them to economize labour; and it is to be hoped that the virgin gem-bearing waters will be washed up-stream. This was the sensible provision of the old Diamantine Regulation. Unfortunately it came too late, when the channels had been choked with rubbish which was hardly worth removing.

CHAPTER XI.

FROM BOM SUCCESSO TO THE CORÔA DO GALLO.

THE SACO OR PORTO DOS BURRINHOS.—INDEPENDENCE DAY.—THE "CACHO-
EIRA DO PICÃO."—THE LAPA DOS URUBÚS.—THE BURITY PALM.—SILENT
BIRDS.

"Cette partie si importante de l'économie publique, en un mot demeure encore
livrée à un état d'abandon que le gouvernement ne peut trop s'empresser de faire
cesser."—(*M. Claude Deschamps, of the French Rivers in* 1834.)

"It is presumed the Brazil will not attempt to dispute the now well-settled
doctrine, that no nation holding the mouth of a river has a right to bar the way
to market of a nation holding (land ?) higher up, or to prevent that nation's trade
and intercourse with whom she will, by a great highway common to both"
(*Lieut. Herndon,* p. 366.)

SATURDAY, *September* 7, 1867.—My letters were soon written,
the trooper Miguel and his mules were dismissed with good characters, and at 9.30 A.M., after embracing our kind host, Dr.
Alexandre, we pushed out of the creek "Bom Successo."

"O Menino," the new broom, swept, as happens for a short
time, uncommonly clean, naming every little break of water or
hole in the bank.* The rocks, sandstone abounding in iron and
laminated blue limestone, were all in confusion. The strike was
to the east, the north-east, the south-east, the west, the north-
west and the north, and sometimes within ten yards the strata
were anticlinal, nearly vertical, and almost horizontal. There
were slabs of clay, with perpendicular fracture dipping towards
the river, and here and there "Cánga" and "Cascalho."

After a few unimportant features,† we left to starboard the

---

* *E.g.* the Corôa do Nenné, so called
after the nickname of a man with a crippled
hand, and the Corôa do Saco, both with the
main channel to the left. Then the Corôa
do Poço do Gordiano and the Corôa do
Cedro, with the Ribeirão do Cedro falling
into the left bank ; these have the thalweg
on the right.

† Córrego do Bom Successo Pequeno on
the right bank, one league by water and
one mile by land from the Fazenda. Then
the Corôa do Saco do Cedro, grassy and tree-
grown, with a break above and below it.
On the right bank the Sitio of Antonio
Alves, with traces of cultivation.

Larangeiras stream and estate, belonging to Colonel Domingos.
Opposite it is the Barro do Maquiné Grande, a little "fishy"
creek of clear water, which has a water-way of five leagues for
canoes, forming a Corôa (do Saco do Maquiné Grande), with a
clear way to the right.* In the Maquiné Fazenda there is, they
say, a cavern which gave fifteen days' work to Dr. Lund, and the
savant found there a "pia" or baptismal font of stalactite, which
would have commanded 400l. in Europe. Shortly after noon we
descended this day's first rapid, the Cachoeira da Capivara, which
has two channels, with a sandbank in the centre. The left is the
deep water-way, but rafts come to grief by dashing against the
bank where the pole cannot touch bottom. We therefore floated
down stern foremost, threw out a cord and hugged the Corôa.
The air was dense with bush-burnings, here producing an
"Indian spring," which corresponds with the "Indian summer"
in the north : mostly Brazilians complain of the smoke, and
declare that it gives them difficulty of breathing. Nothing could
be more picturesque than the long lines of vapour like swathes or
veils, whose undulations overlay the hill-tops, and gradually dispersed in air.†

At 4 P.M. we passed the Rio de Santo Antonio, a pleasant little
stream which admits for two leagues tolerable-sized canoes, whilst
the small dug-outs ascend it about double that distance. It leads to
(Santo Antonio de) Curvello, a town so called after an ecclesiastical colonist ; built upon the Campo, and the last in this region,
it is supposed to demarcate the "Sertão," ‡ or Far West. But
the inhabitants do not readily own to the soft impeachment; the
traveller is always approaching the Sertão, and yet hears that it
is still some days off. He remembers the lands of the tailed
nyam-nyams, which ever fly before the explorer, or, humbler
comparison, the fens of certain English counties which, according

---

* The next holm, Corôa do Palo, which
sent us to the left, is not mentioned by M.
Liais.

† After the Palo are the Porteira, so
named from a creek, and the Corôa das
Mamonciras, with the thalweg to the left ;
neither of them is mentioned by M. Liais.
Then comes the Córrego das Canoas (Ribeirão das Canoas, Liais), exposing on the
right bank a mass of auriferous pudding-
stone, and beyond it the boulders dip 10°
to 30°. Here the Corôa das Canoas blocks
up the right channel. On the left is a

perpendicular bank of brown clay six feet
deep, with red-leaved Copahyba trees growing from it. There is little to notice in the
Porto and Córrego da Anta or in the Porto
do Murici, so called from a small edible
yellow berry.

‡ Southey writes the word after the old
fashion, "Sertam," and declares (ii. 565)
that he does not know its origin. It is
nothing but a contraction of Desertão, a
large wild, and it is much used in Africa
as well as South America.

to the pallid, ague-stricken, web-footed informant, are not honoured by being his dwelling-place.

After passing broken water at the Corôa de Santo Antonio and the Corôa and Corrida das Lages, at 5 P.M. we fixed upon our "dormida." It was a sandbank in a bay called Saco or Porto dos Burrinhos, of the Little Donkeys, and opposite it, on the right, lay Boa Vista, still the property of Colonel Domingos. The moon, that traveller's friend, a companion to the solitary man, like the blazing hearth of Northern climates, rose behind the filmy tree-tops and made us hail the gentle light. We have not the same feeling for the stars, or even the planets, though Jupiter and Venus give more light than does the Crescent in England; they are too distant, too far above us, whilst the Moon is of the earth, earthy, a member of our body physical, the complement of our atom. We did not forget a health to this, the Independence Day of the Brazil. Within the life of a middle-aged man she has risen from colonyhood to the puberty of a mighty Empire, and history records few instances of such rapid and regular progress. This "notanda dies" also opens to the ships of all nations, the Amazons and the Rio de São Francisco; a measure taken by Liberals, but, curious to say, one of the most liberal that any nation can record. In spirit we join with the rejoicings which are taking place on the lower waters of the liberated streams.

September 8.—Pushing off at 6·30 A.M., we passed the Porto do Curvello with a ranch on the left, denoting the high road to Diamantina. The rapid and shallow, known as Saco da Palha, sent us first to the left and then to the right. Again the rocks are quaquaversal, with dip varying from horizontal to vertical. The banks at the beginning of the day were low, but presently they became high and bold; forested hills on the right formed a hollow square. The first rapid was the Cachoeira do Landim,* with its "crown" and shallow; a line of stone, fractured in the centre, stretches nearly across stream, and gives passage to the left. Beyond this point are sundry minor obstructions,† not named

---

* Said to be the name of a fish and a tree. M. Liais writes Landin.

† The Corôa do Jatahy, but little above water, and with a break to the right, shows where Col. Domingos' property ends. Then by the right of the low banks the Corôas do Garrote and do Páu Dourado; by the left of a third, where two sandbanks narrow the bed to fifty yards, and descend the Saco da Varginha or Varzinha. Another little nameless break, the course turning from east to north, and backed by a hill-line wooded to its flat top, and apparently crossing the stream.

by M. Liais. He proposes, however, extensive "ameliorations" of the stream, "tunage," "draguage," canalizing to suppress the useless "chenal," and "attacking" the bank.

After the Varginha, a low sandbank which gave us passage to the left, the Porte do Silverio (P. N.) sent us to the right. Here a reef, at this season very shallow, nearly crosses the stream, and "Marumbés" or iron-coated stone, began to glisten on the bank. Next came the Saco and Cachoeira de Jequitibá, with fields and houses on the left. We landed on the Corôa and inspected this neat mill-dam, a broken ridge of ferruginous rock —possibly derived from the Serras—extending right across from north-north-west to south-south-east. Canoes can creep along the left side, but our ark gallantly plunged down the middle, which a little hammering would easily open. We noticed the magnificent sugar-cane, which exceeds in size that of Bom Successo.

More small troubles * led us to the not very important Cachoeira da Manga. The word denotes a narrow lane, and a square of rough rails leading to the water edge. Cattle are driven in, and the pressure of those behind compels the foremost to set the example of swimming the stream. A clearing ran up the neat hill-slope on the right bank, horses and cows basked on the sands, and men, squatting like Africans under shady trees, shouted warnings of the dreaded Picão, and promised to pilot us if we would wait a day. We expressed our gratitude chaffingly, modifying the puppy pie and the lady in mourning.

Steering to the left of the Tronqueira break, and describing a little circle to the right, at 3 P.M. we entered the Saco do Picão. Here the stream, swinging to the left bank, works round from west to north-east and east. At first a little break extending across nearly home, and well provided with snags, made us present rear and hug the right; the bank was hard and soft argile, quartz-veined, and supporting Cánga, whose strike was east and dip 30° to 35°. Then passing to the left of an "inch" we landed on the right side to lighten the craft and to inspect the formation.

---

* Barra do Breginho, with a turn to the north-east; on right bank, huts and fields with snake fence opposite. The Cachoeira do Saco, a dam of ironstone, with narrow gap to left, and grassy hill in front. The Cachoeira and Corôa dos Tachos (Taxos, M. Liais), with bad break over rock wall to the right, passage on left, but two rocks in the way.

The Picão, or Pickaxe, deserves its ill-fame; it is perhaps the worst obstruction on the Rio das Velhas.* A broad, broken band of jagged serrated teeth dams the stream, besides which rocks and sandbanks extend some two miles above and below it. The material is a very hard blue clay shale, whose laminations easily split apart: it has a metallic ring, it does not effervesce under acids, and it hardens in, without being otherwise affected by, fire; evidently it will be valuable for building. The emerging rocks cause the waters to groan and splash, to dash and swirl by them in little rapids (Corradiças), averaging some nine feet per second. We crept under the right bank, but now drawing sixteen inches, we were soon aground, and required lifting by levers. Passing to the right of a small sandbank below, we had a good back view; the water-fall was between three and four feet, and there would be no difficulty in opening the mid-channel. At 5 P.M. we crossed to the left and nighted on a sandbank, still in the Picão Sack, opposite a hill, and a small cascade which resembled a toy.

Here we enter the land best fitted for emigrants. We are beyond the reach of the great planters who wish to sell square leagues of ground, some good, much bad, and all, of course, at the longest possible price. There are no terrenos devolutos, or Government grounds, but the small moradores ask little. Hereabouts a proprietor is ready to part with four square miles, including a fine large Córrego, for 300$000 to 400$000, less than I paid for my raft. The Geraes, or lands beyond the river, are still cheaper, and generally where water runs in deep channels, land may be purchased at almost a nominal price; the people have no appliances for irrigation, which the steam-engine would manage so efficiently. The views are beautiful, the climate is fine and dry, mild and genial, there is no need of the quinine bottle on the breakfast-table, as in parts of the Mississippi Valley. There are no noxious animals; and, except at certain seasons, few nuisances of mosquitos and that unpleasant family. The river bottom is some four miles broad, and when the roots are grubbed up, it will be easy to use plough or plow, whilst the yield of "corn" and cereals is at least from 50 to 100 per cent.

---

* M. Liais remarks of this Picão (p. 10), "une petite barque vide et à moitié portée par des hommes peut seule passer tout contre la rive droite, et en touchant souvent un fond de pierres."

There is every facility for breeding stock and poultry; besides washing for gold and diamonds, limestone and saltpetre abound, whilst iron is everywhere to be dug. Water communication will soon extend from the Rio de São Francisco below, to the excellent market of Morro Velho in the upper waters. Lastly, the people are hospitable and friendly to strangers; my companion, who had a smattering of engineering, could have commanded employment at any fazenda.

Sept. 9.—The end of the Picão was a shallow break, known as the Portão; it is formed by a ledge projecting from the high right bank of red-stained limestone.* This was followed by a straight reach, with fine bottom lands, wooded hills bounding them to the left. After paddling for about two hours and a half, we descended by the stern "as Pórteiras," the gates, and came to the rapids known as Cancélla de Cima, and Cancélla de Abaixo, the upper and lower barred gate.† These unpleasant gratings were not passed without abundant clamour and fierce addresses, beginning with "Homem de Deus." The river is shallower than ever, we can see the water line below which it has lately shrunk, and evidently the usual rains are wanting in the upper regions. The marvellous dryness of the air continues to curl up the book covers; at sunrise the breath of the morning deadens our fingers, and incapacitates them from writing, though it ranges between 55° and 60° (F.). At noon the mercury rises to 75°, and at 1 P.M. to 85°. Presently a south wind will blow from the Serra Grande or do Espinhaço.

At 11 A.M. the reach bent from north-east to north, and we passed the mouth of the Paraúna River ‡ (Barra do Paraúna), now an old friend. The breadth of this, the most important of influents, is 90 to 105 feet, a mass of sand cumbers the left

---

* Further down was limestone on the right bank, striking to the north-west, and dipping 45°.

† The upper Cancélla is formed by scattered teeth of stone projecting from the banks. We hung upon a detached rock in the centre, and the poor canoe took in much water; levered her off and found passage close along right bank. Rest of run occupied by a ledge stretching from north-west to south-east; touched again and spent a total of twenty minutes before getting into deep water. Another dam from left bank gives free passage to the

right; on opposite side a Barreiro de Gado with huts, sugar-cane, and Jaboticábas. The Cancélla de Abaixo has on the left bank a grating composed of four long walls and detached rocks, the passage is along the right side, where there are two separate stones and a pair of dam lines; here also we struck, and lost twenty-five minutes.

‡ M. Gerber places the Barra da Paraúna in south lat. 18° 50′ 0″. M. Liais in 18° 30′ 19″·9, at fifty-three direct miles from Casa Branca, in 19° 23′ 45″; and eighty-four from Sabará (in south lat. 19° 54′).

jaw, and elsewhere there are stiff banks of brown humus, and white and red clay. The position will make it a great central station when a railway from Rio de Janeiro shall connect with the steam navigation of the São Francisco.

At the Barra do Paraúna began new scenery. Hitherto the mountains have been like crumpled paper; now they assume a kind of regularity, and often lie parallel with the axis of the stream. On the left there is a buttressed calcareous line through which the Rio das Velhas breaks at its confluence with the Paraúna: further south the same ridge is to the right, or east, and flanks the Cipó river on the west. The Rio das Velhas widens to 200 yards; the tortuous stream becomes comparatively straight, with a general direction of north, 11° west, and the slope is greatly diminished.* A "fancy country" showed itself, the blocks of hill drew off, and the banks were gently sloping ledges, with brown drift wood at the water edge; and yellow clay and sand with rocks here and there in higher levels. Large undulating ribbons of tender green, set in sun-burnt flanks, showed the torrent-beds green-lined as those of Somali-land in the rains, and here and there the thicket contrasted with tall scattered trees, the remnants of an old forest. Cattle lay and sunned themselves upon the damp Corôas, and we heard with pleasure the voices of villagers and the barking of dogs.

At 1·30 P.M., we passed the Lapa d'Anta, a formation reminding us of Páu de Cherro. The river runs to the north-east, and its right bank is buttressed by a bold mass of limestone bluff to the west, rising sharply from the sands and clays on both sides, and forming a small bay with a graceful sweep. It is the perpendicular face of a long range, extending from south-east to north-west, and hemming in the river on the east; the feature corresponds with that before noticed. The dip is 25°, exposing only the edges towards the stream: the lower part is a hollow of wavy, blue-tinged strata, whilst the upper half is an overhanging mass of solid matter, looking as if crystallised, stained red by the rusty clay, and curtained with black tongues apparently dyed by the cinders of the burnt soil above. From the summit sloped backwards a brick-coloured hill, with leafless

---

* According to M. Liais, the slope between Trahiras and Paraúna is 0·4355 metre per mile. From the confluence of the latter stream to the débouchure of the Rio das Velhas, it diminishes to 0·2735.

trees, contrasting singularly with the metallic verdure of the banks.

At 1·45 P.M. the river turned from north to west, and we passed a similar formation. Here a cave, the Poço do Surubim or do Loango,* faces south, and shows an arch of blue limestone with soffit-like edges of brick, built as if by art, with their laminations of dark chocolate embedded in a limestone resembling marble. A little below, a sandbank, projecting from the left, contracts the stream to half-size and makes it very deep. The prospect is pleasant, hill piled on hill, and changing colour from brown-red to blue as the lines recede.†

Presently we sighted the Lapa dos Urubús, a limestone bluff like its neighbours ; but rising some eighty feet in height : it is crowned with green trees, and has grey vegetation above. It faces to the west, the river running north to south and the strata are horizontal, except where they had slipped down into the water. On the right bank, and in front, lay a tapering point projecting from a bushy hill, whilst the sand-ledge that banked the stream was tasselled with verdure. A single splendid Jequitibá, with a cauliflower-like head and a wealth of cool verdure, marked the spot.

About 5 P.M. we landed and walked up to the Lapa. Beyond the bank, some fifteen feet high, was a dwarf clearing (Roça), with felled trees and a field of tomatos and Quiábos, or " Quingombos," (Hibiscus esculentus), mixed with the Cordão do Frade.‡ After a few paces we reached a cliff from whose crevices trees sprang and creepers hung down ; here also the arches had a brick-like

---

* According to the people, the Loango is the male of the Surubim ; others declare that the Moleque is the male of the Loango. The fish here supplies the Amazonian codfish, the Pirurucu (Vastus gigas), and the people will learn to salt and export it. It is a kind of sturgeon, scaleless, spotted and marbled, flat-muzzled and whiskered, like the " cats " (Silurus), which drown the negro boys fishing in the Mississippi waters, and ugly as any "devil fish." It is often five feet long, and attains a weight of 128 lbs., yielding two kegs of oil. Several species are mentioned ; for instance, the Surubim de Couro. The people declare it to be a cannibal like the pike ; they net it, and the wild men shoot it with arrows. They split the body, sun-dry it, and sell it

in the Sertão. The meat is excellent, white, firm, and fat. I have never tasted a finer fresh-water fish ; it has, however, the bad name of causing skin disease.

† Here occurs the Ilha Grande which blocks up the right side. Then the Corôa do Clemente with three sandbanks, one tree-grown, the others sandy. Beyond this is another large islet, which must be passed on the right.

‡ Leonotis nepetæfolia. From Ukhete, in Eastern Intertropical Africa, I sent home a specimen of this labiad, which grows wild all over the low damp region of the seaboard. The negroes use it to narcotise fish, and probably it has been introduced into the Brazil by the old Portuguese.

appearance, and the tall organ-pipe Cactus hedged the foot. The cave faced to the south, débris of rock encumbered the entrance, and higher up was a large shield-formed slab, masking a dark gallery some three feet high, and said to extend two miles. Here was a shallow pit whence the saltpetre earth had been taken, and we found nothing within but bats and " horse-bone limestone."

Sept. 10.—The night was cold, a chilly eastern breeze coursed down from the Diamantine mountains, and the " Corrubiana " appeared from afar in fleecy dark-lined clouds. After twenty minutes' work we came to the Cachoeira das Ilhotas, an ugly place,* but easy to be opened, as the crest of the ledge is narrow. The sun waxed hot, the east wind was exceptionally cold and high, and my companions began to suffer. João Pereira was treating a bruised arm with arnica, and was compelled to " lay up ;" a serious matter with a small crew. The other men had for the last two days complained of a sensation of malaise, headache and want of sleep, without any apparent reason. I resolved to begin a new system, and to halt during the greatest heats. Finding the Eliza overweighted to starboard we pulled up a plank and discovered that, in addition to the leak, the carpenter had not taken the trouble to remove his chips. In the Bight of Benin none of us would have escaped fever, and a few would have remained on, or rather in, the banks.

After the Ilhotas we attacked the three Jenipapos. No. 1 is a wooded islet defended by a dangerous snag ; there are rocks in abundance and the current swings towards them. We ran down the left bank of the holm, and crossed water breaking over sunken stones ; here in June, 1866, they wrecked a canoe and implements for sugar-making, en route from Sabará to Januaria.† Jenipapo No. 2, where the stream runs to the north-east, has few difficulties ; there is sufficient water in the mid-stream. After this, for some three miles, we made easting, and gained nothing. Then we crossed the Redemoinho da Beija-mão, the " Whirlpool of

---

* Rocks extend across the stream from right to left, blocking it up in the latter direction. We went to starboard, grounding upon the dexter bank of the Corôa, above the rocks on the right, and rounded its lower end by cordelling. Then we shot through a bad break formed by a rock pier running from north to south, and made the left side to avoid two similar formations, a detached stone and a shallow. The second islet caused us to hug its eastern side to avoid a reef on the right bank of the stream, and we ran the rapid, carefully looking out for ledges below water. This occupied half-an-hour.

† Below it is another break, stones and an islet, crossing the stream from north to south ; further down, the water dances and flows over a newly formed bank ; whilst, lowest of all, there is a break of ironstone.

Hand-Kissing." It is not even a Maëlström, but it may be dangerous to small craft during the floods. The third Jenipapo was a Corôa, which we skirted on the right, the rest of the water-way breaking heavily. Shortly afterwards we passed the Ilha do Hippolito * with a saw of jagged rocks that barred the right side.

At 2 P.M. we resumed work in the teeth of a strong north wind. The right bank showed a bed of quartz-conglomerate four to five feet high, and below it was the dry Córrego do Brejo with its limestone outcrop. At the Váo da Carahyba† there is a ford in the dry season, and the Saco of the same name showed a rock to starboard, not dangerous, for the channel on the left is well marked. Here we followed three sides of a square, and a cut of 1·5 mile would save six. At 5 P.M. we passed the Porto de Arêas, on whose right people were encamped. It was marked by a quaint-looking Angico Mimosa, then leafless, and exposing a smooth rhubarb-yellow bole.‡ Another hour placed us at the Saco da Manga, a sandbank 20 feet high, spangled with the Mangui Hibiscus, and supporting fine rich soil eight feet deep. Here the waters of the Rio das Velhas, probably affected by some influent, were particularly dark and foul, with the peculiar smell of the slimy African river where rain has not washed it. The pilots declared it crystal compared with the waters of the wet season, when the upper washings give it a blood-red hue. At night, however, the evil was mitigated by a strong wind from the " Range of the Spine."

Sept. 11.—The dawn when we set out was clear, but as the horizon waxed yellow, smoke columns began to rise from the water till dispersed by the light breeze which became a strong east wind. At noon the sun was fiery, and the afternoon waxed wintry, but it was a winter in Egypt. It reminded my companion of a "fall day" in Tennessee, when men begin to pick " cutt'n." About eventide clouds like smoke-puffs flitted across the sky and gathered in the north, whilst a purple haze in the west, and a misty moon betokened, said the pilot, not rain but wind.

Sweeping round a corner we saw white sand-drift and tall trees, which showed the Porto da Manga of the Rio Pardo. It drains

---

* M. Liais calls it " de San Hippolýto."
† Also called Caraúba, Carôba (an error), Caraíba, and Carahiba ; we shall find it in quantities upon the Rio de São Francisco, where there are two species, one with pale golden brown, the other with a smaller blossom of pleasant lilac colour.
‡ The guides named it Páu Breu—pitch tree.

the western slopes north of Diamantina. The counterslopes supply the Caéthé-Mirim to the Jequitinhonha. Canoes, after two days, reach its Serra, distant only twelve leagues from the City of Diamonds. The mouth was 140 feet broad, the main stream being 650. The first hour saw us bumping down a shallow formed by a break, and passing a jagged line of limestone slabs with a western strike, and nearly perpendicular, like half-submerged grave-stones. A little below it were limestone blocks, with a south-eastern strike. Again the surface of the land displays extreme irregularity, caused probably by the meeting of different systems of uplands which project their bands from both banks across the stream. It is one of the peculiarities of this Lower Rio das Velhas, and deserves attention.

Presently we shot at the Cachoeira do Gonçalvez,* an ugly place with broken water. Shortly afterwards we struck heavily, and hung for a time upon a sunken rock in midstream, under surface all the year round, and not noted in the plan. Twenty minutes led to a similar accident. On the latter occasion, however, limestone lumps emerged from the water near the bank. These obstacles are dangerous to boats; the Cachoeira must be cut through, and the rocks should be removed. At 9·30 A.M. was crossed the mouth of the Curumatahy River, which heads north of and runs parallel with the Rio Pardo. Here the pretty stream is about 105 feet broad; its right bank is rich with tall trees, and it curves gracefully out of sight.

The Rio das Velhas again alters its aspect. For some time we had seen in front a long grey line, the Serra do Bicudo, so called from a little stream entering the left bank. Now we make a long westerly bend, compelled by the Serra do Curumatahy, a chine rising some 1500 feet above the river-bed, and at this point approaching within 300 yards of the stream. It is prolonged to the north by the Serras do Cabral, do Paulista and da Piedade, whilst opposite them on the left bank are the Serras da Palma and da Tabuá. There is a remarkable correspondence in the lines. The summits are grass-grown, and shrubbery appears in the damper hollows. Here, as elsewhere, more rain falls upon the higher

---

* M. Liais, Cachoeira de Gonçalo. Two separate lines of limestone on the right strike south-east, and dip 75°; all below is rugged, with scatters of rock. We went down on the right, shaving a slab, made for the left side, and then crossed to the east.

than upon the lower levels, but the former readily drain into the
latter. Between the southern chains, which appear to be the
boundaries of the old bed, is an average interval of four miles.
The ranges are composed of gently swelling hills, with a surface
of brown bush from which the timber has been removed, and with
scattered patches and gashed lines of green, denoting water. The
slabs of blue stone, probably lime, are said to form caves and
saltpetre. At the base are bayous and swamps (brejos) lying below
highstream level. The banks show a remarkable difference; on
the right is a fertile calcareous soil, based on a ferruginous argile*
used for whetstones. On the left, where sandstones and laminated clays appear, the vegetation is poor and "scraggy."

At noon we anchored for rest near a bed of conglomerate, six
feet thick, shaded by a noble Jatobá salaaming to the water. The
place is called the Brejo do Burity, and it bears a thin forest of
monocotyledons with a dicotyledonous undergrowth. The word
written by Pizarro and St. Hilaire "Bority," by Martins, Gardner
and Kidder "Buriti," and by the System "Bruti," is a vulgar corruption of the Tupy "Murity."† This Mauritia vinifera is at once
elegant and useful, but I was disappointed with it when recalling
to mind the magnificent Palmyras or Fan-palms of Yoruba. The
people, however, declare that near the river it is an inferior
growth, attaining its full dimensions only in the high and dry
Geraes lands. They could not tell me how far it extends. Most
of them agreed that where the Carnahuba clothes the margins of
the middle São Francisco the "Burity" grows inland. Here it
flourishes isolated and in groups. I saw every size, from the
little ground-fan to the tall column crowned with sphere of leafage.

According to Leblond and Codazzi, a tribe of Guaraunos or
Waraons depended for life upon this palm, where they built their
aërial houses, and whose larvæ are still favourite food with the
"Indians" of the Orinoco. Here the leaves are woven into baskets, and the fronds are cut, rafted down, and sold for fences.
The oily, reddish pulp between the fruit scales and the albumi-
nous substance of the nut‡ is made with sugar into a massa
or lump, and carried bound in leaves to market. The people
relish this "doce," although it is believed that eating the fruit

---

* At 9·50 A.M. we passed in the river
and on the banks, ironstone, apparently
rich.

† Some old travellers have "murichy."

‡ St. Hil. (III. ii. 344) says, "le tronc
est rempli d'une moëlle, dont on fait une
sorte de confiture." All assured me that
it was from the fruit.

stains the skin yellow. The brown-yellow fibre forms strong hammocks, which last longest when the material is greased. On the
Rio de São Francisco they cost from 1$000 to 1$500. The
saccharine juice gives the most highly prized palm wine in the
Brazil, where, curious to say, that of the Cocoa nut, of all the
delicatest, is unknown. It is extracted, after the wasteful negro
fashion, by felling the tree; holes are cut with the axe half a foot
long by three inches deep, at intervals of five or six feet, and they
are soon filled with the reddish liquor. As time advances a more
economical system will be tried. The " Buritizal" suffers much
from the large ant called Içá or Yçá.

At 2 P.M. we left the Jatobá shore, which seemed to be enjoyed
by flies and other pests as well as by ourselves. We made a
straight line of five miles between the parallel range, after which
the narrowness of the right-hand channel drove us to the left of a
Corão. At 3·15 P.M. we passed an island wooded on the north.
The west bank was strewn with very loose Cascalho, and cut by a
limpid stream. Here the bed narrowed to 250 feet. A few " der-
rubadas," or clearings, contained dead trees encumbering the
ground, and little onion plots spoke of population. Half an hour
afterwards a sandbank squeezed the left channel, and drove us
down the right. Here we saw for the first time groups of limestone rocks just above water, and overgrown with the woody
Arinda. The Cachoeira do Riacho das Pedras breaks in the
centre and shows the same features, calcareous blocks bare of
everything but shrub. Lastly we left on the right the Corôa do
Gallo, two bars of limestone almost à fleur d'eau; and at 5·45 P.M.
we anchored on the port bank, a tract of sand thinly covered with
scrub.*

This day we passed over immense wealth, of which, like philo-
sophers, we took no heed. The Rio Pardo, like the Paráuna,
drains highlands rich in diamonds and gold, whilst the bed of the
Rio das Velhas is a natural system of launders. In due time it
will be thought, perhaps, advisable to turn and lay dry certain
bends in this part of the stream, and there are several places
where such an operation suggests itself.

For the last two nights the " Whip-poor-will " and the " Cury-

---

* Opposite this place the map shows a
dwelling-house, " As Porteiras," but from
the stream we did not sight it. Porteira
(originally a porteress,) here means a barred
gate leading to a pasture, &c.

angú" have been silent—they who so often had broken our sleep
with their complaints and responses, delivered from the thickets
along and across the stream. Men are certainly not numerous
enough to destroy them. Perhaps their favourite food abounds in
some places, not in others, and thus they may not inhabit the
banks continuously. Or again, the cold wind is, we may conjec-
ture, uncomfortable enough to interrupt the concert.

CHAPTER XII.

FROM THE CORÔA DO GALLO TO THE ILHA GRANDE.

"CACHOEIRA DA ESCARAMUÇA" (NO. 10, AND FINAL). — THE DELIGHTFUL
TEMPERATURE.—VERMIN.—ECLIPSE OF THE MOON.—THE HOWLING MON-
KEYS HOWL, AND OTHER SIGNS OF AN APPROACHING RAINY SEASON.—
THE JACARÉ, OR BRAZILIAN CROCODILE.—GULLS, AND NOISY BIRDS.—
SERPENTS.—LAST NIGHT ON THE RIO DAS VELHAS.

> ———— o clima doce, o campo ameno
> E entre arvoredo immenso, a fertil 'herva
> Na viçosa extensão do aureo terreno.
> *(Caramurú, vii. 50.)*

THURSDAY, *Sept.* 12, 1867.—We had been idle yesterday. I
had given an inch, and very naturally my men had taken the usual
ell. We began early with the best of resolutions, doomed, however, to be disappointed. Presently slack water prepared us for a
fall, called by M. Liais the "Cachoeira dos Ovos.* Here a mass
of green-clad blocks and a break sent us first to the left and then
down midstream. Half an hour afterwards we reached O Desemboque—the disemboguing.† A little further down an old Morador
put off from the right bank to buy twist-tobacco, which the "Menino" had bought for seven and sold for twenty coppers per yard.
Yet the whole country is admirably fitted for growing the weed.
He gave us a terrible account of a rapid some seven miles down
stream, declaring the fall to be six feet high, and nothing would
persuade him to accompany us. Probably he had never seen it.

Presently appeared on the left the opening of the Rio Lavado,
or Washed River, so called from the diamond diggings in the

---

* The "Menino" named it "Barra das
Pedras," and an old man on the bank
"Cachoeira do Ribeirão," from a little
stream on the right which we passed at
9·30 A.M.

† In Minas Geraes there is a town called
Desemboque. M. Liais writes Desem-
borque and "Embórque" (p. 22), the popular pronunciation ; there is, however, no
such word. Here a shallow tide-rip (ma-
reta) crosses the bed, the effect of rocks
extending from the right bank. We descended, stern foremost, in ten minutes,
and took the right of a small Corôa.

upper bed.   The gap, 150 feet broad, appeared to be choked with
green.   We easily shot a small break garnished with three lumps
of stone, and went to the left of a Corôa and its shallow.   Now a
reach and banks, regular and artificial dykes, backed by a fine
mass of blue Serra, prepared us for the Cachoeira de Escaramuça,
the tenth and last serious obstacle on the Rio das Velhas.

This rapid is formed by a broken wall extending nearly across
stream from north-west to south-east.   The hard clay is capped
with iron, and the shapeless rocks are tilted up nearly vertically.
In the centre is the main drop, about three feet high, and here
the channel would easily be opened.[*]   We went half way down
the shallow thalweg, close to the eastern bank, and after six
minutes we made fast near a patch of bright green " water grass,"
hardly sweet enough to be good forage, whilst the pilot went
ahead in the tender to prospect.   Under the shady trees the rush
and bubble of the cool waters made pleasant music, and it was
interesting to see the old man balancing himself like a rope-dancer
upon his hollow log, tossed by the tide-rip.

Below the principal fall were three channels.   That lying to the
right of the Corôa proved too shallow.   Above the sand-bar was a
bad broken passage, rejected because of the rocks to leeward.   Between them and the gravel-islet lay the clear way.   The river was now
at its lowest, and the drift timber showed that it had lately fallen
two inches.   The crew was obliged to clear away the rock-frag-
ments, and the Eliza was led like a vicious mare down the hand-
made channel.   On the Corôa we found for the first time the
bivalve shells of a "river mussel,"[†] which extends all down the
Rio de São Francisco, and which is valued for fish bait.

After working nearly an hour we made for the left bank, and
anchored near the mouth of a small marsh drain, " S. Gonçálo das
Tabocas " (of the wild bamboos).   Here the men changed their
dripping clothes, and guarded against the rheumatics with a dram.
At 2·20 P.M. we resumed work, passed sundry Corôas,[‡] and ran
under the Serra do Paulista.   At 4·30 P.M. we attacked the Cacho-

---

[*] M. Liais proposes to open the right
channel, but this portion would, I venture
to think, soon be filled up.

[†] It is the No. 1 of my small collection.
According to the pilots this mussel, when
alive, keeps in deep water, and only shells
are found in the shallows.

[‡] The first was a small Corôa with a
break and snags ; the right bank a little
below it showed heaps of black stone, in
which sand, and frequently the blue calcareous matter, reappeared.   The next Corôa
was close to the Serra, which in the map
is placed one mile too far east.

eira das Prisoẽs—of the prisons. It is formed by a Corôa of large pebbles between which spring tufts of grass. On the northern end grew a clump of the largest trees yet seen. The right channel being too narrow, we took the left, and bumped along the islet, leaving the break to port. It was not easy to escape a snag in the middle, where there are also many rocks. The Mandim fish croaked like a frog and grunted like a pig under our bows.

This day's sun had been burning hot, and till 1 P.M. we had no breeze. As we descend the atmosphere undergoes a notable change, like the air of the Mediterranean after the English Channel. Nothing can be more delightful than this sensation; one feels thawed; the "snow gets out of the eyes," the "ice leaves the bones," and man is restored to the passive enjoyment of life in the medium where he was first born to live. Hence our seamen, it is well known, prefer the West African Station, despite its fevers and dysenteries. A "spell of cold" easily explains the preference.

Nor can we complain of heat, remembering that we are in S. lat. 17°, about the parallel of Mocha in Southern Arabia. Here we have 85° (F.), there 105°. The climate is tempered by the large area of sea compared with land, by the abundance of water causing a regular ventilation, by the height above sea-level, by the hours of darkness being nearly equal to those of light, and generally by the shape of the continent. At times, however, especially under the tree-shade, the vermin bite viciously. Of the larger nuisances, I have not yet seen during my Brazilian sojourn the centipede, or any but spirit specimens of the lacraia, or scorpion, although Koster was stung by one, and in Patagonia the latter is plentiful as in El Hejaz. Hence the term is sometimes applied to the Bicho Cabelludo, or hairy caterpillar, called by the indigenes Taturana. The Carrapâto tick and the jigger, except in huts, are rare. We did not suffer from the Berne or blow-fly, nor from the Marimbombo, the "Jack Spaniard" of trappers. The borrachudo (Culex penetrans) which greatly affects cool and wooded Serras, at times gives trouble. The bite draws a point of blood which must be pressed out, and the place rubbed with ammonia, otherwise the itching becomes intolerable. I never travel without a large supply of "smelling salts," which are equally valuable against a snake or a headache. In this arid atmosphere the mutúca or motúca (which Southey writes "mutuça") gadfly is rare. The Mosquito,

generally called mosquito pernilongo,* but here muriçóca or muri-
sóca (Morisoca, Koster), at times pipes a small song, in "la sharp,"
say the musically eared. The "bar," however, is as little necessary as is the "fever-guard." The insect is not a large variety,
like the Vincudo of the coast, especially of the Mangrove rivers,
and its threat is worse than its bite. In February and March,
when the waters recede, and the banks, like those of an African
river, are dressed in mire, the infliction is said to be severe. The
most troublesome is the diminutive dark sand-fly, known as Mu-
cuim (Muquim, St. Hil.) or polvora. The Maruim or Morúim
(Maroim, Koster; Miruim, St. Hil.; Merohy, Gardner), burns
like a "blister" of fire; it produces swellings, especially around
the eyes, even in those who do not suffer from the Mosquito, and
where swarms are found it is as well to wear gloves and a gauze
veil connecting head and body gear. The Carapana and a smaller
variety, the Puim, which delight in the Assacú (Hura Brasiliensis),
also bite by day.

At 5·45 P.M., after much labour, short and sharp, we were not
sorry to find on the left bank a clearing known as the Curralinho.
A little above was the Córrego do Negro, with a white-tasselled
Ingazeira† drooping over the water. A black morador sold us a
gourd full of eggs at the rate of five per "dump," copper or
penny. Here we saw fine sugar, castor plants 15 feet high, and
magnificent cotton. It was a fine study of wild life. The screams
of the wild fowl told us of a lakelet on the right bank, and as
the after glow deepened, flights of wild duck and the splendid rose-
tinted Colheireira‡ winged their way across the stream. The
moon, nearly at the full, and almost obscuring Jupiter, rose
majestically above the misty wall, the Serra da Piedade, which
bounded the view to the left. The shadow of the vegetation upon
the far side, as the lunar disc tipped the tallest trees, was nearly
as well painted upon the mirrory waters as in the soft blue air.
The river seemed to sleep, and over its depths brooded unbroken
silence, except when a fish sprang to its prey. The stars and

---

* Mosquito, both in Spanish and Portuguese S. America, is properly speaking a
"little fly," namely a sand-fly, and the
name which we have perverted is thoroughly
appropriate.

† Not Angaseiro as Halfeld has it. The
name, Ingá or Engá, is applied to Mimosas
of various species, some bearing an edible
legumen.

‡ The "spoonbill," so called from its
chief peculiarity. The zoological name,
"Platalea Ayaya" or "Ajaja," is evidently
derived from the Tupy Ay' áya.

planets rose with no glimmering indistinct beams, as they appear upon the horizon in northern lands; the rays strike the eye at once in the full blaze of their beauty. At times a cold breath came from the highlands to the north-east, soon to be followed by a warm and violent gust from the north, which swept harmlessly over our sheltered raft. Then recommenced the persistent clamour of the " Curyangú " and the complaint of " Whip-poor-will," whilst in the distance the wolves bayed their homage to the Queen of Night. What a contrast to the hum of civilisation and the glaring of the gas!

September 13.—The morning was warm—65° F.—and we were en route with the rising of the " fall-sun," whose smoke-stained disk was harmless as in England. Presently we passed the Piedade River, which heads far to the north-east.* Under its influence the Rio das Velhas spreads out into a bay widening to 1500 feet and half a mile—my companion was reminded of the Yazoo River. The flat benches and ledges of the banks, fifteen to twenty-five feet high, show by their regularity the action of water. Half a mile below the Piedade we found the Cachoeira dos Dourados,† with rocks on the left; the channel to the east is shallow, and a bottom of heavy pebbles causes a break. Below the Coroa we poled across to the western side, shaving two large trees in the stream.

At 7·15 A.M. we passed the Córrego de São Gonçalo,‡ which takes its name from an old village and chapelry on the upper course. After making another channel by removing loose stones and safely cordelling down a difficulty,§ we came to another " Cachoeira do Desemboque," which M. Liais calls the most dangerous point on the Lower Rio das Velhas.‖ It is a complicated feature; at the north is a gravel islet covered with trees;

---

* The mouth is 110 feet wide, and the left jaw is garnished with a green patch and fine trees; the stream is said to be full of fish, and, though shallow, it gives passage to canoes as far as its Serra.

† The Dourado, or gilt fish, the Aurata of Dr. Levy, so called from its red yellow belly and fins which flash in the sun, is one of the Salmonidæ, found in salt water and in streams where it cannot escape to the sea. It resembles a trout in shape of the body, not the head, and it grows to a length of two to four feet. It readily swallows bait and devours small fry. The

people consider it one of the best fish for the table, and the head and belly are the parts preferred.

‡ M. Liais has called it Córrego de Maria Grande.

§ Below the Córrego was the " meio brabo (half fierce) Cachoeira das Taboquinhas "—of the little bamboos. Then a long mass of black rock forms two distinct ledges, the northern stretching from southeast to north-west almost across the stream.

‖ Here M. Liais has placed on the right bank a tall block of hill, which does not exist.

a white and sandy holm also well clothed, and below it a common sand-bar. The brawling river-channel on the right has not water enough for canoes. Here magnificent masses of green bulge out towards the stream, and are set off by large bunches of rusty-red yellow flowers, resembling from afar the autumnal and maturing leafage of the sugar maple (Acer saccharinus). This Paú Jahu,* when seen singly, is by no means beautiful; the Sertanejos make tea of the blossoms, and the ashes are used for soap. We took the left of the islet down a thalweg with a small sand-bank and two breaks in the centre. The second was the more dangerous; a rock below the surface threw back the flood in white foam. We then poled to the north of the Corôa and down the centre, inclining to the right.

Followed a confusion of small sand bars,† while in front rose the "Serra do Brejo," trending from east-north-east to west-north-west. The height is from 1300 to 1500 feet, and there are two distances, the nearer forested, whilst the farther is dyed blue with air. We halted for an hour at 1·30 P.M., when a high northerly wind set strong in our teeth; it lasted till 4 P.M., when it fell to the deadest calm. These breezes greatly retard progress, as the men seem to disdain the shelter of the bank. A charming reach then appeared, a long perspective of corresponding sides, some ninety feet high, into which ran large blocks of stratified and weathered stone. Below the small " Corôa da Carióca,"‡ (of the white man's house,) the Rio das Pedras opens on the left a mouth of ninety feet from jaw to jaw. It comes from a distance of ten leagues, but at this season it is dry; such indeed is the case with all but the very largest drains.

The Corôa-cum-Ilha do Cahir d'Aguia was the largest we had yet seen; it took us fifteen minutes to run by it, and in England it would have been a pretty little estate. The narrow right hand channel is garnished with splendid forest-trees, faced on the left

---

* The pilots called it Marmelo do Mato, or wild quince. The Jahú is also the name of a large Silurus, not found in the Rio das Velhas, but abundant in the Rio de São Francisco, the Upper Paraguay, the Tiété, and other streams.

† The first was a shallow break, the "Cachoeira da Cannella;" it is just below the "Corôa do Curral," a deep strong current with the passage on the left. The "Cachoeira do Cotovêlo" (of the elbow)

subtends a long island, and the "Corôa do Cantinho" (of the little corner) is a double islet, with dark rusty pebbles to the south and tall trees to the north.

‡ From Carýba, a "Carib," a white man, a Portuguese, and "Oca," a house. "Carióca" was often applied to a small fort, and hence the name of the suburb of Rio de Janeiro. This Corôa has many snags on the right, the swiftness of the stream to the left sweeps them away.

by second-growth and scrub.* About 5 P.M. we anchored near the left bank, at the " Porto da Palma," a peculiar formation. Projecting into the stream, and flaked and caked with the mud of the last flood, was a natural pier 150 yards long by twenty deep, with a dip of 5° and a westerly strike. The substance is the " Pedra de Amolar," an argillaceous schist of greenish colour, sometimes bare, more often capped with ironstone ; the cleavage is in all directions, the subaërial portion is very fragile, and the laminations vary from wafer thickness to a foot deep. A little below on the right bank there is a sister formation. We picked up specimens of this clay shale ; as whetstones they were too easily broken.

The river plain on the left bank is baked to white mud and sprinkled with silt, showing that it is regularly inundated ; small drains bear scattered lines of trees, and the rest of the vegetation is mostly bitterish water-grass, which will not feed cattle without salt. To the south-west the land, as the forest shows, is beyond the reach of water ; here the soil must improve. The flat had lately been burnt, and the shrubby trees, well warmed, had put forth the tenderest green leafage in lieu of the scorched brown tatters that hung loosely to the twig tops.

This evening was the perfection of climate, fresh yet balmy. The boys fished successfully ; everything bit voraciously, even at the bird-bait. Five douradinhos† and eight mandins soon lay crimped upon the ground, and when the line, nearly the thickness of a little finger, was left in the water, it was cut, the pilot said, by a piranha. Again the noise of water-fowl told us that a lakelet was not distant. Clouds high in air flitted over the moon's full disk, which threw across the water a pillar of tremulous fire, and crested with red the ripples that rose from the inky surface swirling under the further bank. The mobile physiognomy of this river is not the least of its charms. Its expression is changeful as that of the human face. Yesternight it was still and shallow as a mountain tarn, now it is swift and deep, covering the backwaters with flecks and curds of foam.

Presently the eclipse came on, and the dark shadow of our globe creeping slowly over the disk of the old " harvest moon," was

---

* The left bank showed tolerable soil in the beginning of this day's work, but the improvement was only temporary. As a rock it was not nearly so fertile as the other side.

† Considered to be a small species of the dourado.

shown by reflection in the new moon's arms; the gibbus of the crescent, however, faced to the south. There were none of the sinister appearances, rather appalling than imposing, which accompany solar obscuration, the lurid copper-coloured air, the slinking of beasts and the silence of birds, and in man the feeling that even the sun is not above or beyond change. Here the light slowly waned, the various voices of frogs and night birds came from swamp and forest, bats flitted about, fireflies lit up the copse, and the fish splashed merrily to catch the gentle breeze. As might be expected, the human beings there present hardly noticed the phenomenon by looking upwards; a comet would not have roused their attention.* Then the glorious satellite climbing the zenith finally emerged from the shadow, and again shed silvery light and gladness over the nether world. By way of anti-climax we "turned in."

September 14.—We set out at 6 A.M. in warm and perfectly still air; foam was floating in lines down stream, and curdling near the banks where the deep water lies. An hour's work took us to the Ilha da Maravilha, where the Córrego do Lameirão † enters the left bank. On the opposite side appeared a good "improvement," the soil was excellent, and a fence of stakes and poles had been run down to the waterside. Presently we heard, for the first time, from a tall Jatobá tree, whose fruits are its delight, the hoarse roar of the Guariba monkey (Mycetes ursinus, or Stentor). It is here known by the general words bugío, and barbado, the bearded; the French colonists call it alouate. John Mawe declares that it snores so loud when sleeping, that it astonishes travellers; the enlargement of the larynx into a square bony box which causes the disproportionate noise, is now familiar to naturalists. This brown monkey was eaten by the Indians, and in wilder parts Brazilians do not disdain it. The pilot mentioned a similar species with a long fine black coat, which may be the Mycetes Beelzebub. He declared that the roaring of the guariba was

* Mr. Buckle, whose first volume had the good fortune to be designated by a popular writer "a *farrago* of energetic nonsense and error," remarks (i. 345) "there probably never has been an ignorant nation whose superstition has not been excited by eclipses." Possibly in the New World, where the operations of nature are on so grand a scale, man is steeled against appearances which in other countries would stimulate his imagination. Who that has ever inhabited an earthquake country would think of dreading an eclipse, unless at least it be connected in the popular mind with earthquakes?

† "Of the big mud;" the pilot gave this name to an opening in the right bank.

a sign of the Rainy Season drawing near, and noted a variety of other small symptoms, such as the trooping of butterflies in moist places, the louder frog-concerts, the hum and chirp of the Cicada, the biting of the sand flies, and the song of the Sabiá, that Prince of the Merubidæ. During the last three days also, the soft and balmy atmosphere had been disturbed by gusts of wind, vapours here lay upon the ground, there accumulated into clouds, and distant sheet lightning flashed from the mists massing round the horizon. The smoke of the prairie-fires rose in columns, and they might have been mistaken for the fumes of a steamer; by night those that were near glowed like live coals, whilst the more distant gleamed blue. We prepared for an Ember week of equinoctial gales, but we hoped to be far down the São Francisco River before the beginning of the wet summer, which usually dates from the middle of October. As will appear, we had deceived ourselves.

About 10 A.M. we passed on the right bank the Ribeirão do Corrente; a small stream, which greatly swells during the inundations, was trickling down it: the line is not navigable, but the waters abound in fish, and these places will act as preserves when life is driven from the main line by steamers. The embouchure is marked by a columnar conical mass, which suggests an enormous cypress formed by vines and creepers swarming up a broken tree-shaft. Here a dog swimming across the stream showed little apprehension of the "Jacaré" (Crocodilus sclerops), and the people declare that those of the lakes are dangerous, whilst the river-caymans * are not. Lately, however, a woman was carried off in the Ribeira de Iguape by this congener of the dreaded African crocodile. It is said to prefer its meat "high," as does its big brother, and before deglutition to break the bones of its victims by blows with its ponderous head. According to Koster, the wild people eat it, but the negroes will not touch the meat; even the Gabams (Negroes of the Gaboon), who are believed to be cannibals. Both on the Rio das Velhas and the São Francisco we often saw the Jacaré protruding its snout from the water, basking in the mud, or lurking amongst the drift wood. No specimen exceeded five feet in length; in the Apuré and the equinoctial

* In old French, Caymand and Caymande are equivalent to "faineant;" perhaps the early travellers found the huge lizard unwilling to move. It becomes unwieldy with age, but in youth it is very agile.

rivers it grows to four or five times that size. The negroes, it is
well known, use the crocodile gall in their philters and poisons;
the molars of the Jacaré are here hung round the neck as talis-
mans against disease. The musky smell of the meat must deter
any one but an "Indian" from using it, and the people ignore
the alligator-skin boots which Texas invented.

A lumpy hill, grassy above and forested below, and stretching
from north-east to south-west, strikes the stream at this point,
and bends it from a straight course to the south-west and the
north-east; this "sack" is seven miles long instead of one.
Presently we passed a large Fazenda on the right bank of perpendicular clay, some thirty-five feet high; it belongs to Dr. Luis
Francisco Otto of Guaicuhy, and we begin to acknowledge the
odour of civilization. After a few obstructions,* we rested at
noon on the left bank, sheltered from the strong north wind;
here was a mass of bluish stone, which appeared to be finely
laminated calcaire when it was only clay shale.

Resuming our way, we passed to port the Córrego das Pedras
do Burity,† where the great bend terminates, and two nameless
influents—I mention them, because they are "Corrégos de
Morada," where men have settled, and which afford a good "situa-
ção," giving value to the lands adjacent. At 4 P.M. hove in sight
a tall blue wall of mountains, denoting the line of the Rio de
São Francisco; the crew disputed about the name,‡ and also
about a couple of Córregos further down.§

At 5 P.M. we made fast to the right bank of the Ilha da Tabua,
which the pilots called Ilha Grande. It has a large Corôa to the
south, with a mound of stiff clay, tree-grown and root-compacted,
extending from south-east to north-west. The left arm of the
river is here greenish in the centre, and beautifully clear under
the banks; on the other side we saw a farm with a line of

---

* A sunken sand-bank (Areão) which
must be passed on the left, a Corôa in the
bend called Saco do Jequi, and a double
tide-rip inclosing smooth water; this is
formed by a beach (praia) on the right,
which narrows the stream to 120 feet.

† "Córrego grande dos Buritis." Liais.

‡ One named it Serra do Jemipapo, and
another the Serra da Tabuá; it may have
been the Serra da Porteira (Liais) on the
right or eastern bank of the junction.

§ In the map "Córrego da Gamelleira"

and "Córrego do Tamburil." The wild
fig here attains a great size, and sometimes
six stems spring up together. The Tamburil, pronounced Tamburi (M. Liais "Tam-
bury"), also called Vinhatico do Campo, is
a tall hardwood tree. The "Menino"
insisted that the "Tamburil" influent
should be called the "Gamelleira," and
that it is "de morada," not navigable
but coming from afar. The mouth is eighty
feet wide from jaw to jaw.

noble trees, whilst the north is a tangle of wood, thicket, and grass.

For the first time we found the Corôa well stocked with birds.* The Urubú scavenger, regardless of the rifle, expanded his wings to the sun, and looked as if he wore a silver back. Small Charadriadæ hopped gleesomely about the sands, together with Manuelsinho da Corôa—little Emanuel of the Sandbar—a Scolopax with red-stockinged stilts, much resemblng our sandpiper. The South American plover (Vanneau d'Amerique, Vanellus cayennensis, Neuw.), also with red stockings and pretty variegated plume, followed the cattle tracks. Spanish America calls it after its cry, Tero-Tero, the Portuguese prefer Quero-Quero (I want! I want!) and Espanta boiada, "Startle Cattle:"† its manners are those of the peewit, it haunts marshes and pastures, it seems never to sleep, and it is a great plague to the sportsman. In remarkable contrast with its unpleasant vivacity, is the solemn Acara, or heron with the long thin legs supporting a body always delicately white and clean. A tern very like the Sterna hirundo, looking snow-white against the slatey blue sky, fluttered in the lower air with the rising and falling flight of the butterfly. The Gaivota, or gull, which the Tupys term Atyaty, or Cará-caraí, dark-backed and red-billed, reminded my companion of those which show communication between Memphis and the Mexican Gulf, one of the colonies which I saw upon the Tanganyika Lake. The whole flock rose and with circlings and swoopings followed and seemed determined to fight the dog Negra, occasionally varying the exercise by feinting to assault the men. They were enraged at our intruding upon their private property, and with proverbial stupidity they told by screams the secrets of their ménage. We retaliated by taking their eggs,‡ which were about the size of a plover's, with "splotches" of light and dark chocolate brown upon a dirty cream-coloured ground. They revenged themselves by a persistent "corrobory" round our camp-fire, which effectually banished

---

* The number, however, gradually increased below the Paraúna River.

† Thus Sr. Ladisláo dos Santos Titará sings,—

Vão quero quero pelo ar soltando.

‡ So on the lower Purús, in July the eggs of the Gaivota may be picked up by scores from the nests, round holes, four inches across and three deep, in the sand-banks, where an upper coating of mud prevents the drifting of the wind; "the eggs, three or four in each nest, are of a dirty light green or brown, with patches as of dried blood; when fresh they are very good eating and much like puffins' eggs." Ascent of the River Purús, by W. Chandless. Journal Royal Geo. Soc., vol. xxxvi. 1866.

sleep, and they were viciously ready by early dawn to " see the last " of us with taunts and execrations.

The " Menino " found upon the sands the parallel lines which might easily have been mistaken for cart-ruts ; he declared it to be the sign of the dreaded Sucuriú,* or Watersnake, whilst Chico Diniz declared that the straightness of the trail showed a small Jacaré. This hideous boa mostly haunts stagnant waters, occasionally visiting rivers ; it is amphibious, and when not disturbed by man and prairie fires, it attains the enormous length of thirty feet. I heard of one that measured sixty, and swallowed a bullock ; in old travellers we read of men sitting down upon a fallen tree-trunk, which presently began—like the whale with the fire on its back—to change location. The " Indians " eat the Sucuriú which, like most serpents, is savoury and wholesome food ; the civilized confine themselves to eels. The skin used to be tanned for boots and housings, now it is kept chiefly as a curiosity.

At Maquiné, a morador threw into the river, before I could secure it, a fine specimen of the Surucucú, or Çurucucú, first mentioned by Marcgraf. It is the Lachesis mutus of Dandin, the Crotalus mutus of Linnæus, the Bothrops Surucucú of Spix and Martius, the Xenodon rhabdocephalus of my friend, Dr. Otho Wucherer (Zool. Soc. London, Nov. 12, 1861), and the " great viper " of Cayenne and Surinam, which is supposed to cause death in six hours. The length of this trigonocephal varies from three to eight and even to nine feet ; its skin is of a dirty tawny yellow, with dark brown lozenges on the back, and the broad head gives it, to the connoisseur, a peculiarly vicious appearance. It is reported to be attracted by fire, but rarely to injure travellers. There are two species of this snake, the less common being the " Sururucú bico de jacá."

The other serpents of which the people spoke were the fol-

---

* The Boa Anacondo of Dandin (the Boa Murina of Mart., Cunectes murinus). " Sucuriú," properly " Sucury," is derived from "Suu " beast, and "cury" or "curu" a snorer, a snorter, alluding to its sibilant powers. According to Prince Max. (ii. 172) this boa is called " Sucuriú " in Minas, and " Sucuriuba " on the Rio Belmonte. Pizarro prefers " Sucruyu." Some write " Sucuruju " and even "Sucuriuh," and pronounce " Sucuriú." It is also called "Cobra de Veado " because supposed to be fond of venison, and Spix and Martius heard from M. Duarte Nogueira that it has attacked a man on horseback, and has even swallowed an ox. A Brazilian gentleman assured me that in Maranham he had seen the terrible reptile swimming across the stream with a pair of horns protruding from its mouth.

lowing. The rattle-snake (Crotalus horridus), is known as the Cascavel (not Cascavella, as some write), a "hawksbell," and the Tupys called it "Maracá," a rattle, or boicininga, from "boia," or "boya," a serpent, and "cininga," a chocalho, or bell. It is well proportioned, in length between four and eight feet, and brown grey with lozenges of lighter and darker colour. It prefers stony and hilly ground, where it can easily sun itself, and has a kind of domestic habit of making a home. It is very lazy and harmless, except when troubled; hence, probably, its fame for listening most willingly to the voice of the serpent-charmer. The rattles* soon give warning, and it may be killed with a switch; cattle are often poisoned by it, but I have not heard in the Brazil of a man dying by its bite. Possibly the dampness of the climate may modify the venom. The fiercest of the lance-headed vipers, and emphatically declared to attack mankind, like the Cobra de Capello of the Guinea Coast, is the Jararocá (Cophias or Viper atrox; Bothrops Neuwiedii of Spix and Martius, alias Crespidocephalus atrox). It is of a dirty dark yellow, turning to brown-black about the tail, and although Koster gives it nine feet, it seldom exceeds five feet in length, and the Jararacussu is the same reptile when full grown and old. The Caninina often mentioned by old writers, is a Coluber not much dreaded, and the papo-ovo or egg-eater much resembles it. The Cobra Coral is so named by the people from its resemblance to a necklace of mixed corals; the term, however, is applied to four, five, or more animals of different species. The common Coral, Elaps corallinus, called Coluber fulvus by Linnæus, who saw it when the beautiful colours were tarnished by alcohol, has black, carmine-red, and greenish-white transversal rings upon a smooth thin body. All declare, both in books and vivâ voce, that it is as venomous as it is charming; but the fangs, though formed for offence, are so placed as to be almost useless. Another Coral (Coluber venustissimus), is also ringed with tricolor ornaments, but the head and gape are larger than that before-mentioned. A third ringed snake is the Coluber formosus, with an orange-coloured head, and not venomous. Lastly, there is the Cobra Cipó, or whipsnake (Coluber

---

* Dr. Renault of Barbacena declares that the rattle (sonnette), is perpendicular in the male and horizontal in the female.

bicarinatus, the Cypô of Koster), with a line of carinated scales
on each side : it is often confounded with the Cobra Verde, a
fine, green, harmless Coluber. I have killed it in a tree despite
the prayers of the bystanders, who declared that it can project
itself like an arrow. The same tale is told of the Cananina, which
is mentioned as a " flying snake " by Koster.

When first visiting the Brazil, travellers come prepared to
meet serpents on every path, their minds are brimful of beasts,
every spider is deadly, they suspect the intentions of the cockroach, and a thorn-prick suggests a scorpion. Even the unfortunate Macaco fly, the African Millipede (piolho de Cobra),
the Amphisbæna or Slow-worm ("Mãi das Sámbas"), the
innocuous "Dryophis," and the Gitaranaboia * are capable of
dealing sudden death. Presently they find out that the reptiles have retreated before man, either to the seclusion of the
maritime regions, or into the Far West. As in Africa, so here,
" snake " means something more or less fatal. I presume that
man's aversion to this harmless and maligned animal is partly
traditional, derived from the old Hebrew myth, and, to a certain
extent, instinctive ; the brightness of the eye, upon which Mr.
Luccock could not look, and the form of the head, a curious
resemblance to humanity, being the most remarkable points. I
have heard, even amongst the educated, of an inherited horror of
the snake, but this must rank with the tales of the Serpent kings,
and with the " Indian " fancy that a man when bitten must not
look at a woman.

The Brazilians inherit from the old inhabitants † a sensible
way of treating snake bites, but their system admits of improvement. The savages applied above the wound a ligature, which

---

* This insect, of which the traveller will
often hear, is described as about two inches
long, with an oblong body, a snake-shaped
head one third of its total length, and
wings like those of the tree cricket (Cigarra),
but much longer. The proboscis folds
under the abdomen like the blade of a pen-
knife ; this stylet is supposed to be thrust
forth like a bayonet when the insect flies
straight as an arrow, and as it is always
blind it victimises everything which comes
in its way.

† And from the Africans. I could not,
however, find any traces of the "Mandi-
gueiro" or serpent charmer, who, according

to Koster, is the West Indian Obeah. But
the word is evidently a corruption of Man-
dingo, the old and incorrect form of
Mandenza, a Semi-Semitic Moslem race,
well known at "Sã Leone." Wonderful
tales are told of these "Curadores de
Cobra," how they could handle the most
venomous reptiles, cure the patient (curado
de cobras) by wrapping a tamed snake
round his head and shoulders, or by re-
citing magical words, or by the use of
"contas verdes," literally "green beads,'
which were probably nothing but the blue
Popo bead of which every West African
traveller has left an account.

prevents the blood reaching the heart for some time; the civilized bind it so tightly that mortification of the limb has followed. Both indulge in a butcher-like style of surgery, which has been imitated by the scientific man.* They almost always administer as sudorifics spirituous drinks in large quantities, and this is the secret of the cure; the action of the heart is restored, the venom is expelled, and the brain returns to its normal functions. When the patient, who mostly complains of a "sinking" sensation, as in cholera, becomes intoxicated he is safe. On the other hand they mix with the alcohol what is either harmless, as lemon juice, or spirit in which a Cobra Coral has been macerated, or what is positively injurious, as mercurials. There are many simples in general use, such as the Herva Cobreira, the Aristolochia, the leaves of the Plumieria obovata, and the grease of the Teyu, tree-lizard,† whilst Aves and Paternosters do the rest. "On dit que les sauvages guérissent très bien les morsures des serpens, et l'on m'a même assuré que parmi eux personne ne meurt de cet accident."‡ Evidently the civilized man ought not to die unless he delay too long to apply ammonia, eau de luce, or the "whisky-cure."

Our last night on the Rio das Velhas recalled to mind the words of an eloquent Brazilian writer. " I cast my eyes now on the stream spanned by a line of fire reflected from the planet Jupiter, then on the banks whose beautiful woods concealed the rich champaigns. The river, a natural line of navigation, despised by and despising art, rich in a thousand kinds of produce, fertilizing in its sinuous course millions of acres, was full of all but human life; to its silent banks here and there a canoe was tied, and from its waters rose the log which the solitary fisherman makes his perch; while at rare intervals a

---

* Thus M. Sellow records treatment by scarification, repeated burning with gunpowder, and peppering with Cantharides. Labat, to mention no others, scarifies the wound. Koster observes, "le rum est aussi administré jusqu'à produire l'ivresse."

† The Teiú or Teyu (Lacerta Teguixin, Linn.), is black spotted with yellow, and, including the tail, four feet long. Yves D'Evreux writes Tyvu, Marcgraf Toinguaçu, M. Denis Tiú (Tupinambis monitor), and declares with St. Hil. that the white, savoury, and delicate meat is eaten by Brazilians in good circumstances. This is certainly not the case in the Sertão of these days. Koster mentions the Tijaaçu, which he believes to be the Teguixin; the Calango, a smaller variety also edible; the vibra, and the lagartixa, a house and wall lizard, a vivacious little animal which destroys flies and other insects. Some travellers have confounded the Teiú with the Jacaré, as the old Greek who wrote the Periplus did at Zanzibar. The good missioner (Yves) specifies the Taroüire as a grand lizard, but his editor corrects him, and declares the Taranyra to be smaller than the Tiú.

‡ Prince Max. ii. 294.

dwelling-place, and clearings that ignore civilized agriculture, dotted the forest-shore. Such misery and so much want in the Old World!—here such neglected wealth, and so much that can make life happy! Lands that will fructify every manner of plant and grain cast into their bosom, shoals of fish to feed the poor, a wealth of precious stones and ores, a channel easily connecting with the outer world! But the age shall come, and the day has dawned, when men shall flock to these unknown regions, when gardens, quays, and works of art, shall adorn the river side, when town and village shall whiten the plain, and when the voices of a happy people shall be heard where the profound solitude and silence are now broken only by the moan of the dove, by the scream of the night-bird, and by the baying of the wild dog."

So be it !

CHAPTER XIII.

TO AND AT GUAICUHY.

"A descripção das scenas de natureza deleita, a dos costumes instrue."
"Aquelle que só deleita toma se superficial, o que só instrue, aborrecivel;
casemos pois estas duas qualidades."—*A. G. Teixeira e Souza.*

A HOUSE on the left bank kept up during the night a red
fire, which shone through the dark trees, another evidence that
we were approaching a centre of settlement. After a few days of
traveller's life and liberty, of existence in the open air, of sleep
under the soft blue skies, of days without neck-ties, the sensation of returning to "Society" is by no means pleasant; all
have felt, although, perhaps, all will not own the unamiable
effort which it has cost them. The idea of entering a town
after a spell on the Prairie or on the River, is distasteful to
me as to any Bedouin of the purer breed, who must stuff his
nostrils with cotton to exclude the noxious atmosphere. I
looked forward with little pleasure to breaking up my crew,
and to entering Guaicuhy.

The first of Ember Week (Sunday, September 15) showed
a warm cloudy morning with a north wind, contradictory signs.
We passed on the left the Córrego da Tabua, it comes from
the Serra of that name, a continuation of the Palma range;
about two miles from the mouth is an Arraialsinho, or little
village. Presently rose before us the peaky Serra do Jenipapo.
The uniform river-banks would in Europe be called a forest;
here they seemed utterly civilized, with their Coqueiro palms,
their huts and vegetable-plots, and their scatters of old and new

clearings. The river widened out and became somewhat shallow; the sole obstacle was a sunken rock known as the Páu Jahú.

We " cleaned ourselves "—literally not funnily—and prepared for delivering letters of introduction, which, being directed to absentees, all proved useless. About 10 A.M. we made fast at the Porto da Villa do Guaicuhy, the port being a rough clay bank, covered with thicket, through which a path is cut to the upper settlement. Presently we received a visit from the Delegate of Police, Sr. Leandro Hermeto da Silva, and sundry friends; he kindly detached a sergeant to find us a lodging at the Porto da Manga, a few hundred yards down stream, and close to the junction of the two great rivers, das Velhas and de São Francisco. We were soon established in the house of Major Cypriano Medeiro Lima, who had offered us its hospitality at Diamantina. It was in the usual style, mud and wattle walls, containing a well-ventilated room which boasted of a table, a dark closet with a pair of " catres," or cots, one with a bottom of cow-hide, the other with leathern thongs. A passage nearly blocked up by the big water-pot led to a kitchen distinguished by thin stones upon the ground, and to a little railed compound well calculated for the accommodation of beggars, pigs, and dogs.

Here a mature old age ends the stream which we have accompanied from its babyhood for the last three months: this, however, is not a Thanatos, it is a Mokshi, an absorption. It was impossible to contemplate without enthusiasm the meeting of the two mighty waters which here lay mapped. The " River of the Old Squaws " sweeps gracefully round from north-east to nearly due west, and flowing down a straight reach about 550 feet broad,* merges into the São Francisco, which flows from

---

* M. Liais gives 167 metres. The figures of the junction are as follows : the Rio das Velhas discharges 209 metres per second, and lies above sea-level, 2,365 palms (Halfield), or 567 metres (Gerber), or 432 (Liais at the confluence). I made the Manga 1774 feet high (B. P. 209°·40, temp. 45°). Before the confluence the São Francisco is 359 metres broad, more than doubling the Rio das Velhas, and the debit is 446 cubic metres. The limited discharge is 655 cubic metres per minute.

The Barra or mouth of the Rio das Velhas, south lat. 17° 11′ 54′, and west

long. (Rio) 1° 43′ 35″, may be considered almost in a straight line of prolongation of Rio de Janeiro, Barbacena, and Sabará. The distance from the arc of the great circle uniting these points is only five geographical leagues to the west, although the old maps placed it far to the east. The deviation from the direct line prolonged from Rio de Janeiro to the Barra do Rio das Velhas, is only 3800 metres, about half a Brazilian league, or $\frac{1}{172}$nd of the total distance, 656 kilometres, or 5° 55′ 31 ″ ·4 (355 geographical miles).

the east to receive it. The right bank of the Rio das Velhas
is of stiff clay standing almost upright. On the other side is
a little Chácara with the plots of Castor-shrub which stretch in
blue-green tufts towards the water, backed by a clump of
oranges and bananas. Beyond it, at the point projecting into
the united rivers, is a matted forest of wild figs, Páu Jahú, and
other wild growth.

I remained at the Manga from the 15th to the 18th of September; the house, which had been long unoccupied, was well
tenanted by the Bicho do pé, and two of them chose to lodge with
me. It is a beast of many names, Pulex penetrans, P. subintrans,
or P. minimus. The old French Missionary Yves D'Evreux
(1613—14), calls it le Thon, and the modern Gauls speak of
"des biches"*—thus the neo-Latin tongues borrow from one
another, only changing the terminal vowels. I have also seen
Brûlot and "Pou de Pharaon," although Pharaoh was never in
America. The Tupys knew it as "Tumbýra." The Spaniards
chose Nigua and Chigua,† from which again the French took
Chique, and the term has descended to us in various forms:
Chigre, Cheger, Chegre, Chegoe, Chigo, Chigoe, Chigger,‡ and
finally the Jigger, thus immortalised by the Negro minstrel:

> Rose, Rose, lubly Rose,
> I wish I may be jiggered if I don't lub Rose.

This nuisance especially affects coffee-stores and deserted
abodes: § the old travellers bitterly complained of it, and carried
camphor in their boots, being careful never to go barefoot. "All
persons of whatever rank," says Southey, speaking of Santa
Catharina, the island (iii. 861), "carefully wash their feet every
night, as the best preservative against the Chiguas,"—which it is
not. A traditional naturalist, wishing to carry home a live speci-

* "Bicho" in Portuguese is a very comprehensive word, as Sir Charles Napier said of Hindostani; it applies to everything, from a flea to an elephant, and even to a steam-engine (bicho de fogo, bicho feio). Koster pleasantly relates how, being a Protestant, he was called in the out-stations "Bicho."

† "Chica" is also used, and M. F. Denis, the editor of Yves D'Evreux (Notes, p. 416) writes "Niga."

‡ The "Chigger" or "red bug" of the Southern States of the Union, is, I believe, a kind of tick which, like the Carrapato, affects the woods. It does not hatch its young in the body, but the result is a painful pimple.

§ According to Koster (ii. xix.), it is not found in the plains of the Northern Sertaõ, and some people in parts badly infested have been so much preferred by the insect, that they were compelled to leave the country.

men, would not be operated upon, induced mortification, and became a "martyr to science." I have often seen boys with their toes dotted over, as if pepper had been sprinkled upon them, but no death has been recorded, and I have heard that careless negroes have lost their feet by amputation.

The Jigger, seen under a microscope, has the appearance of a small flea with well developed body, and of somewhat lighter colour. It crawls more quickly, but does not jump so well as the ordinary pulex; the popular belief is that the male is never found. It burrows under the nails of the hands and feet, especially the latter; I have extracted as many as six in one day, but never from the fingers. The sole is also a favourite place; in fact the Bicho colonizes wherever the skin is thick—hence its preference for negroes. Its proper habitat is between the cuticle and the flesh, into which it does not penetrate, and where there is not lodging room it falls off after drawing blood. Having ensconced itself bodily, the jigger proceeds to increase and multiply; the small dark point develops to the size of a pea, and can move no more. The light-coloured bag is enormously distended with eggs of a slightly yellow tint, and after producing her fine family the parent departs this life.

The small livid point which appears about the nails is generally accompanied by a certain amount of titillation which old stagers enjoy; they describe it as sui generis, and make it almost deserve the name of a new pleasure. Men with tender skins easily feel the bite, and remove the biter before it can penetrate. They then send for a negro, always the best practitioner, and he proceeds to extract the intruder with a pin in preference to a needle. Should the sack be burst, and the fragments not be all extracted, the place festers, and a bad sore is the result; some sufferers have had to wear slippers, and have walked lame for weeks. The wound is finally cicatrized with some light alkali, even snuff and cigar ashes are used, and a little arnica completes the cure.

If any place bear the stamp of greatness affixed by Nature's hand, it is this Junction. It is the half-way house on the mighty riverine valley; it has, or rather it can have, water-traffic with Sabará, Diamantina, Curvello, Pitangui, Pará (or Patafugio), Dôres de Indaiá, Campo Grande, Paracatú, São Romão, and the other settlements on the São Francisco River. It links together the Provinces of Goyaz, Pernambuco, Bahia, and Minas, and before many years the steamer and the railway will connect it with the Capital of the Empire. I shall ink more paper than enough for the present settlements; thus, when my forecast of their future greatness shall have been justified, the traveller may compare his Present with my Past, and therein find another standard for measuring the march of Progress as it advances, and must advance with giant strides, in the Land of the Southern Cross.

In early colonial times the Junction of the rivers and the settlement near it were called Barra de Guaicuhy, and formed an old Julgado, or Chef-lieu de Justice, extinct about fifty years ago. The later generations translated the Tupy name into Barra do Rio das Velhas. The district and the municipality were created in 1861 (Provincial Law, No. 1,112 of Oct. 16) by taking in part of Montes Claros, São Romão, Paracatú, Curvello, and Diamantina, and the principal town took the name of Villa de Guaicuhy. Afterwards were annexed to it Mumbuca and the new districts of Estrema, Pirapora, and São Gonçalo das Tabocas, and now it is divided into four, namely, Guaicuhy, São Gonçalo, Pirapora, and Estrema. The population is stated to exceed 15,000 souls, with 1200 voters and seventeen electors; the latter seldom exercise their functions, as the College sits at Montes Claros, distant 120 to 200 miles of vile road from their several homes.

The settlement is divided into two Bairros, or Quarters. Near the confluence is the (Arraial da) Manga, or the Cattle-ford, popularly called the Port. The upper village is the Villa, formerly the Arraial da Porteira, so called from a neighbouring range, also an old name. The municipality has a single parish, the "Freguezia de Nª Sª de Bom Successo e Almas da Villa de Guaicuhy."

The Manga is a wretched decaying village, apparently doomed to destruction. It is perched upon an almost upright bank of white-yellow clay, twenty-nine feet six inches high, and the walls of the tenements show a water-mark of more than six feet; thus the total of the rise is between thirty-five and thirty-six feet, with a weight which nothing can withstand. The river, as usual with large streams, flows upon a ridge, and swings towards the north side, which readily melts away; its course will be arrested only by the Serrinha da Manga, or Muritibá, a long low lump of hill to the north. The southern bank projects into the São Francisco a long tongue of sand, with hardly five inches of water at this season.

The Manga bank is painful climbing, as that of Angolan Kui-sambi, and its rude attempts at steps, when greased by rain are safe only to the semi-prehensile feet of the natives. The only conspicuous building, upon whose tall, gaunt, sloping roof of tiles the traveller's eye first lights, is the Bom Jesus de Matosinhos; it fronts the meeting of the waters, or south, with a little westing, and it now stands almost at the edge of the precipice. Built of ashlar and lime, it shows that in Colonial times the place knew better days; as usual it is half-finished, a "work of Santa Engracia." The southern entrance has never been roofed, the sacristy to the east is bare scantling, and the belfry is the normal gallows of three timbers. Pilasters and pulpits of cut stone are destined to remain in embryo, and a neat arch of masonry intended to mark the high altar to the north, now the body of the temple, is foul with weeds. Beyond the Bom Jesus is a small rum-distillery, and further down stream the "bush."

Formerly the Manga had two thoroughfares, but in 1865 the inundation swept away the most convenient portion, and only part of "Water Street" shows a double line of blocks and huts, numbering twenty-four. They are built upon flags of hard blue sandstone, resembling lime, sometimes capped with iron, or showing junction with reddish gneiss. The new thoroughfare to the south, and running parallel with the former, has thirty-three tenements which look upon a road ankle-deep in sand. These lodgings contrast badly in point of comfort with Dahome, or Abeokuta, in Egba-land: they are unwhitewashed cages of wattle and dab, roofed with half-baked tiles. All have ground-floors of tamped earth, except the Sobradinho,* belonging to Sr. João Pereira do Carmo, merchant, and Juiz de Paz. In the Brazil this official has conciliating powers, intended to·spare appeal to the Juiz Municipal. But in country places the servants of old Father Antic, the Law, not unfrequently recall to mind the Scotch saying about a far cry to Loch Awe.

Most of the houses have back-yards, green with bananas, Cuietés or Calabash trees (Crescentia Cujete or Cuyctc) and

---

* The Meio Sobrado is a single-storied house upon a raised platform of masonry. The Sobradinho is a one-storied house with a single room above it, and the Sobrado is a two-storied dwelling—a casa nobre when well made.

oranges, which are exported down stream. The Settlement abounds in manioc, and as wheat was not to be found, we laid in a store of polvilho or Tipioca cakes (roscas de Tipioca)* and fubá-meal, which is very expensive on the Upper São Francisco. As in Africa, the housewives would not sell their eggs. Turkeys thrive here, and cost 2$000 a head. About half that sum is paid for fowls and for Guinea-fowls, which are exceptionally held to be good food. The people do not readily part with their provisions, and they are perniciously frugal. A month's work at manioc gives them bread for a year. Moreover, much more is to be had by barter than for money. All determined that we were merchants, and offered cent. per cent. for tobacco. Had we known this I should have invested heavily in the article, and thus made myself a something intelligible. A fatted bullock costs 30$000, a cow 15$000, a pig from 10$000 to 16$000, and fine goats and sheep, mostly four-horned, 2$000. Fish is, of course, cheap. A fresh Curumata, weighing 4 lbs., is worth a halfpenny, and a salted Surubim of 32 lbs., from 3$000 to 6$000. The high value of the latter is owing to the price of salt, which must be imported from the lower river, and the plate of 4 lbs. or 5 lbs. fluctuates between 0$800 and 1$320. Washerwomen and sewing women gave their services at the cheapest possible rate.

At this season the Manga is tolerably healthy, but between January and June, agues, typhus, and malignant marsh-fevers (carneiradas) decimate the inhabitants. Many are chronic invalids, paralytic, or suffering from ophthalmia and goître, which below Guaicuhy will cease to offend the eye. The climate has won for itself an enduring bad name;† but the blame attaches

---

* Our "tapioca" is a mere corruption. I bought,—

Half a quarta of Manioc flour	1$000
4 lbs. of toucinho lard . .	1$280
32 lbs. of carne seca (sun-dried beef)	3$840
Total	6$120

† "Le long du Rio San Francisco, à l'époque où le fleuve baisse, le pays est affligé d'épidémies qui enlèvent beaucoup de monde et deviennent surtout très dangereuses pour les étrangers, ainsi que pour les voyageurs qui ne sont pas acclimatés" (Prince Max. iii. 185). This is repeated by many a writer, and is sensibly modified by Lieut. Herndon (p. 326). "The mere traveller passes these places without danger. It is the enthusiast in science, who spends weeks and months in collecting curious objects of natural history, or the trader, careless of consequences in the pursuit of dollars, who suffers from the Sezoens." As a rule on the São Francisco the fevers, though at times of malignant type, are mostly "chills," and the people, where unable to procure the much valued Quinine, treat them with simples ; such as Sal Amargo, the antifebrile Quina, the purgative Fidegoso, the bitter root Cipó de mil-homens or de Jarrinha (a diaphoretic and diuretic Aristolochia).

more to the dirty and dissolute habits of the people than to the maligned river.* Drainage is absolutely unknown, and the worst sites are preferred, because they are the most handy. The houses are impure to the last degree. The pig lives in the parlour, and "intramural sepulture" here survives. The diet,—fish and manioc, manioc and fish,—assists the work of dirt; hence the sallow unwholesome look and the listless languor of the people. They drink to excess new rum, the "Kill-John" of the Mediterranean. On Sunday evening hardly a soul was sober, and two of my men, the "Menino" and Agostinho, could hardly stand. Having little else to do, their libertinism is extreme.‧ They sit up half the night chatting and smoking, playing and singing. Of course they are unfit for work till nearly noon on the next day. Hence too often poverty, misery, and churlishness.

The inhabitants are all more or less coloured, and as the yellow skin denotes the Brahman, so here a light-tinted face is invariably a token of rank. The genus Vadio abounds, and as these idlers are not above a little stealing, we removed, by advice, the iron grating of our raft's galley. On common days many of the men are absent at their roças or are fishing with seines (Puças),† and with long hand-lines. The street and a half shows here and there a vagrant stretched upon a bench or on a mat to protect him from the sand. Rarely a great man passes, with wooden box-stirrups and ambling nag. The animals are like those of Pernambuco, small for want of breeding, but showing original good blood in the shape and carriage of the head. At times a Caipéra, mostly a vagueiro or cattle driver, rides in leather-dressed cap-à-pié, showing that he is a denizen of a thorny land.‡ The slave boys

* Dirty not in person but in lodgings. St. Hil. (III. ii. 37) remarks: "En général, c'est là une des qualités qui distinguent les Brésiliens; quelque pauvres qu'ils soient leurs chaumières ne sont presque jamais sales, et s'ils ne possèdent que deux chemises, celle qu'ils portent est toujours blanche." He doubtless spoke as he found matters, but he wrote much from memory. My experience amongst the poor showed me that they reverse the practice of the Netherlanders, amongst whom I have seen a woman whose arms required a bath-brick, diligently scouring a door-step white as snow.

† The Puça is a bag-net of reeds which two men drag along the bottom.

‡ These leathers are best made at Janu-

aria, on the Rio de São Francisco. A whole suit costs from 5$000 to 25$000, and it is far superior, softer and more durable, than what a London tailor supplies for £5. The preference is given to the skins of deer, Veado, Sassuapara, Catingeiro and Mateiro; an inferior kind is made of the Capivara, here called Caititú. Bullocks' brains are principally used to soften the leather, which becomes like casimir; this is a trick doubtless inherited from the savages of the land. The full suit consists of the Chapéo, a billy-cock hat, sometimes flapped behind like a sou' wester, the Gibão or jerkin, a short jacket opening in front and with pockets outside, the Guardapeito, an oblong piece of skin extending from throat

sit upon the cruppers of their lean garrons as the youth of Egypt bestride their donkeys. On ass-back the seat is correct, not on horseback. Nothing else is to be seen but birds, beasts, and naked lads. The dogs and pigs are apparently in a state of chronic civil war, and the only gymnastic of the citizen and the citoyenne consists of "sticking" them.

Amongst these half-breeds respectable men are invariably civil and obliging. Churlishness increases with the deepening tint of the skin, and at times, when very dark, it indulges in the peculiar negro swagger which speaks of a not unintentional rudeness. When, however, the men are sober, they show nothing of the ruffianism so common amongst the European uneducated. A stranger would often look upon their manners as offensive, whereas the offence proceeds not from intentional ill-will, but from a total want of tact, incapability of discerning the decorous, and absence of perception that they are giving offence. Men come to the door, lean against the post, stare like the Ophidiæ, stare like the gods of Greece and Rome, with eyes which never wink. They care not whether the man in the den is eating, shaving, or bathing; they intrude conversation, and they make *vivâ voce* personal comments and remarks, as the Central Africans would do. In fact the

Realm of Bocchus by the Blackland Sea

is the best of patience-teachers. You there learn, and must learn, to endure what the Englishman hates, perhaps, most. The women enter uninvited, cigarette in mouth, and sit down for the first time like old friends. We have a pretty neighbour, much resembling the "Yaller Gal of New Orleans." The Sⁿᵃ Miner-vina of Salgado loved, said the tongue of médisance, a soldier, not wisely and not too well. Like the rest of her sex in this region she carried one shoulder always bare, and she asked for everything, valuable or valueless, which met her sight. The

to stomach, with a hole through which the head is passed, and acting waistcoat, and the Perneiras or tights, which reach the ancles. Over these boots are drawn on the feet, and protected by closely fitted sole-less shoes, like the under slippers of Egypt.

I soon adopted leather. Brazilian travel, especially in the interior, wears out a pair of overalls per month. Where the land,

however, is not very thorny, the suit may be limited to the breeches or even to leg-gings; some backwoodsmen here economize the "seat." A modern author justly praises the material for long lasting, but he probably never tried what he describes as "frais et léger." It is, as most of Master Shoetie's brethren know, heavy and cumbrous, hot in hot weather, cold in cold wet in wet.

smallest trifle was thanklessly received, because better than nothing. The women are here tolerably independent of the men. I often saw them paddling themselves and their children across the streams.

We took an early opportunity of visiting the Serrinha, behind or north of the Manga. Beyond the fifty yards of river-ridge lies a bayou-bed, mud-flaked and in parts still green; this partly explains the fevers. On the damp margin grew a circle of Crioulina trees, regular and domed like enormous oranges, with thick trunks two feet high, and leafage like the myrtle of tenderest pistachio-green; the perfume of the flower resembles vanilla, and the small red berry is eaten by children. They contrast strongly with the "Carrascos" and Cerrados of the broken waterless ground further from the stream. This vegetation is European rather than tropical in want of variety, and it presented anything but a gay prospect, this depth of winter in the heat of the dog days. Many were leafless, like hazels in our winter; some were dead, killed, according to the people, by the heat of the sun; others said that frost was the cause. The ground was rich in the black "formiga douda," or mad ant, which loves the orange tree; it is so called because it moons about as if mad or drunk. Wens of termite nests* throttled the branches, and we were once pursued by a swarm of furious Marimbondos or tree-wasps. This nuisance must be abated by breeding birds; we found few of the feathered race, and ornamental rather than useful, with brilliant tints they lighted up the dull and arid view. Passing a few outlying hovels, each of which sent forth its barking cur, we began the ascent. Here the land shows, where denuded, red and yellow sandstone, new, shaly, and regularly stratified; perhaps it is the "Old Red," discovered in the Serra da Porteira by Dr. Virgil von Helmreichen, the same who detected granite in the limestone near Gongo Soco.† The dry grass was still burning in parts, for the future benefit of the few cows, and the surface was cut by wet-weather rivulets. On the higher levels, well swept by the cool breeze, houses might be built beyond the range of malaria, but there

---

* The nest of the termes arborum is called "panella," a pot.

† Similarly on the Amazons River, older observers believed the slaty rock and hard sandstone seen on the banks at Manaos to be Trias or Old Red: Prof. Agassiz (p. 199) determined both to form part of the "great drift formation."

is no surface water, and none but a madman would now dream of laying down pipes.

The view from the summit delighted us. To the north the riverine valley of the joint streams was broader than the eye could estimate, and the least width was nine miles. To the east is the crescent-shaped Serra da Porteira,* a long tongue of raised land, convex towards the stream. Southwards the horizon was broken by the high blue lines of the Serras do Rompe-dia and do Saco Redondo. A little north of west stood the Serra do Itacolumi,† forming with the Jenipapo and the Varginha to the south-west another half-moon, whose bulge faced the river. The Jenipapo is said to bear a plateau on its head, and to abound in gold. These western mountains have gaunt forms, as if broken by volcanos, and there are two pyramids connected by natural curtains, which make magnificent land marks. Below the peaks there are gradings of horizontal lines, evidently formed under water. The surface bore the growth of the great and arid plains called Campos Geraes, and resembling the upheaved "levels" of England and the "carses" of Scotland. Here it was dull and grey, there the trees were donning their spring dresses of liveliest green.‡

Between these limits of the stream in days of yore, the Rio de São Francisco winds up through its verdant avenue from the south-east, spreading out into bays 1800 feet broad. Above the thin confluence-point of trees and sand, its noble tributary, the Rio das Velhas, serpentines from the south-south-east, and shows a silvery lake on the left bank. Grand are the curves described upon the lacustrine lowlands, the "straths" and "dales" of our country, whose vast.extent smokes like a battle-field with prairie fires. During the rains the flats must become a broken line of lakes. Below us lies the shallow line of village,

---

* M. Halfeld calls the northern part of this Serra "da barra da Manga," and connects it to the south with the Serra do Rompe-dia. South of the confluence he places the Serras da Tabua and do Truichete.

† Down stream, near the town of Remanso, there is on the left bank a "Serra dos Columis," and a hill called "Itacolumita" is at the junction of the Rio Preto with the Rio Grande.

‡ The extreme breadth of the riverine valley as determined by its tributaries, lies between the Serra Grande or do Espinhaço on the right (east) and the highlands that divide Minas Geraes from Goyaz, under the names Serra dos Pilões, da Tiririca, dos Aráras and do Paraná (called by St. Hil. Serra de S. Francisco e do Tocantins). Thus its extreme breadth would be 240 geographical miles from Rio de Janeiro to W. long. 4° (Rio).

and scattered near the junction are plots of bright and luxuriant sugar-cane.

I did not neglect to inspect the Villa de Guaicuhy, distant from church to church about three-quarters of a mile. The path wound along the right bank of the Rio das Velhas, which is only partially subject to inundations; their limit is denoted by green grass and thickly foliaged Almacegueiros (gum trees); the prettiest feature is the Páu de Arco de Flor Roxa—the red-flowered Bowdarque. This Bignonia, rich with mauve-coloured trumpets, is used as an anti-syphilitic, and the cerne or heart is made to do the duty of lignum guaiacum. In places there is good ground for cotton grown annually, and "topping" would turn its fibre to lint; here the comparative aridity of the soil would save the trouble of cutting the tap root. The people say that there is too much sand and too little water for coffee; the "Cafezal" is an exception, and the best are in the Fazendas of Rompe-dia, Bejaflôr, Canabrava, Mumbuca. We crossed a dwarf ridge of the usual shaly sandstone, and a fiuman now dry; beyond it lay Campo ground, dotted with a few cattle.* Two bulls eyed us curiously, but the novelist's pet animal is here unknown.

Presently we crossed by stepping over stones the normal bridge, the small Córrego da Porteira, which drains the crescent-shaped Serra of the same name; other streams can be added to it, and thus there will be a sufficient water supply for the future city. Passing the Quartel or barracks, a more substantial house than usual, we issued into the square, where the superiority of the site at once became apparent. The floods reach only to the lower portion; the upper part slopes gradually up to the skirts of the stony hill, and affords a beautiful view of the double distances which buttress the riverine plain. At present the settlement consists only of the square, and the square has a total of forty-five tenements, not including the church. At present it supports itself by exporting provisions, and it

* In the true cattle-breeding countries, such as Texas and the Argentine Republics, a few head turned out to graze and completely neglected multiply exceedingly in the shortest possible space. Here, as in the southern part of the São Paulo Province, they do not, and the cause is hardly apparent. The climate is excellent, and the surface of the ground is favourable, while forage, possibly not of the best description, abounds. On the other hand the animals cannot live without salt, and want of communication, by adding 400 to 500 per cent. to the price, greatly limits the supply.

imports from Joazeiro salt and dry goods, and from Januaria saltpetre, hides, and sole leather. The post reaches it twice a month, on the 7th and the 27th.

The vicar, Rev. P^e Francisco da Motta, was confessing at Desembrigo; I was sorry not to meet him, as all spoke highly of his local information. The excellent Delegate insisted upon giving us coffee and sponge cake (pão de ló); my companion bought at his store a bit of cotton marked J. Bramley Moore; full of starch, leucom, and dextrin, it contrasted badly with the substantial home-made produce of Minas. Our friend led us to the village school, which could easily be traced by the sound. The Brazilians have facetiously described the *rivá voce* system, borrowed from the Arabs.* It should not, however, be condemned precipitately; it assists in forming pronunciation, it fixes the subject upon the memory, and it teaches abstraction of thought. My system of learning foreign tongues has long been to "read out loud," and mentally to repeat whatever is said to me. The process is tedious, but it masters the language in three months.

The fault of every old settlement in the Brazil, beginning with Rio de Janeiro, is the narrowness of the streets, and after a time it can hardly be corrected. We advised the Delegate to lay out the wide open space in regular parallelograms, with thoroughfares at least 100 yards broad, and thus to make ready for the days when, pace the manes of Sir John Shelley, tramways will become universal. We visited the church in charge of a Sacristan, born about 1796. Founded some 150 years ago, by the piety of an old philanthropist, the Rev. P^c Nicoláu Pereira de Barros, it faces the fair view to the setting sun. The stone front is pierced with three windows, a door, and what by courtesy may be called a rose light, and the material is taipa, armed with mail of broken pottery wherever the rain strikes it. The bells depend from the normal gallows outside, and of the two Sacristies one is in ruins. Inside there is an organ-loft, and the two plain wooden pulpits resemble magnified claret chests. The High Altar bears the Patroness supported by São Miguel and by N^a S^a Maẽ dos

---

* "Ouve se um concerto infernal e monotono, uma especie de canto descompassado e confuso, composto de gritos de uma modulação especial. Grita o mestre, grita o discipulo, gritão os monitores, todos gritão, e finalmente ninguem aprende."

Homens; it has been gilt, but I detected a bird's nest in a cosy corner. On the left are two side chapels, one of S^{to} Antonio, unfinished still, the other of Santa Anna, somewhat in the pier style of Bahia, and gilt by a devotee in the olden day, João da Rocha Guerreiro. Opposite Santa Anna is N^a S^a do Carmo in newer fashion, with pillars and capitals, the gift of Joaquim José Caetano Brandão. The fourth is completely modern, columns resting on consoles, the liberality of a Genoese, Antonio da Costa. The worst part of the Matriz was its floor; the nave was paved with loose boards, and the sanctuary with coffins and brass tacks, forming dates and initials. The sacristy had the huge boxes de rigueur, the waterless fountain, a spout projecting from a human face, and the stool and sieve confessional.

Sr. Leandro lent me the last papers from Ouro Preto, and the Presidential annual reports, together with the original description of the São Francisco by M. Halfeld. He had travelled little, and ignored even Rio de Janeiro, yet he had collected a variety of information; his thirst for knowledge was unlimited, and he often spent half the night in study. He was great upon the education question, and as a moderate politician he deplored the excesses to which zeal and interest led, appropriately quoting the fable of the old man, his son and the ass, to show how difficult it was to please even his own party. He wrote for me a variety of introductory letters to his friends on the Great River; in the Brazil generally the handwriting would have charmed Lord Palmerston, but the Delegate's caligraphy was positively copperplate. We had every reason for being grateful to Sr. Leandro, and I embrace the first opportunity of expressing to him my best acknowledgments.

CHAPTER XIV.

TO THE RAPIDS OF THE PIRAPORA.

> . . . And streams as if created for his use,
> Pursue the track of his directing wand
> Sinuous or straight, now rapid and now slow,
> Now murmuring soft, now roaring in cascades.
>
> *Cowper.*

WE were strongly advised to visit the Rapids of the Pirapóra, which are said to be, after the Casca d'Anta at the beginning, and the Paulo Affonso at the end, the important feature upon the Rio de São Francisco. The word means a "fish leap,"[*] and is applied to places on more than one Brazilian river; it has, however, many significations. On the Tiété, in São Paulo, the people translate it "Sign of fish," making "Póra" a corruption of "Bora."[†] With a flush of joy I found myself upon the bosom of this glorious stream of the future, whose dimensions hereabouts average 700 feet. I had seen nothing that could be compared with it since my visit to the African Congo. In due time the banks will be levéed, the floods will be controlled, the bayous will be filled up, and the great artery will deserve to be styled a "cœlo gratissimus amnis."

The author of the Noticias do Brazil (1589) informs us that

---

[*] Pira, or pyra, a fish, and pora, salto, a leap. Thus Colonel Accioli explains it, "lugar onde o peixe salta." The word must not be written with St. Hil. (III. ii. 213), "Piraporá."

[†] The dictionaries explain pyra-pora by "fish-inhabitant, a great fish which lives in the open sea—that is to say, a whale." "Bora," contracted from "Bor véra," is a verbal desinence corresponding with the Hindostani -wala in such expressions as "Canheu-bora," which a Hindú would render "Fujne-wala."

the once numerous and now extinct tribes living near this river, the Caétés, the Tupinambas, the Tapuyas, the Tupiáes, the Amorpiras, the Ubirajaras, and the Amazonas—of course there were Amazons—knew it as "O Pará," the sea. The old Portuguese explorers went down the coast with the Romish Calendar in hand, and thus the Rio de São Francisco (de Borja) derived its name from the Jesuit saint presiding over the 10th of October.* So Varnhagen assigns the honour to the little squadron of five caravels which, commanded by João da Nova, and bearing on board as pilot the cosmographer Vespucci,† sailed from Lisbon about the middle of May, 1501. It must not be confounded with the little Rio de São Francisco in the Province of Santa Catherina, a port also described by the author of the Noticias (chap. 66); and it is as well not to suggest California by giving to it the Spanish form San Francisco, instead of the Portuguese São or San Francisco.‡ The river soon attracted the attention of those dwelling on the seaboard; like the Nile and the Congo, it floods during the dry season, and *vice versâ*—sufficient to excite, in those days, the marvel-faculty.§ Adventurers who determined to

---

* Thus we find the Promontory of São Roque first visited August 16; Cape St. Augustim, Aug. 28; Rio de São Miguel, Sept. 29; Rio de São Jeronymo, Sept. 30; Rio de São Francisco, Oct. 10; Rio das Virgens, Oct. 21; Rio de Santa Lusia (the Rio Doce?), Dec. 15; Cape St. Thomé, Dec. 21; São Salvador da Bahia, Dec. 25; Rio de Janeiro, Jan. 1, 1502; Angra dos Reis (Epiphany), Jan. 6; Island of S. Sebastião, Jan. 20; Rio or Porto da São Vicente (São Paulo), Jan. 21.

Frei Gaspar Madre de Deus would attribute the naming of São Vicente to the fleet of Martim Affonso de Souza, who touched there on his return from the Rio da Prata, Jan. 22, 1532. But the port is mentioned under the name of its saint in the Diary of M. Affonso's brother, Pedro Lopes de Souza, before the squadron in which he commanded a ship reached it. It is, moreover, found in the map of Ruysch, 1508 (Varnhagen, i. 425).

† Sr. Varnhagen (i. 27) ably rehabilitates the name of Amerigo Vespucci, the god-father against whom for many years America, and even Europe, have been so furiously raging. He quotes the Phisices Compendium: Salamantice, 1520 (eight years after Vespucci's death), "Prima est Asia, secunda Africa, et tertia Europa . . . . addenda tamen veteribus incognita

America a Vesputio invente quæ occidentum versus," &c. Columbus did not complain of him, and the fortunate Genoese died convinced that he had discovered the Eastern portion of the "Indies," to which Castella added the term "Western." The historian sensibly remarks (i. 27), "And the designation of 'West Indies' would best perpetuate for us the work of Columbus and his genius in perseveringly working out a great idea. It will ever remind human nature of the respect due to genius, even where it greatly errs, inasmuch as these errors often lead to the discovery of truth, which in the exact sciences is reached by setting out at times from gratuitous hypotheses."

‡ This inadvertently has been made by stranger authors from Southey to Agassiz. I know only one who has avoided it, Lieut. Netscher, "Les Hollandais au Brésil," 1853.

§ The same is the case with the Paraguassú of the Bahian Mediterranean; in fact, with all the streams which in these latitudes rise west of the sea-fringing uplands; they flood during the dry season, and they shrink when the coast rains set in. The reason is simply that the dry season of the coast is the rainy season of the interior.

solve this great mystery, and who probably had heard of the then abundant "Brazil wood," and mines of gold and silver, ascended as far as the Great Rapids in early days. The "protomartyr" was one Sebastião Alvares, of Porto Seguro, who was sent to explore by the second governor of the captaincy of Pernambuco, Luiz de Brito de Almeida, who succeeded Duarte Coelho de Albuquerque.* After four years of travel he and his twenty men, an insufficient force, were massacred—there has been many a "Bloody Run" in these regions. Presently João Coelho de Souza ascended more than one hundred leagues above the Rapids, and published a Rôteiro, now curious.

Two new men were hired to guide us in the "tender" canoe, which they described as very "violenta e banzeira," crank and kittle. We eyed curiously the contrasts of the new stream with that which we had lately left. Here the water was of a transparent green, like the mighty Zaire; it is said to be "heavier," when drunk, than that of the Rio das Velhas; the influents, often so deeply embedded as to be now useless, were clear, especially when they drained little bayous on the sides. The water seemed to break even from the stiff clay, which was in places caving in. The Corôas were either mere sandbanks, lines of gravel or lumps of boulder, or clothed with the Arindá, which in places grows twenty feet high. Cattle, here the chief produce, made them their favourite haunts. The barreiro, or salt lick, cribbled the sides, but we lost the aluminous white rash which distinguished the Rio das Velhas. The banks were cut and graded into steps by the receding floods, and where not broken by "riachos," they were above high-water mark. In places there were heaps of decayed leaves crushed and pressed together; they formed layers often 3—4 feet deep. At noon we passed on the left bank ledges caked over with hard "Cánga;" water trickled from it upon the loose Cascalho and the felspathic clay of the São João Mine. This is a true diamantine formation. A natural pier projected from the right side, hard clay deeply tinged with iron; and the violence of the floods was shown by a tree root, weighing at least

---

* Here the Noticias para a Historia e a Geographia das Naçoês ultramarinas (March 1, 1859), which has a chapter (No. 20), "On the Greatness of the Rio de São Francisco and its Sources," seems to be incorrect. Luiz de Brito de Almeida about the end of 1573 governed the Captaincy of Bahia. D. Coelho de Albuquerque (the second Donatory, not to be confounded with Duarte Coelho the First) became in 1560, third governor of Pernambuco.

a ton, and lodged in the fork of a fig, whose gigantic limbs were distorted by the burden.

All this region is of the greatest beauty and fertility; when the Rio das Velhas shall have been opened it will become the garden of the land. On the banks were many clearings and small sugar plots, with which the owners are ready to part. Beds of melons show that the fruit has now grown to be a favourite, and will, presently, become the daily bread;* the Mangui Hibiscus and the Castor shrub here stand thirty feet, and everywhere we saw the broad-leaved Brazilian tobacco growing half wild; the people prefer to pay heavily for the gifts of Baependy and Pomba. In the patches of cultivation the women had stuck, as in Harar-land, a cow's horn on an upright stick, to keep off the evil eye—por olho da gente. Fishermen and boys appeared at times, and the negroes and negresses washed by the waterside; here there is no cause to fear the crocodile or the slaver. Before the banks were sloped cuttings of sugar-cane, ready for planting in October if the rains be early, if not in November. A fish hung lifeless, hooked to the stern of a small canoe, whose beak was the wedge-formed projection used in Africa as a handle; and the turkey buzzards were hard at work upon a dead terrapin (Kagado), which infatuated humanity in these regions will not eat.

During the ascent we hugged the left bank as closely as possible; the descent was, till struck by the storm, viâ the "fio de agua," or mid-stream, crossing to the headlands and points round which the current swings. The distance was said to be five leagues, and if so each league must represent six and a half geographical miles.† After nearly nine hours of hard work, we doubled a wooded projection from the left bank, and sighted the Cachoeira of the Pirapora. The break is now at its worst; like most others, it is easier to pass during the rains, and the more water upon it the better.

The Pirapora differs from anything that we have yet viewed; it is a superior article in quality as well as in quantity. This is, in fact, partly a true fall, divided into two sections; but we have

---

* The fruit is of two kinds, the melancía, or water melon, and the melão, musk melon. The former is a great favourite with the barquemen, who seem to have its name ever in their mouths. Yet they declare that it gives them "dumb chills," and the same is the belief in the Southern United States: few will touch the fruit when working in the sun.

† It is about six leagues west of the Rio das Velhas.

come a long way to see a small sight, and we tremble to think what Paulo Affonso may really be. On the western bank rises a lumpy hill, the Curral da Pirapora—some day it will be built over—at whose foot is a narrow stony beach. The course of the Rio de São Francisco is here from south to north, and the rocky mass crosses it in ledges and scattered blocks, mostly disposed diagonally. There are evidently several breaks, and southwards the dark blue of the swift gliding river, backed by the light azure of the Saco Redondo range, contrast with the boiling raging flood that forms the "foreground."

Glad to stretch our cramped limbs we landed at the Porto da Pirapora, on the right or eastern bank, and proceeded to inspect the Cachoeira from above. The path led through " Barandão," a caricature of the Arraial da Manga ; its principal features were huge seines and large fish, split, hung on gallows to sun dry. The people do not export this produce, but sell it only to passing mule troops. Finding that we did not trade, and suspecting us of being agents of Government, they were scantily civil, but they offered for purchase their refuse " desmonte "—sand without diamonds. The dogs were even more churlish than their masters. Had we had tobacco and other small matters for barter we might have been received in another way.

At first we walked over loose sand ; the rest of the right bank is a flooring of rock, which probably extends far under the eastern bank. The natural course of the water is to this side, and canoes prefer it during the floods. M. Liais opines that canalization would here be easy ; it is hard however to predicate this until careful piercings shall have been made. M. Halfeld proposes sluice gates, moreover, which the French authority does not consider necessary.* There is no danger of the Brazil undertaking any such work in the present generation.†

The stone platform is composed of slabs, some forty feet long, and mostly narrow : the cleavage is perpendicular with the stream

---

* M. Liais makes the length of the Pirapora obstacle a total of one kilometre, and the difference of level 3·55 metres. This would give a velocity of only 3 to 4 metres or yards per second.

† The estimates for opening forty leagues are as follows :—

Canalizing up the Pirapora	.	.	.	1,400 : 000$000
To the Cachoeira Grande	.	.	.	4,100 : 000$000
To the Porto das Melancias	.	.	.	3,200 : 000$000
Total	.	.	.	8,700 : 000$000 say £870,000.

and the water-turned pot-holes and channels, cut a yard and more in depth, show the effect of floods. The substance is generally a hard compact gneiss (grauwacker sandstein, gris traumatico) of light purple tinge, dotted with specks of mica glistening white. We found also sandstones and impure calcaire which effervesced but little under acids. From this point we could easily distinguish the two main steps separated by about 700 yards, a length which makes the slope of the rock planes appear very gentle. The upper rapid, six feet high, seemed more formidable than the lower of about seven feet. Near the right bank these form catadupas, or true falls; they are also garnished with escadinhas (little ladders), miniature cascades in gerbs and jets, rushing furiously down small narrow tortuous channels, between the teeth of jagged stone-saws, and tumbling over dwarf buttresses. Thus the total height between the upper and the lower "smooths" is thirteen feet; above the break the stream narrows to 1800 feet, whilst below, at the Porto da Pirapóra, where the serpentine arms, after crossing and dividing between the boulders, unite, the bed broadens to 3500. During the dries the fair way, if it may so be called, is a thin sheet of water near the western bank; no ajôjo, however, can pass, canoes must be unladen and towed up, and without a good pilot there is imminent risk. At the present season it is broken by outcrops of rock, and during the floods it has dangerous whirlpools.

The Pirapóra is a serious obstacle. It is not insurmountable, but it would cost more money, and take a longer time to remove, than all the most serious obstructions upon the Rio das Velhas. No work could be carried on in the rainy season, and the inundations would damage the labour done during the dries. Hands would have to be sent here at a great expense, and even on this most wealthy soil imported provisions would be required. Above it also the Rio de São Francisco becomes a mass of rapids, and when you clear one you are within hearing of another. Canoes ascend with difficulty to the mouth of the Abaeté.[*] M. Liais accurately surveyed as high as the embouchure of the Paraopéba, and he found that no expense would clear more than a hundred leagues of its course.

Returning to the Porto, we visited the diamond diggings, which

---

[*] Etymologically, the true man (ábá, man, and été, veritable), or hero. This is the stream which produced the celebrated diamond in 1792.

are of some antiquity; formerly gold was washed, but this industry has now ended. The gem, which comes, perhaps, from afar, is found in the Cascalho arrested by the rocks. Most probably the Caixão or hollow at the foot of each fall would yield a better supply. About a dozen men raising " desmonte " from a pot-hole (panella) between two boulders deeply channelled out by the joint action of sand, gravel, and water. For small and valueless stones they asked per vintem (two grains) from 12 $ 000 to 14 $ 000, something above the London prices.

This part of the São Francisco should be eminently diamantiferous. On the east it drains the Cerro, which we have already visited. To the west it receives the washings of the Rio Bambuhy (of old Bamboi), which falls in south of the city Dôres do Indaiá. Beyond it is the Rio Indaiá, or Andaiá, where in May, 1800, Dr. Couto's party took from one hole forty-two stones. Further north is the Ribeirão do Borrachudo, which also gave one gem; and its neighbour is the Abaété, draining the old Sertão Diamantino. These four streams, to mention no others, issue from the eastern flanks of the great chain, whose western counterlopes supply the diamonds of Bagagem. Further north is the Serra da Gamelleira and the valley of the Somno, an eastern branch of the well-known Paracatú. I will allude to these rich diamantine deposits as we pass them.

During the last night a raw south wind had set in from the mountains, and told us that rain had fallen there. It was the beginning of the wet season, but the people called the showers Chuvas da Queimada—of the bush-burning. Prairie fires are popularly said everywhere to bring down water; they sublimate a vast mass of humidity, the heat and steam rise, a cool draught supplies their place, and thus the atmosphere cannot support the condensation. In the temperate parts of North America, during the fall of the leaf, the tree-trunks restore to the ground the juices which spring injected into the wood-pores, and hence the phenomenon of streams swelling without a drop of rain. Here, however, though the dry season was just ended, the vegetation is assuming its vernal green.

As we began the descent lightning flashed from the east, the south, and presently from all the horizon, followed by low grumblings of thunder. To the right hand appeared the Olha de Boi, or bull's eye; it is not, however, the white patch under

the black arch of the African tornado. Here the sign is a little
section of distant rainbow glistening in all its colours against the
slaty grey background of the discharging cloud, and showing that
a gale will blow up from the falling shower. Mostly we shall see
it in the east, meaning therefore in the afternoon, and when it is
accompanied by wind that sinks the thermometer 8° (F.), we shall
expect a patter of rain, and a storm like a charge of cavalry.
The people call it either simply or "com rabo de gallo"—accompanied by cirrus. Presently our cranky canoe was struck by
the gale (rajada de vento), one of the especial dangers of the Saõ
Francisco. The east wind was heard roaring from afar; and, as
it came down upon the stream, white waves rose after a few
minutes, subsiding as easily when the gale had blown itself out.
In July, 1867, a white squall of the shortest possible duration
carried off the tiles from the roofs of Guaicuhy.

Our men preferred the leeward bank upon which the blast
broke, leaving the water below comparatively dead, and thus they
escaped the risk of falling trees. The surface of the central
channel being now blocked by the fierce wind, the side current, a
backwater during our ascent, bore us swiftly down. It was very
dark at 7·30 P.M., when we climbed the steep and slippery bank
of the Manga. Shortly the thunder growled angrily overhead,
and heavy rain fell, fortunately upon a tight roof. This was the
first wet weather which we had experienced since July 21, and it
began a season desolate as a fête-day in England.

At the Manga we saw for the first time the "Barca,"* which
reminded my companion of the Mississippi "yawl." It has been
introduced only during the last forty years; before that time all
the work was done by ajôjos and canoes. The shape is probably
taken from the Douro, but here the form is more in Dutch style,
round and spoon-shaped to suit the stream; it wants also the
immense Portuguese caudal fin, though by no means without a
large and powerful rudder. The planks are of the best woods in
the country, Cedro and Vinhatico, the keel is of Aroeira, and the
huge ribs (costellas or cavernas), together with the heavy cross-
pieces and gangways, are of the stout, tough Rosca. The average
length may be 45 feet by 14 broad, drawing 3, 4 to 5 feet when

* Barco is the general term for large
craft, whilst Barca is larger. In this point
Portuguese agrees with the Italian, which
makes the feminine major than the mascu-
line; for instance, "trivella," the augmentative form of "trivello." Some
authorities, however, make Barca the
smaller.

loaded, and carrying some 400 arrobas, reckoned by rapaduras or sugar cakes, each about 4 lbs. At Salgado was built the Nª Sª da Conçeição da Praia, now broken up; she was 81 feet long and 6 feet in the water. These large craft are always flat-bottomed (de prato) to work off shoals. Keels are dangerous, as they cause upsets when the current carries them to the shallows. The bows and stern are raised, as in the old caravel, and the cargo is matted over or covered with hides in the centre, leaving a narrow trampway of plank at each side. Above the Paulo Affonso the toldo or standing awning is unwisely placed in the stern so as to catch every puff of wind. The lower riverines prefer the cabin in the bows, and diminish its dimensions. It is made in tunnel shape, resembling the surf-boats of the Guinea coast, and it is worthy of imitation by the dwellers on the upper stream. The stern-cabin, which from 8 feet long sometimes takes up a quarter of the length, is of solid planking, in the poorer sort arched and matted with fronds of the Indaiá or the Carnahúba palm, or even with common grass; the ends hang over both sides so as to carry off the rain. A rich trader assumes some fine name, as the "Baroneza de Minas," displays a flag with a "Santa Maria," and has doors and glass windows. His cabin, which is also his shop, is fitted up with shelves for goods; he comfortably swings a hammock, and he disdains to sit at table without a table-cloth.

The crew of a moderate-sized craft may be ten men, the extremes being six and fourteen. The pilot stands or sits at his rudder on the raised stern. His men, dressed in white kilts, and at times in tattered shirts, with hats of leather or straw, have hard work. Their poles, 21 to 23 feet long, are much heavier than those of the ajôjo, and like the Bedouin lance, require a practised hand. They also work huge oars like sweeps, one man pulling whilst another pushes. During the floods they must creep up at the rate of two leagues per diem, wearing, as they say, holes in their chests, and exposed to all the insects of the shore; hence as a rule they make only one trip per annum, and at the beginning of the rains they return home to cultivate for themselves or for others.

I was surprised at the absence of sails; they were seen at only two places, Pilão Arcado, and "Joazeiro;" and even there they were limited to ferries crossing the stream. The people declared that the channel, besides being studded with snags, was too tortuous. This, however, is very far from being the case. They also feared sudden gusts (Pés de Vento or Redemoinhos), which would cause accidents. The chief reason is, doubtless, ignorance. On the Lower São Francisco, where the sea-breeze from the south-east sets in regularly at 9 A.M., every barca goes up under sail and at steamer pace.

The Upper São Francisco has its regular trade winds, which vary with night and day, and still more with the seasons. The east, sometimes veering towards the north, is called the Vento Geral,* and it often acts as a " soldier's wind," useful both ways. By night in the lower parts of the stream it is followed by a Terral or land-breeze from the west. Of this " vent traversier " also the barca-men declare that with canvas their boats would be driven out of the channel.† During the four rainy months, which of course are different in the different sections of the river, and which as a rule follow the southing and the northing sun, the trade shifts to south with westing, and thus blows down stream. The regularity of the meteor suits admirably not only for sailing but for all manner of simple and economical machinery.

In this portion of the Brazil, where the simplest labour-saving contrivances are unknown, they have never heard of the " horse-boat," now so common upon the streams of Continental Europe, and still used in the United States. The machinery might easily be adapted to the rafts and boats. A platform some seven feet long, and raised at an angle of 20° to 31°, faces the stern, and the animal is taught to walk up it. It is composed of some forty-two slabs, each four inches square, and the hard, unelastic woods of the country would supply the best material. Connected by vertical joints of iron, which work loosely upon one another, forming an endless band or chain, the platform is fastened to an " idler " axle in the fore flooring and aft to the tranverse tree which works the paddles. This portion is made fast to strong uprights, and the diameter of the working wheel is about 3 : 1 of the axle.

---

* The regular east wind of the Amazons is also known as "Vento Geral" (Mr. Bates, i. 213).

† "The fault of the vessels navigating the Amazon is the breadth of beam and the want of sail. I am confident that a clipper-built vessel, sloop, or rather ketch-rigged, with a large mainsail, topsail, topgallant-sail, and studding-sails—the last three fitted to set going up before the wind, and to strike, masts and all, so as to beat down with the current under mainsail, jib and jigger—would make good passages between Pará and Egoás " (Lieut. Herndon, 262).

Thus it would be easy to get over thirty miles per diem with a tithe of the present toil.

At the Manga I dismissed with the highest recommendations to future travellers my good old pilot, Chico Diniz and his stout-hearted companion, João Pereira. The expense was 190$000, but on the Rio das Velhas wages are now at a fancy price ; on the São Francisco there is a regular demand and supply. Joaquim volunteered to accompany me, but he was short-sighted and soft-bodied. The "Menino" agreed to remain with me on condition of being supplied with a return passage from Joazeiro. On the great stream barquemen do not leave their beat; it is the custom to engage them per travessía or trip, of which, as will be seen, there are eleven. I hired the cousins Manuel Casimiro de Oliveira and Justino Francisco da Conceição ; both were very dark, and the latter, 6 feet 3 inches long, reminded me of Long Guled the Somal. They were well acquainted with the water, civil and obliging, but they lacked the pluck and bottom of the Highlander crew.

As a rule the worst hands offer themselves to the stranger, and thus he may find himself in great trouble. All men are here more or less amphibious ; the canoe, as they say, is their horse. The real barqueiro is a type as peculiar as the bargee of olden days in England ; he is also a free-born man ; few traders like to employ slaves. More handy than a sailor with us, like the African, he is perfectly acquainted with all the small industries necessary to his comfort ; he can build his house or his dug-out, and make his tiles or his clothes—arts which among the civilised demand division of labour. Thus he is mostly inferior to those of his own class in more advanced lands where society has split up into thin strata. Here, as elsewhere, it is wonderful how little foul language is used. The same has been remarked of the North American backwoodsmen, and the aborigines of both countries know, we are told, neither swearing nor abuse, "bad man" being the worst reproach. The good specimen is quiet, intelligent, tolerably hardy, and perfectly respectful to his Patrão, the proprietor or hirer of the boat. He usually eschews drink altogether, fearing the drunken quarrels to which it leads. The worst lot is rough as its own barque, and desperately addicted to strong waters and women, to the nightly Samba and Pagodi, the local "orgie." My last gang will be a good specimen of the bad.

All are headstrong, a race of "autonomoi," who will have their own way, and who do not like to be directed or contradicted. I was advised to carry plenty of spirits and tobacco to prevent them jumping ashore at every house. They have enormous appetites, which come, they say, from the shaking of the barca. This is probably an "Indian" derivation; the savages, we are told, would sacrifice everything for food, and ate with the voracity of jaguars. Although they know that it is injurious, the barqueiros delight, like the Peruvians, in rapadura or Chancaca sugar; I have seen a man eat 2 lbs. of it at a sitting. They have the usual Portuguese and tropical horror of fresh milk; on the other hand, the soured form, here called Coalhada, and in Hindostan "Dahí," has a high reputation; it certainly is antibilious. The rest of the diet is Jacúba, which has been before mentioned, sun-dried meat, water melons, and beans * with lard. Almost all smoke, a few take snuff, and very few chew.

A characteristic of the barqueiro is his aptitude for mild slang-ing and chaffing, the latter being a practice abhorrent to the Brazilian mind in general. "O Senhor e muito caçuador"—a great joker—means that you are not pleasant. He has also the habit of the Hindu palanquin-bearer carrying a "griffin," and will, if impudent, extemporise songs about his patron. The lan-guage renders the rhyme easy, but the stranger is astonished by the facility with which men and women squatting on their heels† answer one another in Amabæan verse, made without a moment's thought. Although we have had an Ettrick Shepherd, many deride the pastorals wherein the swains prefer poetry to prose. They should hear the barqueiro of the São Francisco River capping verses with his "young woman," and making songs about everything in general. Similarly the opera is held to be fictitious and unreal because emotions and passions are expressed in music; but the negroes of Central Africa show by chaunting when their sorrow is deepest, and the South American Botocudos evince excitement by singing instead of speaking. "Ils ne parlent plus; ils chantent," says the traveller.

---

* This is an excellent food, not only for cattle (70 per cent. of nourishment to 60 of oats). The principal species of these Papilionaceæ are Feijão Preto (Phaseolus derasus), Feijão Carrapato (P. tumidus and sphæricus), and Feijão Mulatinho (P. vul-garis). There are many others.

† This position is usual in the wild parts of the Brazil. The eye familiar with it in Eastern lands is struck by it when the squatter wears the garb of the West.

Naturally the subject matter is mostly amorous. The barqueiro delights in screaming "a largas guelas," at the top of his voice, some such verse as

> Hontem ví uma dama
> Por meu rispeito chorar.

He eternally praises the Côr de Canella or brunette of these regions, and he is severe upon those of the sex who dare to deceive the poor mule-trooper or boatman.

> Mulher que engana tropeiro
> Merece couro dobrado
> Coitadinho tropeiro coitado ! (chorus).

He thus directs Mariquinha to put the kettle on :—

> Bota o frango na panella
> Quanda vejo cousa boa
> Não posso deixar perder.
> > O Pilota (chorus).

Some of the songs still haunt my ears, especially one which much resembled "Sam 'All." The more and the louder they sing the better for the journey; it seems to revive them as the bell does the mule.

The superstitions of the barqueiro are as numerous as his chaunts. He believes firmly in the Duende or Goayajára, wizard and witch, the Lobishomem or loup-garou of Portugal, the Angai, the Anhanga,* the Alma or ghost, the Esqueleto or skeleton apparition, the Gallo Preto, or bad priest turned into a black cock, and the Capetinha or imp. They have curious tales about the Cavallo de Agua and other fabulous animals. This beast is the size of a small colt, round-hoofed, red-haired, and fond of browsing on the banks. The "Menino" declares that he saw it in a poção or kieve below the Cachoeira dos Geraes in the Rio das

---

* Angai in the "Tesoro de la lingua guarani" is translated "the evil spirit," also called Giropary, Jurupari, and Jerupari. I presume that it was really applied to that injured man or to some ghost that had made itself notoriously unpopular. Anhanga is Anglicised "phantom" (phantasma) from "anho," alive, and "anga," ghost, soul, spirit : thus it means soul only — soul without body. Of course, "soul" and "spirit" are civilised terms applied to a barbarous idea. They denote subjectivities which may be reduced to the totality of central and nervous action. The Alma is like Dr. Johnson's, or rather Mr. Cave's "ghost"—"something of a shadowy being." Nobrega and Anchieta wrote Anhanga, Yves D'Evreux, Aignan, Barrière, Anaanh ; and other forms used upon the Continent and the Islands are Uracan (hurricane ?), Hyorocan, Amignao, and Amignan (F. Denis).

Velhas, and that a youth fired at it. Perhaps it may be the Lamantin, so well known in the Amazonas waters, but I am not aware that the Peixe boi (Manatus Amazonicus) has been found here. The Cachorrinha d'Agua or water-pup has a white coat and a golden star upon the forehead; whoever sees it will command all the gifts of fortune. The Minhocão or large worm, the Midgard, the Great Sea Serpent, the Dabbat-el-Arz of the Arabs, plays a part as important as the Dragon in China. It is 120 feet long by 2 in diameter, barrel shaped, scaleless, bronze-coloured, and provided with a very small mustached mouth. The Minhocão is a perfect "Worm of Wantley" in point of anthropophagy. St. Hilaire (III. ii. 133) heard of it in the Lagôa Feia of Goyaz. At first he believed it to be the Gymnotus Carapa, then a gigantic Lepidosiren. Col. Accioli (p. 8) holds it to be an extinct monster. Castelnau (ii. 53) speaks of it in the Araguaya. It was 30 to 40 metres in length, and the terrible voice resounded for many leagues. Halfeld (Relatorio, 119) mentions that his men mistook for it a tree trunk, and thinks it fabulous. Farther down we shall pass a part of the bank which has been injured by the Big Worm, and many educated men have not made up their minds upon the subject. The superstition is evidently of " Indian " origin.*

All these legends have a taint of the Tupy, grotesque savage who best adorned his person by spreading upon a coat of gum the hashed plumery of gaudy birds, in fact who invented tarring and feathering by applying it to himself; experimentum in corpore vili. Classical, and worthy to rank with the Sea Fairy Tales, however, is the Mãe d'Agua, a spirit, a naiad, a mermaid who aspires to be a mer-matron, and who inhabits the depths of Brazilian rivers. Of perfect form, utterly disdaining the fish-tail, and clothed only in hair glittering like threads of gold, she is also a siren. Her eyes exercise irresistible fascination, and none can withstand the attraction of her voice. She is fond of boys, like most of her sex, when arrived at a certain age, and she seduces beautiful boatmen. Unlike the churlish Undines and Melusines

---

* Thus Lieut. Herndon (Chapter 8), speaking of the Lake Country of the Upper Amazons, remarks, "Many of these lakes are, according to the traditions of the Indians, guarded by an immense serpent, which is able to raise such a tempest in the lake as to swamp their canoes, when it immediately swallows the people. It is called in the Lengua Inga (Inca), 'Yacu Mama,' or mother of the waters; and the Indians never enter a lake with which they are not familiar that they do not set up an obstreperous clamour with their horns, which the snake is said to answer, thus giving them warning of its presence."

of Europe, when she proposes a change, she dismisses her lovers with great wealth.   Gonçalves Dias, the poet, has made of her a malevolent pixy, a Lurelei, whose object is to drown youth; but he takes away none of her charms.

> Olha a bella creatura
> Que dentro d'agua se vê!

CHAPTER XV.

THE RIO DE SÃO FRANCISCO.

" One of the best gifts of nature, in so grand a channel of communication,
seems here wilfully thrown away."—*Darwin, Naturalist's Voyage*, Chap. vii.

LIEUTENANT MAURY is undoubtedly correct when he remarks
that the valleys of the Amazons and the Mississippi are commercial complements of each other, one supplying what the other
lacks in the great commercial round. The geographical homo-
logy of the riverine formations in the Northern and Southern
divisions has also been remarked by many writers. The Amazons
represents the comparatively diminutive Laurentian system.[*]
The Rio de la Plata is the Mississippi, the Paraguay is the
Missouri, and the Paraná is the Ohio, whilst the Pilcomayo, the
Bermejo, and the Salado are the Plata, the Arkansas, and the
Red River.

The Rio de São Francisco has been trivially compared with
the Mississippi and with the Nile. It presents an analogy with

[*] The valleys of the Amazons and the Paraguay Rivers can easily be connected like
those of the St. Lawrence and the Mississippi.

the African Niger, but none with those of North America. One of many, it rises in the South, flows to the North with easting, and near the end of its course it bends eastwards and disembogues into the Atlantic. It is the external segment of sundry similar sections of circles, bounded by basins draining north to the Amazons, and west and south-west to the Parana-Plata: the included arcs are the great Jequitinhonha and the Doce rivers. Further South is the Parahyba do Sul, and South again the Ribeira de Iguape.* Except the latter all these streams burst through the barriers which more or less developed subtend this part of the South American, as they do the corresponding portion of the African, seaboard.

The oldest traditions (Noticias do Brazil, 1589) derived from the savages, make the São Francisco rise in a " great and famous lake which it would be very desirable to discover." Luccock (p. 53Q) remarks, " in the St. Francisco and the Paraná we beheld the drains of an immense internal lake, bounded on the east by the Serros Frio and Mantiqueira, on the south by that of Maracaná, and on the west by those which separate the Paraná from the Paraguay, or lie beyond those streams. The waters of this ancient elevated sea have burst their barriers in S. lat. 15° and 20°, and are still wearing their channels deeper at the Falls of Pirapóra in the north, and Setequedas in the south; just as the Lakes Erie and Ontario, in North America, will, in all probability, be drained by wearing away the impediments which now form the Falls of Niagara." M. Halfeld (Relatorio, p. 108) is inclined to think that the Serras of " Ibyapába," † and the Itacutiára, Bréjo and Itacaratu, with the minor features near the Monte Escuro were of old the walls of an extensive " salt-water sea." He drains it off through the Rapids of Itaparica (317 leagues ‡) which burst and formed the great future Paulo Affonso. Salinas abound upon its line, the grits and calcareous

* This stream rise s to the east or seaward side of the great Serra do Mar, which in the southern province of São Paulo bends away from the shore. The etymology is "yg," water, "cua," belt, and "ipé," a place where. I reserve the Ribeira for a future volume.

† Sr. J. de Alencar prefers to write "Ibyapaba." Vieira translates it "Terra aparada," and Martins explains it by "Iby," land, and "pabe," all. "Iby" is often corrupted: thus the name of the celebrated Pytiguara tribe was originally Iby-tiva-cua-jara, the lords of the land-valley. According to M. Brunet, of Bahia, the height of the range does not exceed 2200 metres. Mr. Keith Johnstone has adopted Ibiapaba, and Gardner informs us that the Portuguese name is Serra Vermelha. M. Halfeld writes Hippiapaba.

‡ M. Halfeld's first league was at the Pirapora, and he places the junction of the Rio das Velhas.

marls contain an abundance of salt (chloride of sodium), and Chilian saltpetre (nitrate of soda),* and as in the Valley of the Indus the Sal da terra effloresces during the dry season. I may add that the presence of iodine would explain the absence of goître, and the fact that the cocoa-nut palm flourishes at such abnormal distances from the ocean.

The main source of the Rio de São Francisco is in the eastern versant of the Serra da Canastra, the great central platform of Minas Geraes, between S. lat. 20° and 20° 30′ and W. long. 3° (Rio de Janeiro). "From the gap of a perpendicular rock more than 1000 feet high," says the Baron von Eschwege, "bursts the principal nascent of the São Francisco." The spot was visited by St. Hilaire (III. i. 184), and "tore from him a cry of admiration." He gives to the Casca d'Anta Cascade 667 feet of altitude, and he remarks "qu'on se tâche de se représenter la réunion de tout ce qui charme dans la Nature ; le plus beau ciel, des roches élevés, une cascade majestueuse, les eaux les plus limpides, la verdure la plus fraiche, enfin des bois vierges qui présentent toutes les formes de la végétation des tropiques."

The waters of the young river sweep from west to east for a distance of about fifty-five and a half leagues, and are mere mountain torrents. Before receiving the Paraopéba the breadth of the united stream is 140 metres, and the maximum depth 3·25 metres, with a discharge of 130 cubic metres per second. The direction is then from south to north with the Serra Grande or Espinhaço on the east, and the Mata da Corda forming the western wall. From the Paraopéba to the Pirapora Rapids the line has been surveyed, it inclines at first to the west and then to the east, the distance being forty and four-fifths geographical leagues (226,845 metres). From the Pirapora to the Cachoeira do Sobradinho, a distance of 239—240 leagues, the whole line is ready for a steamer, and including the Rio das Velhas, a total of 508 leagues can be made transitable with little difficulty. Below the Sobradinho there are twenty-nine clear leagues, followed by forty-four which, though dangerous, are transitable to rafts and canoes. From Varzéa Redonda, twenty-five to twenty-six leagues are unnavigable, and in this section occur the Great Rapids of

---

* See Chap. 19, where the nitrate of potassa will be mentioned.

Paulo Affonso. Finally, below the line of rapids, forty-two leagues, upon which steamers now ply, connect the Lower Rio de São Francisco with the ocean. It is here unnecessary to enter into details of direction * or distance, as we shall float down the whole way.

The Annuaire du Bureau des Longitudes de France assigns to the São Francisco the fourth rank amongst the streams of South America. It follows the Amazons (5400 kilometres †), until lately held to be by far the largest river in the world ; ‡ the Paraná-Plata (3440 ks.) and the Tocantins (2300). But M. Liais has shown that the São Francisco has been wrongly assumed to represent 2100 kilometres : from its source to the mouth of the Rio das Velhas is 800 kilometres, and 2100 from that point to the sea : the total, therefore, will be 2900. § Thus the cosmic rank of our stream will be seventeen or eighteen. ‖ In Europe it is surpassed only by the Volga ; in Asia by the Yenissei, the Yang-tse-Kiang, the Hoang-ho, the Oby, the Lena, the Amour, and the Meï-Kong ; in Africa by the Nile, the Niger, the Zam-

---

* We may briefly remark that it runs north, with a little westing as far as the Urucuia R. (30th league from the mouth of the Rio das Velhas), north-north-east to the Bom Jesus da Lapa (106th league) ; north, with a little westing to the Villa da Barra (162nd league). This meridional course is pleasant to the traveller, who always regrets when he must east or west, and thus catch the sun. Then begins the long north-eastern bend, whose apex is Cabrobó or Quebobó (278th league). Thence the stream curves to the south-south-east, and finally to the south-east.

† Lieutenant Herndon assigns˙ to the Ucayali-Amazon an uninterrupted navigation of 3360 miles. He estimates in round numbers the fluvial lines of the valley for large vessels at about 6000 miles, and he supposes that, including the numerous small streams, the length would swell to 10,000 (p. 280).

‡ The Nile is rapidly rising in point of length. My friend Mr. A. G. Findlay, the geographer, says (June 3, 1867)— "If the source be near the Muxinga Range . . . . . the total course will be 3500 geographic or 4050 British miles, almost unparalleled by any other river."

§ Professor D. T. Ansted (p. 34, Elementary Course of Geology, Mineralogy and Physical Geography) gives the São Francisco a total

direct length of 1000 British statute miles ; and of 1600, including windings, whilst he sets down the area of drainage at 250,000 square miles. Sir John Herschel (Physical Geography, p. 188) says— "The basin of the San Francisco includes the district (?) of Minas Geraes, the great source of the mineral wealth of Brazil. It includes an area of 187,200 square geographical miles in length to its source in the Sierra da Matta da Corda (?)." It is regrettable when any but professed geographers write geography. Mr. Gerber makes the total of the two hydrographic basins in the Province of Minas to contain 20,000 square leagues (180,000 square geographical miles), and amongst these he places first the São Francisco, to which he assigns 8800 square leagues, or 79,200 square geographical miles.

‖ M. Liais assigns to it the 16th place. But at present it is very difficult to calculate the area of the Zambesi and the Congo Rivers. Assuming the former to rise in E. long. 16° and to debouch in E. long. 36° (and to extend between lat. 8° and 18°), with the mean length of a degree of 58·472, we have a greater direct course than the São Francisco. The Congo is not to be estimated in the present state of geographical knowledge : it will probably prove itself equal to the Niger.

besi (?) and the Congo (?) ; in America by the Amazons, the
Mississippi, the Paraná-Plata, the St. Louis, St. Lawrence, and
the Mackenzie.

A late expedition has decided that the basins of the Piauhy
and the Amazons are identical, and that both are like the Mississippi, cretaceous formations. Neither Professor Agassiz nor
Mr. Orestes St. John found marine deposits, but these may
have escaped the notice of a flying survey. They judged that
both were of fresh-water origin. During the cosmic winter the
glaciers had moved down to the valleys, without, however, plough-
ing their soles, or leaving those " glacial inscriptions," furrows,
striæ, and burnishings which characterise ice-action. When the
frozen masses were raised by thaws, the triturations were deposited at the bottom, and now form the underlying distinctly
stratified sandstones and the loose sands. Upon these rest the
clay formations, laminated, stratified, cross-stratified, and unstratified, with lines and waves of coarse gravel and pebbles,
whose material is quartz, often highly ferruginous. Capping the
whole is the sandy and once pasty clay, red with ochre, and
common to the Brazil and intertropical Africa. It overspreads
the undulating surface of denuded sandstone, following all its
inequalities, and filling up its furrows and depressions. The
breaking up of the geological winter, and the final disappearance
of the ice, formed a vast fresh-water lake. This, after a somewhat
complicated history, finally burst its seaward dyke, effected denudation on a gigantic scale, and wore the land down to its
rock-core, except where the strata were hard enough to resist.
Professor Agassiz found distinct moraines, and shows that instead of forming a Delta, the mouth of the Amazons has suffered
extensively from the encroachments of the ocean. In the case of
the São Francisco the river builds up faster than the sea can
destroy ; and the denudation of the coast is not to be compared
with that further north. Its Delta does not equal in size those
of the Nile, the Niger, and the Zambesi, but it is distinctly
traceable.

M. Halfeld (Relatorio, p. 172) opines that the grès, or sandstone
grit, is the characteristic formation of the Rio de São Francisco.
The stream rises, as has been seen, from the great central platform of Minas : its material is the Itacolumite or granular
laminated sandstone, which seems to compose the central and

the newer portions of the continent.* Some would compare
these deposits with the vast Silurian beds of North America. At
present characteristic proofs are wanting. M. Chusin (Bulletin
de l'Académie de Bruxelles, viii. 5) found the print of a
univalve in the modern red grits of Minas Geraes. Travellers
and miners, however, are both agreed that hitherto the Brazil,
and even Southern America generally, resembles Africa in
the difficulty of finding organised fossil bodies, and thus it is
difficult to decide the geologic age of the immense grit deposits
in the Eastern and Northern plateau. This Itacolumite reappears at Bom Jardim (138th league) and runs down stream
alternating with coast granite.

Below the gneiss and schist of the Pirápora, we find sand and
sandstones here brown, there of a deep ochre, often highly ferruginous, rarely stratified, and more or less nodulous and porous.
This formation resembles the coast "drift," and once covered
continuously all the river valley; it is still superficial except
where the flood-mud has accumulated upon it, and in parts it
shows intervening layers of clay. It is also broken by outcrops
of hard, blue mountain-limestone, and by argilliferous or hydraulic limestones, compact or stratiform, and abounding in
silex.

Further down stream are close sandstones resembling ferruginous quartzite and covered with a polished crust, either
chemical or mechanical. The rocks are blackened to the colour
of dark coke in places where the floods have less polishing
power and the presence of the mirrory glaze upon the brown,
yellow or red rock, sandstone, granite and syenite, readily gave
the high-water mark. In many parts it resembled magnetic
iron, and I tried it upon the needle without any effect. The
coating did not exceed wafer-thickness, and in places where the
softer material had yielded, glazed sheets, and surfaces partially
glazed, stood up detached. The people term these tinted rocks
Pedras de Marumbé, evidently believing them to be ironstones.
The glaze, however, is of three kinds; the darkest purple, which

* The same grit was found by Castelnau
on the Tocantins River, and on his route
from Goyaz to Cuyabá, in Mato Grosso.
Near Santa Cruz of Minas Geraes he also
mentions erratic blocks of a granite which
does not exist in the neighbourhood. That
traveller records the absence of fossils, and
believes that as a rule the low-lying and
hot portions of the South American continent are of much older date than the
Highlands offsetting from the Cordilleras,
and whose formations are placed regularly
as those of Europe.

appears black in the shade, another is plumbago-black (pedra negra), while the third is a warm red-yellow, probably pure ferruginous matter deposited upon boulders whose inner colour is the same (pedra cabocla).* On the São Francisco the further we went down the deeper became the tint, and the denser the glaze, till in places above and below and about the Great Rapids, the monstrous masses looked like castings of solid metal. This would suggest that it is the work of the stream, but it is difficult to decide whether the waters carried it in solution, or whether their friction had drawn it from the interior to the surface. Analyses by Berzelius and Charles Konig made it to consist of oxides of manganese and iron.† The specimens from the Atures proved to contain, besides oxide of manganese, carbon and super-carburetted iron, but they blackened the paper in which they were wrapped. Such is not the case here, nor do the people attribute to them any noxious influence upon the atmosphere.

The subject was, I believe, first discussed by Humboldt.‡ He found that "whenever the Orinoco, between the missions of Carichana and of Santa Barbara, periodically washes the granitic rocks, they become smooth black, and as if coated with plumbago." On the Congo River I observed the thin shining black crust, strikingly resembling the coatings of meteoric stones, to begin at Boma, just below the narrows of the Zaire, and to extend up to the Yellalah or Great Rapids, in fact where the stream is most turbulent. Here it was first observed by the expedition of 1816 under Captain Tuckey, and the specimens were described by M. Konig.§ In 1832 Mr. Darwin found near Bahia, where a rivulet entered the sea, and where the surf and tidal waves supply the polishing power of cataracts, coatings of a rich brown like those of the São Francisco, and he justly remarks that "hand specimens fail to give a just idea of these burnished stones which glitter in the sun's rays." He could assign no

---

* I never heard the people say, as on the Orinoco, that "the rocks are burnt" (or carbonised) "by the rays of the sun," or that "the rocks are black where the waters are white."

† I have sent to Europe specimens of these curious rock-incrustations from the São Francisco River. During the few months since they were removed, the glaze has become comparatively dull, and looks as though it required renewal.

‡ Personal Narrative, vol. ii. chap. 20 : Bohn's Scientific Library, London, 1852.

§ That geologist (Appendix to Captain Tuckey's Expedition, No. 6) argued from the primitive rock-formations of the lower Zaire the probability that the "mountains of Pernambuco, Rio, and other adjacent parts of South America, were primevally connected with the opposite chains that traverse the plains of Congo and Loango."

reason why these coatings of metallic oxides always remained of nearly the same thickness. During his second expedition Dr. Livingstone (chap. ii; Zambesi and its Tributaries) remarks of the rocks of the Kibrabasa Rapids, that "they were covered with a thin black glaze, as if highly polished and coated with lamp-black varnish." This was apparently deposited while the river was in flood, for it covers only those rocks which lie between the highest water-mark and a line about four feet above the lowest. This appearance has also been remarked upon the Cataracts of the Nile. *

In the river valley, running parallel with the glazed rocks, are detached hills rising abruptly from the level surface, and divided from one another by low spaces. † Some of these piers, which appear to be pinned down, as if they were segments of dykes to control the stream, and to keep it from wandering, are composed of almost pure magnetic iron; ‡ we ascended several of them, and

---

* M. Rozière pointed out to Humboldt that the primitive rocks of the little cataracts of Syene display, like those of the Orinoco, a glossy surface of a blackish grey or almost leaden colour.

† For the first few leagues below the mouth of the Rio das Velhas, the São Francisco runs between containing walls. Thence to Urubú in the 127th league, it is bounded by the scarps of ridges which divide the secondary river-valleys. The detached hills backed by "denudation mountains" appear below Urubú.

‡ This vast iron formation is not noticed by M. J. A. Monlevade, who in 1854 addressed Sr. Diogo de Vasconcellos, then President of Minas Geraes. He declares the Province to be peculiarly adapted for the industry, having a healthy temperate climate, a vast expanse of virgin forest to supply charcoal, and waterfalls which will everywhere facilitate the application of machinery. The united deposits contain more iron than the whole of Europe, considering the richness of the gangue, which gives 76 per cent. of pure metal. It is principally martite, or magnetic ore almost always accompanied by Jacutinga, oxydulated iron, or protoxide of iron, with layers of manganese and titanium in the sandy state. The analysis by Dr. Percy of the micaceous Itabirite gives 68·08 per cent. of metal thus distributed—Sesquioxide of iron, 97·25; peroxide of manganese, 0·14; lime, 0·34; residue, silica, &c., 1·88; a trace of magnesium and no phosphoric acid : total, 99·61. Overlying the rich ores is often Cánga, or hydrate of iron, worked in Europe by air furnaces : it is here neglected, because its yield is only from 25 to 35 per cent. There are besides huge scatters of mineral, five principal ranges lying at a mean distance of eighteen leagues east and west of one another on a line perpendicular to their trend. The richest diggings are associated with gold, which occurs for the most part in the lower hills, slopes, and valleys. The metalliferous strata strike from north-north-east to south-south-west, inclining to the east : the breadth is one-eighth to one quarter of a league, and the depth is unknown.

No. 1. Cordillera, beginning from the east, extends from near Sacramento, Municipality of Santa Barbara, Parish of Prata, crosses the Piracicava River viâ S. Domingos and Jequitibá, covers a vast surface near the Ribeirão de Cocaes-Grande, and after twelve leagues, is lost in the forests. The land is everywhere wooded on both versants, the soil is fertile, and water abounds.

No. 2 is ten leagues long. It rises in the farm of Professor Abrèu, 3½ leagues above the village of S. Miguel, and it forms the left-hand wall of the Piracicava River. "Morro Aguado" (Agudo ?), its culmination, fronts the foundry of M. Monlevade, and crosses his grounds for a whole league.

No. 3, twelve leagues long, appears in the "Capão," south of Ouro Preto, is rich to the west of that city, prolongs itself viâ Santa Anna and Antonio Pereira, forms

I reserve a further notice. The low lands are finely laminated sands and clays with regular cleavage, where sun-burnt and air-baked, and patched with a variety of colours, white and black, blue and grey, pink and yellow, crimson and orange. The iron-dotted levels are backed by ranges of denudation mountains, which, from the stream, appear to be concave. Their smooth table tops and terraces show that they were once continuous walls, now isolated by weathering on a vast scale, and being still degraded by tropical rains and suns. The superior hardness of their ferruginous sandstone saved them from being worn down to the low alluvial levels, and the laminated formations at their base.

The great granitic formation of the coast reappeared about the 238th league, and continued with interruptions to the Rapids of Paulo Affonso, where it passed into syenite. Approaching this feature, and due south of the Araripe plateau, where Dr. Gardner found, on argillaceous ground, the stone-cased fishes of the cretaceous system, the end of the secondary epoch, appeared signs of a remarkable correspondence with the Amazons. On both sides of the river were arenaceous buttresses suggesting gault. The coarser materials had invariably settled in the lowest levels, and above were the fine grits known to the people as "Pedras de Amolar," or whetstones. In this part he found agates and an abundance of flint, with the coticular sandstone re-occurring about Paulo Affonso and the Porto das Piranhas. On the lower S. Francisco, after passing the rapids, about Talhado (332nd league) in Alagôas, I saw the same sandstone overlying granite and underlying limestone. Near the town of Propiá (367th league) there is an outcrop of lime, and extensive deposits of modern calcaire are met with on the lower courses of the short broad streams which cut the coast line.

the Morro d'Agua Quente and the cross chain of the Caráça, and is lost opposite the mine of the Guarda Mór Innocencio.

No. 4, twenty leagues long, begins at the south of the Caráça half a league from Capanéma, and extending north *viâ* Cachoeira Morro Vermelho, Roça (Rossa) Grande, Gongo Soco, Cocaes, Brucutú, and the Serra da Conceição, forms the peak of the Northern Itabira.

No. 5, eighteen leagues long, begins south of Itabira do Campo, which is composed of pure oxide of iron, accompanies

the great Cordillera to Curral d'El-Rei, crosses the Rio das Velhas at Sabará, forms the Piedade Range, and probably reappears far north at Gaspar Rodrigues, Candonga, in the Serra Negra and in the Grão Mogor —all places very rich in iron.

Evidently, says M. Monlevade, it wants nothing but roads, which will save 7 $ 000 out of 8 $ 000, and an import duty on foreign iron of 25 per cent. A few model establishments would soon give an impetus to the trade.

St. Hilaire (I. ii., 14), when describing the course of the São Francisco, had remarked, " La rive gauche, plus élevée que la droite, est généralement moins exposée aux débordemens." Col. Accioli (p. 14) seems to confirm this observation, which was probably only local. The great river, however, flows on a meridian, and the result of the compound motion produced by its northern course and the earth's revolution from west to east, tends theoretically to withdraw the weight of water from the left or western side, and to throw it against the right or eastern. Thus it has been remarked, that on long lines of railways running north and south, the wear is on the eastern rail. Practically I did not find that this theory, which has been extensively discussed in Russia, affected the São Francisco.

This stream is not a "holy river," caret quia vate sacro, but its future will be more honorable than the past of the Ganges or the Indus. The valley and the high dry Geraes which limit it on both sides contain all the elements of prosperity required by an empire. The population is now calculated at 1,500,000 to 2,000,000, probably nearer the latter than the former, and it can support 20,000,000 of souls. As was said of the Upper Amazon, " here the sugar-cane and the pine-apple may be seen growing by a spectator, standing in the barley-field and the potato patch." The uplands can breed in any quantities black cattle, horses, mules, sheep, pigs and goats, while there will be no difficulty in acclimatising the camel. Of mineral wealth, besides diamonds and opals (?), agates, gold and iron, we find mentioned by M. E. de la Martinière* and others, platina, argentiferous galena, mercury, copper (near the Sete Lagôas), antimony, arsenic, manganese, cobalt and various pyrites. Salt and saltpetre, sulphur and alum have been found in large deposits. Of building materials we notice marble, freestone and slate, lime generally dispersed and hydraulic cement; silex, grindstones and potter's-clay are also abundant. The land is admirably fitted for the silk-worm, and for the cultivation of cotton, which will some day rival its immense fisheries.† The basin of the São Francisco is terres-

---

* Official Letter. Annexe N. to Presidential Relation of 1867.

† The names of the fish not occurring in the following pages, but mentioned by M. Halfeld, and referred to by the people, are :—The scaly :

1. Camurupim (?), short and thickset.

2. Camurim, mirim, and assú (large and small), white with dark stripes on both sides.

3. Tubarana, dourada (yellow), and branca, a large fish, lean up-stream, but much admired below the rapids.

4. Bagre de Ouro. (?)

trial not aquatic, and it is completely isolated by cataracts near the source and above the mouth. The fishes, therefore, which have Amazonian names will probably be found to be distinct. The localisation of species lately found, even to a greater extent than he expected, by Professor Agassiz, who remarked that the main artery of the great northern basin was broken up into distinct families, will be the case here. The riverines, who have never attempted classification, or distribution, or limitation, can generally tell whether a fish is or is not caught below certain grounds. The naturalist who shall attempt the ichthyology of the São Francisco will have before him the task of years. The stupendous results obtained by Professor Agassiz, the revolution of ichthyology of which he speaks, were effected by an immense collaboration, public and private, as far as collection extends. That savant may be said to have been assisted by the forces of the empire.

The hop, and to a certain extent the vine, will flourish. Amongst the cereals it produces a wealth of maize and rice, whilst barley, rye, and probably wheat, will succeed in the Geraes. Most of the fruits and vegetables that belong to the sub-tropical and the temperate regions may be introduced. A sugar plantation lasts ten years, although the cane is most inefficiently treated. Coffee grows admirably; tea, congonhas (or maté), and the favourite of North-Western Brazil, the guaraná (Paullinia sorbites) will succeed in low, hot, humid spots. The tobacco is some of the best in the Empire : salsaparilla and cochineal-cactus, aloes

5. Robalo, a kind of pike common in the streams of the Brazil.

6. Pacamon and Pacamon de Couro, which, says M. Halfeld, is a soft fish that lives in mud. Gardner describes the Pocomó as an ugly black fish, about two feet long and covered with hard scales ; it keeps near the bottom, is easily netted, and makes good bait, but is rarely eaten. The Pacamum of the Amazons is described as of a bright canary colour, and weighing 10 lbs.

7. Sardinha.

8. Sarapó.

9. Sibeirá or Aragú.

10. Cará.

11. Pirampeba, white and black, a small flat fish with teeth like needles.

12. Lombia, about one foot long.

13. Sudiá.

The smooth-skinned are—

1. Niquim.

2. Cumbá.

3. Prepetinga.

I heard also of the Tamburé, about one foot long and held to be good eating, and the Piguri and Lambari, small fishes from which oil is extracted, on the Upper Paraguay River. The Shark (Tubarão, Squalus tubero, Linn.) has carried off people near the mouth, and they speak of another large fish, the "Meru," probably a Squalus, which some say is anthropophagous, and others not : it is also found at the mouths of the short disconnected tidal rivers which drain into the east coast. Of course the Manatu or Sea-Cow, that representative of the Dinotherium, and the Porpoise of the Amazons, are wanting in the upper waters of the São Francisco.

and vanilla grow wild.   The lumber trade is susceptible of a vast development; the Aroeira, the Braúna, the Candêa, the Peroba, the Canella, and the fine hard-woods of the Brazil generally, await exploitation.   Oil-plants and tanning barks, basts and fibres, drugs and gums, as the Jetahy-copal, the Balm of Peru, the Copahyba and the Asafœtida, are yielded in abundance, and the same may be said of beeswax and of the Carnaúba wax, which is converted into candles at Rio de Janeiro.   The dyes are abundant, from indigo to the Páu Amarello, and of cabinet woods a long list is headed by the Jacarandá and the Brazilian cedar. In the presence of such vast and unexploited wealth awaiting the distressed classes of Europe we may exclaim with Goethe, "Who says there is nothing for the poor and vile save poverty and crime?"

We will now consider the Rio de São Francisco in another most important light, as a line of communication linking the maritime and sub-maritime regions with the Far West, the north with the south, facilitating commerce and colonization, obviating scarcity by giving an issue to the surplus of the central regions, especially when the irregular seasons of the coast injure agriculture, or when the seaboard may be blockaded.   And thus will be completed the strategic circle which the Empire, if it would preserve its integrity, now greatly needs.   I may here premise that the streams of the Brazil between the Amazons and the Plata are, like those of the great African peninsula, to be distributed under two heads.   The many are short and direct, rather estuaries than rivers, surface-draining the ranges which subtend the coast.   The few are the long and indirect, like the São Francisco and the included arcs before specified.   The former are of limited value, the latter may be extensively utilized.

The Brazil is emphatically the land of great, but as yet "unimproved," rivers.   They have, however, gained for themselves a bad name; * and water communication has been deplorably neglected as in British India.   Capital for railways being pro-

* I came to the Brazil prepared to believe and to regret with Mr. Kidder that, "notwithstanding the number and vastness of the rivers flowing through the northern and western portions of the Empire, and finally mingling their waters with the Amazon and the La Plata, there is not one, besides the Amazon, emptying into the Atlantic along the whole Brazilian coast, which is navigable any considerable way from its mouth inland." But actual inspection soon showed that the lower beds of many streams can be joined by short railways with the upper lines, which are naturally adapted for communication, and which have been completely passed over.

curable at heavy interest from England, the various modes of communication have been performed in the reverse order of their merit. Water communication, a vast and economic power, which should have been first undertaken, will be the last; roads have been limited to the use of the mule or the pack-bullock; and the Empire is threatened with a railway system of marvellous ineptitude. In Europe, Italy is perhaps the only country which prospected before breaking its ground. Here the want of a Topographical Commission on a large scale has made the Pernambuco threaten to run into the Bahian Railway at Joazeiro, and the D. Pedro Segundo cut across the Mauá line, and prepare a campaign against the Cantagallo and the Santos and Jundiahy. I shall reserve this important subject for future consideration.

Communication by the Valley of the São Francisco is still in embryo. Dr. Mello Franco, Imperial Deputy, drew attention about 1851 to the importance of the Rio das Velhas. As has been seen, this stream drains the northern versant of the Minas Plateau, whose culminating point is the Itacolumi. Its eastern valley wall is the Serra Grande or do Espinhaço; and westward it is divided by a long spine of many names from the Valley of the São Francisco. More tortuous than the latter, its declivity, as far as the junction, is less, being an average slope of $0^m \cdot 3941$ per kilometre, to $0 \cdot 4890$. During the months of high water the whole river is naturally navigable, and exceptional rises would be dangerous for only a few days. In March, 1852, a respectable Portuguese trader, Manuel Joaquim Gonçalez, whom I met at Januaria, floated down the Rio das Velhas with three ajôjos, of which one was lost. In 1862, when the Councillor José Bento da Cunha Figueirado was President of Minas, the Imperial Government ordered a survey under M. Liais and two assistants, Lt. Eduardo José de Moraes and Sr. Ladislao de Souza Mello Netto; and their admirable plans of the Rio das Velhas and the Upper São Francisco are now well known to Europe.

This Commission preferred the Rio das Velhas as the line of communication with the Empire, and apparently for the best reasons. * The opening of the Upper Rio de São Francisco

---

* Thus the riverines truly observe " O Rio de São Francisco faz barra (falls into) o Rio das Velhas." The discharge of the former at the confluence is 446 cubic metres per second, of the latter only 209.

But these proportions do not last long. At the Porto das Andorinhas, sixty-two leagues above the junction, the debit of the São Francisco is but fifty-nine cubic metres, and the Rio das Velhas has the same

would be a gigantic work for which the country is not yet prepared; the Pirapóra Rapids alone would cost more to remove than all the most important obstacles on the Rio das Velhas. In the thirty-four leagues above this point, the São Francisco has as many "Cachoeiras" as the whole of its rival between Sabará and its mouth. The ridges traversing the latter are mostly friable and shaly; the bars rarely exceed six to seven yards at the summit, whilst many obstacles are merely detached rocks or sand-bars. In the former the material is of the hardest gneiss and sandstone, and spread out horizontally sometimes forty to fifty metres. For a description of other obstacles, such as the nine terrible leagues, so fatal to human life, about the Porto dos Passarinhos, the reader will refer to M. Liais. Trade, moreover, has preferred the former between the mouth of the Paraopéba River; from above the confluence hardly a dozen ajôjos descend per annum, whilst many boatmen, fearing for their lives, refuse to hire themselves. The small towns are sparsely scattered; and during the rains, when Carneiradas drive the inhabitants into the interior, the banks are almost deserted.[*]

On the other hand, it has been shown that a meridian, with a small deviation, connects the metropolis of the Empire with the line of the Rio das Velhas. Sabará is only sixty-four direct leagues from Rio de Janeiro; the analogous point on the S. Francisco would be ninety leagues—a weighty consideration when looking to a Railway. This proximity, combined with superiority of climate, will recommend it to colonists. Finally, it is connected with more important places, such as Diamantina and Curvello.

M. Liais also decides, I believe rightly, in preferring water to land communications. Here again, as in British India, village intercommunication has found no place in the system of public works. "Nature's roads," the vilest paths made by the foot, and never bearing the impression of the cart-wheel, run down both banks of the Rio das Velhas and the São Francisco. Both are bad, but usually one is worse than the other. Even in the dry season the canoe is preferred, and during the rains these lines are inevitably closed. There would be great difficulties in

---

volume at 111 leagues from its embouchure. The reason is that the former receives more affluents in the lower, the latter in the upper course.

[*] All agree upon the subject of these fevers, yet the plane of the Upper São Francisco is higher than that of the Rio das Velhas.

making, and even greater in preserving, a rolling road; and the expense from Sabará to Joazeiro (244 leagues) would not be less than 12,200:000$000 (say £1,220,000), whilst the high tolls would do away with all the benefit. A similar objection would apply to tow-paths for tracking boats.

M. Liais divides the obstacles of the Rio das Velhas into five varieties—stone-piers or detached rocks; whirlpools, with vertical axes; sand-bars and shallow sharp curves, snags and timber encumbering the bed. While greatly admiring his plans, I cannot agree with his system intended "pour assainir la rivière:" he wants to make of this wild stream a Seine or a Rhone; and my experience of India and the United States suggests far more attention to economy. He is too fond of mines and blasting applied to soft stone, of "suppressing" boulder-piers, or marking every rock, and even shoal, where an accident can happen. Here "un petit travail de canalization" is no joke, yet he would suppress channels; to prevent "échouage," alter the stream-bed; change its direction, rectify every abrupt détour, and canalize even the shallows: doubtless the first flood would restore the "status quo ante." Often, too, he would obstruct one half of the channel and canalize the other, a precarious work. I have alluded to his plans of draguage and tunage, either simple or "avec enrochements;" the removal of the Rapids will render these costly works useless by increasing the current, and by narrowing the bed where it spreads out in the dry season. He wishes to "nettoyer" the stream of floating wood, which of course will stick where it has stood. To obviate the deposit of sands from the gold washings of and about Sabará, he would compel proprietors to dig tanks, through which the muddy streams would pass and deposit their burdens before entering the river. But in the present condition of the Brazil such precautions would be impossible; nor would the profits derived from gold-digging enable, as he supposes, mine-owners to make the necessary disbursements. He would establish a water-police to prevent trees being thrown into the stream; the policemen would probably be the first to throw them. Finally, the key-note of his estimates is that the channel should be made independent of pilots, and offer no risk even to a mismanaged steamer. I need hardly characterize these as works of supererogation.*

---

* The Brazil is already but too well inclined towards "monumental works." "Les ouvriers Mineiros," says St. Hil. (I. i. 394) "s'ils mettent de la lenteur

A considerable portion of the labour could be carried on only at dead low water—that is to say, for three or four months in the year. Half water would suffice for another part. During the floods (enchentes) from November to March nothing could be done. About April there is often a small inundation called Enchente de Paschoa, which would limit the season to six months. Thus the swelling of the S. Francisco system is almost synchronous with that of the Amazons, which begins in November, and lasts till May or June, the greater extent of time being the result of its superior dimensions. Both streams have the preliminary freshets, which will presently be described; and in both the oscillations are known by the name of "repiquete." During the retiring of the waters (vasantes) sickness must be expected amongst unacclimatized workmen seduced from distant parts by a rise of wages.

The following is the estimate proposed by M. Liais:—

200:000$000	Between Sabará and Macahúbas, to admit in the dry season a vessel drawing 0ᵐ·60 (deeper draughts would require a great increase of outlay). Canalization of four places and "suppression" of rock.
1,730:000$000	Between Macahúbas and Jequitibá, draught 1ᵐ·25. Draguage, suppression of a ford, rectification of Poço Feio, and removing rocks.
195:000$000	Between Jequitiba and Paraúna. This is one of the worst sections. For same draught.
480:000$000	Between Paraúna and embouchure of the Rio das Velhas, the finest part of the course; draught 1ᵐ·50.*

Total 2,605:000$000 (say £260,000) between Sabará and the mouth, 120 leagues.

The following are the figures for opening the Upper Rio de São Francisco:—

1,400:000$000	opening the Pirapóra Rapids.
4,100:000$000	from Pirapóra to Cachoeira Grande included,
3,200:000$000	from Cachoeira Grande to Porto das Melancías.

Total 8,700:000$000 (say £870,000) between Pirapóra and the Paraopéba River, 41 leagues.

dans leur travail, au moins ils donnent beaucoup d'attention à leurs ouvrages, et je crois même qu'ils les finissent plus que ne feraient les ouvriers européens."

* I need hardly observe that such a draught is wholly uncalled for. In 1849, according to M. Claudel, on the high Seine, empty boats drew on an average

We will now proceed to the Rio de São Francisco.

M. Halfeld has made a detailed plan rather than a map; it wants meridians, parallels, and the astronomical determination of eight or ten points before it can be considered correct. The letterpress describes every league of the stream; but as the distances are not checked by instruments, it is evident that one league must often run into the other. And as any amount of paper has been expended, it is much to be regretted that a place was not given to enlarged plans of the Rapids and the obstructed parts. This is one of the chief merits of M. Liais' publication. The German engineer, with true Teutonic industry, probably chained down the whole distance, and thus also he must have ascertained the breadth; when the stream is very wide, no figures are given. Moreover, he was engaged in this gigantic labour for the space of only two years, which would be insufficient accurately to lay down the topography of the complicated thirty-one leagues between Boa Vista (269 leagues), and Surubabe (300 leagues).

From the details of a "désobstruction," which would convert this enormous bed into a clear channel—a kind of canal—like the Rhine or the Rhone, M. Halfeld proposes a total of 1,089:000$000 (say £108,900). A considerable portion of this expense is mere waste; removing rocks, building dams, applying fascines (which suggest the proverbial tide and pitch-fork), clearing of snags and timber, sloping banks, erecting quays and other improvements—all these may be reserved for the days when steam navigation shall have begun. I may observe that a total of 12:900$000 (say £1290) has been devoted to the stream between Porto das Piranhas and the Villa de Piassabussú, a line upon which steamers have plied since August 1867, without expending a farthing. Strong objection must also be raised to any attempt at a canal fifty palms broad at bottom, and extending seventy-two leagues (206 geographical miles) between Boa Vista and the Porto das Piranhas, the present terminus of steam navigation. This can hardly succeed; the land is alternately sandy and stony, deeply flooded during the rains, and subject to enor-

---

0ᵐ·27; on the Loire and Moselle 0ᵐ·22. Steamers on the various streams of France and Germany drew, say MM. Mathias and Callon, between a minimum of 0ᵐ·36 (Ville d'Orleans, on the Loire) to a maxi-mum of 1ᵐ·23 (Bretagne, Basse Loire). In the United States we find flat-bottomed steamers drawing 22 inches, and a metre suffices for sea-going craft.

mous evaporation in the dry season. Evidently a line of light
rails will be the true system of communication.

Compared with the two preceding estimates, M. de la
Martinière is economical. Their united sum for the Rio das
Velhas and the Rio de São Francisco is £368,900. He reduces
it to 2,000:000$000 (say £200,000); and for this amount, besides
clearing the channel, he builds bridges and workshops, boats,
slips, and five tug-steamers. But he runs only between Sabará
and Joazeiro. Other writers adopt the estimates of M. Liais for
the désobstruction of the Rio das Velhas, adding 2,400:000$000
(£240,000) for clearing the channel between the Sobradinho and
Varzea Redonda; and 12,000:000$000 (£1,120,000) for a road
round the difficulty of Paulo Affonso. This estimate represents
a total expenditure of 17,000:000$000 (1,700,000) for a naviga-
tion of 476 leagues (1428 miles).

I will now propose my own estimates, simply premising that
the plan is not professional, and that I do not intend applying to
the Brazilian Government for the privilege of carrying them
out :—

£55,000	for the Rio das Velhas.
40,000	to remove the Sobradinho Rapid and the obstructions above Joazeiro.
108,000	Railways and locomotives past the Great Rapids between Varzéa Redonda and the Porto das Piranhas, thirty-six miles (at £3000 per mile, gauge 2 feet to 2 feet 6 inches).
£203,000	

With respect to the first charge, £4000 for twenty tons of
blasting powder, which, however, might possibly be made cheaper
upon the spot. The machinery would amount, transport included, to £15,500—viz., two big sledge hammers, and two
smaller ditto; and two picks working in slot or cradle, with a
slot-joint adjustable to the piston, £1000; drags for the
Rapids, £2000; and five-horse-power engine mounted on a
raft, £2500; first-class steam-tug, with donkey-engines to follow
and assist in working, £10,000. The wages and support
of the working hands may be set down at £30,000; and the
remainder for "contingencies," which in these lands demand a
large margin.

The second item I take from M. Halfeld, who proposes to

expend upon the correction of the São Francisco channel (240th league) to Joazeiro (247th) the sum of 416:320$000 (say £41,632). This is the highest possible estimate; the work is the only absolute necessity between the Rapids of Pirapóra (league 1), and the Villa da Boa Vista (269th); and as will be seen when we reach the place, Nature is doing there her own engineering.

From the Villa da Boa Vista to the Porto das Piranhas, seventy to seventy-two leagues (216 miles), the São Francisco can hardly be called navigable. Rafts like my own, and canoes traverse even in the dry season the first thirty-four leagues between Boa Vista and Varzea Redonda, but with a thousand perils. The remaining thirty-eight leagues (114 miles) between Varzea Redonda and Porto das Piranhas are absolutely unmanageable. The minimum of railway required will be £342,000; the maximum, £648,000. If a marginal tramway be preferred, the expense will be reduced to one-half; a cart road would cost about one-third. I rejoice to hear that the Government of His Imperial Majesty has sent a well-known German engineer, M. Karl Krauss, to ascertain the levels which can connect the Lower with the Upper São Francisco.

As the great riverine valley becomes settled, the rapid drainage will tend to increase the floods and corresponding droughts. It will then be necessary to build dams on the main artery and the tributaries, solid piers projecting from either shore throwing a strong current into the centre, and creating sufficient depth of water for navigation. Thus, combined with the removal of the Cachoeiras, the lower valleys will be secured from inundations. Again, the droughts of winter can be avoided by deriving supplies from artificial lakes and reservoirs constructed on the secondary streams. This plan has been proposed for the Mississippi, whose area of drainage is a million and a quarter of square miles, and whose navigable lines are ten thousand miles. Such bold and magnificent schemes have been proposed and partly carried out in the New World,[*] whilst the engineers of Europe have had a chronic fear of "meddling" with great rivers, and have propounded the theory that these were made to make canals. It is only a question of time when the Brazil will follow the example of the United States.

---

[*] Ellet "On the Ohio and Mississippi Rivers," Philadelphia, 1853.

Steam exploitation of the Rio das Velhas is upon the point of commencing. On June 25, 1867, the President of Minas Geraes, Councillor Joaquim Saldanha Marinho, entered into a contract with Sr. Henrique Dumont, C.E. The Provincial Government bound itself to pay before June 30, 1867, the sum of 4:000$000 (£400); before July 15, 33:000$000 (£3300); 19:000$000 (£1900), when a tug-steamer of not less than twenty-five horse power should reach Rio de Janeiro, and make up a total of 75:500$000 (£7550) after the vessel's first satisfactory trial-trip. Counting from June 25, 1869, the engineer was to have for ten years the use of the steamer, after which it is to be handed in good condition to the Provincial Government. The latter also undertakes to solicit admission free from duty of all imported articles, such as steamer, boats, tools, and machinery required for clearing the channel; or should the application fail, to take upon itself the expenditure. The désobstructions of the bed were to be carried out according to the estimates of M. Liais; and the report was that £160,000 would at once be devoted to the work.

M. Dumont, on the other hand, bound himself, under penalty, to place within two years after date of signing, a steam-tug at Sabará. The vessel to make per mensem two passages, going and coming (viagens redondas) over the portion of the channel which would permit, and at the rate of ten leagues per day. The passage money to be 1$000 per league; and for goods, 0$100,* while Government employés were to pay only for rations. The contractor to keep the steamer in good order, and to be responsible for its injury or loss (except by act of God, or unavoidable accident), till it should belong to the Provincial Government. The stream between Sabará and Jaguára to be reformed, according to the plans of M. Liais; and to be rendered navigable, as the public purse shall permit, to its confluence with the Rio de São Francisco.†

* The public at once began to complain of these conditions. From Sabará to Jaguára the passenger will pay 20$000; and each arroba (32 lbs.) of merchandize 2$000. But the same distance may be done for 4$000 by a mule carrying six to seven arrobas. Time of course is never taken into consideration.

† M. Liais calculates that a poling-boat drawing three palms (2 feet 1·8 inches), with a crew of 10 men working 8 hours per diem, and spending 15 days between Sabará and the mouth of the Rio das Velhas, would carry 4000 arrobas (50 to 60 tons). At present this would be done by 340 mules and 42 men in 36 days. The ascent of the boat would demand treble the time and double the crew, yet it would have a great advantage over transport by animals.

On the other hand, a small steamer of 20-horse power, burning wood, which is everywhere plentiful, would tug the same load, working twelve hours a day, in five days

Sr. Dumont lost no time. In March, 1868, he brought from Bordeaux to Rio de Janeiro the sections of the "Conselheiro Saldanha" and "Monsenhor Augusto." The steamers are of forty and twenty-horse power, and their speed will be about eight miles an hour, upon a draught of ten inches. About the beginning of the next year they will begin operations upon the Rio das Velhas. I have already alluded to the horse boat, with inclined planes working paddle-wheels, and it is to be hoped that this improvement will soon follow the appearance of steamers.

As early as 1865 His Excellency the Councillor Manoel Pinto de Souza Dantas, then President of the Bahian Province, proposed to place a steamer upon the Rio de São Francisco. The little "Dantas," ninety feet by fourteen, and of about ninety-four tons, was built by Mr. Hayden at the Ponta d'Arêa Works, opposite Rio de Janeiro. The plates and machinery had been taken to pieces, and were sent numbered, with a model and detailed drawings, by land to Joazeiro. The road, however, was found unfit for wheel vehicles; of 346 bullocks sixty had died in the shortest time, and there had been an equal loss of horses. It is regrettable that the fine timber of the Rio de São Francisco had not been preferred to iron plates, and that local jealousies, of which I shall have more to say, had delayed the execution of a great project.

Of late years there has been a revival of an idea first suggested, I believe, in 1825 by a certain Colonel Joaquim de Almeida, and which, since 1832, had fallen into oblivion. This is to erect the valley of the São Francisco into the twenty-first province of the Empire.* The main object is to remedy the social, commercial, and political evils which arise from the isolation of the settlements; these are often 150 leagues distant from their provincial capitals. The only objection of which I am aware is the trifling increase of expenditure; it would, however, soon reimburse itself.

---

down, and in eight days up stream, with five hands for the tug and two or three for the boat. The expense of descending, including commander and engineer, would be 100$000; of ascending, 160$000. Doubling this sum for time lost in taking in and discharging cargo, and adding per trip 100$000 for wear and tear of machinery, we have a total outlay of 600$000 for each descent, and 900$000 for the return. Thus the arroba should pay a maximum of 0$150

from Sabará to Guaicuhy, and 0$225 from Guaicuhy to Sabara.

* "I find that most of the gentlemen of the lower Province are disposed to sneer at the action of the Government in erecting the Comarca of the Rio Negro into a province; but I think the step was a wise one. . . . If the country is to be improved at all, it is to be done in this way" (Lieut. Herndon, 329).

Foreigners, who are accustomed to view the Brazil with the most superficial glance, have represented to me the evils of increasing an official staff already far too large. They seem not to be aware that the highly constitutional government, which has been well described as a republic under the disguise of an empire, requires to be strengthened as much as it legally can be, and that good " appointments " (as they are called in India) form the readiest and most practical mode of strengthening it. And if the Brazil cleave to number twenty, she may borrow from her northern sister, the United States, an admirable system " of territories " which are there States, and would be here Provinces, in statu pupillari, educating for self-rule.

On the Rio de São Francisco, where the subject of No. 21 is perpetually ventilated, every city, town, and village is prepared and resolved to be the capital. The great rivals are Januaria in the south, and to the north Joazeiro; both would, I believe, remain as they are than accept a subordinate position. The desiderata for a chief settlement are many : a central site, facility of communication with the seaboard and the interior, a healthy climate, and, if possible, rich and fertile lands. As will be seen, I would award the palm to Bom Jardim, or to Xique Xique.

The new province or territory might embrace the whole valley of the São Francisco River. The south would borrow largely from Minas, the Serra de Grão Mogor, Minas Novas, Montes Claros and Formigas, on the east; to the west the valleys of the streams Paracutú, das Egoás, Urucuia, Rio Pardo, and Carunhanha. From Bahia it would take the western watersheds of the Serra das Almas and the Chapada Diamantina, and from Pernambuco the western river valley north of Carunhanha. It would extend to the Rapids of Paulo Affonso, and communicate with the sea by a railway or a tramway, and the steam navigation now upon the lower river. And when population and wealth shall increase, it may admit of further subdivision into a southern territory, with Januaria for capital, and a northern, in which Joazeiro would command. Each of these would own about 500 miles of river, and both are more worthy of provincial honours than the unimportant Provinces of Alagôas and Sergipe, which are crushed like dwarfs between the two giants Pernambuco and Bahia.

The direct distance between Rio de Janeiro and Sabará is 3° 12′ 39″ or 192 geographical miles, and the usual calculation for the length of railway lines is 276 miles. Of this, however, a portion has been covered by the D. Pedro Segundo. For steamer navigation we have down the Rio das Velhas 366 miles, and down the Rio de São Francisco, from the mouth of the Rio das Velhas to the Villa da Boa Vista, 792 miles, perfectly clear, save at one point. From Boa Vista to the Porto das Piranhas, the railway or tramway will run for 216 miles, and from the Porto das Piranhas to the mouth of the São Francisco, in south latitude 10° 27′ 4″, and west longitude (G.) 36° 21′ 41″, there are 129 miles of good navigation.

Thus we have the segment of an immense circle, whose arc numbers 1779 geographical miles, exceeding the average breadth of Russia. Of these by railroad are only 492, the rest (1287) being water communication, which is usually considered to be ten times cheaper.

Communication even by steamer will not create population, except by attracting colonists; on the other hand, it will, like the railway, benefit the country by collecting and centralising the now scattered homesteads. This route of nearly 1800 miles, connecting the heart of the Brazil with its head, the metropolis, and placing its richest Provinces in direct communication with the outer world, will be the most important step yet taken. The opening of the Rio de São Francisco will not only benefit directly the Provinces of Minas Geraes, Bahia, Pernambuco, Alagôas, and Sergipe, and indirectly those of Goyaz, of Mato Grosso, of Piauhy, and of Ceará—it will contribute potently to maintain the integrity of the Empire.

CHAPTER XVI.

FROM GUAICUHY TO SÃO ROMÃO.

FIRST TRAVESSIA, 24 LEAGUES.*

ASPECTS OF THE RIVER.—ESTREMA VILLAGE.—GAME.—THE OTTER.—THE
CASHEW SHOWERS.—REACH SÃO ROMAO.—ITS HISTORY.—GIANT FIG-TREES.—
ACTUAL STATE OF THE TOWN.—A GOOD TIME COMING.

> Montanhas vimos, campos mil patentes,
> E hum terreno nas margens tão extensa,
> Que poderá elle só neste hemisferio
> Formar com tanto povo hum vasto imperio.
> <div align="right"><i>Cara.</i>, 6, 27.</div>

THE Pirapóra had been on the São Francisco my terminus *ad
quem*, and now it was *a quo*—the rest of the voyage lying down
stream. The weather was still surly from the effects of the last
night's scold, but the air was transparent, cleaned of atoms,
spores, and molecules, whilst increased humidity, as in England,
rendered it still clearer. The books no longer curled with
drought, as in the Rio das Velhas, and an increased reference to
the quinine bottle was judged advisable. The Vento Geral, or
Eastern Trade, set in, but we were evidently at the break of the
rainy season.

Wednesday, September 18, 1867.—Ember Day. Of course delays
were numerous; the new crew had to shake hands with the
villagers. It was noon before the Eliza was poled off from the
bank of Guaicuhy, and turned "head downwards" into the Great
Stream.[†] We left on the right the Ilha do Engenho, upon which
people were congregated; canoes were made fast to the alluvial

---

* The word Travessia is written by
Koster (i. iv.) Traversia, and translated by
M. Jay "Traversée." It is probably a
local form of Travéssa, a "passage." In
Spanish South America "Traversía" is
applied to a land stage. This Travessia, or
trip, begins normally at Pirapóra, and thus

numbers thirty leagues. I have heard the
boatman, when we crossed the stream under
difficulties, call it a "Travessa braba."

† "Navegar cabeça abaixo," in the dialect of the river, is opposed to "Cabeça
acima," going up-stream.

banks, rising in regular steps or grades; this side of the island is sandy, and fir-trees rise from the banks. The Ilha do Boi led to the Barra do Jatobá, a stream coming from the west, and this we shall find to be the rule of almost all the great affluents. Its waters, called "seizoentes," "sesonarias," "pestiferas," breed, they say, chills. A little below it were detached rocks, Pedras do Agato; * these the pilots did not expect to pass, as the head wind, especially during the afternoon, often waxes fierce there—it did not offer let or hindrance. Passing "A Barreira,"† where there was a clearing and a few huts on the right bank, we saw large deposits of the iron-revetted amygdaloid "Cánga." Beyond it was the mouth of the Jequitahy‡ stream, breaking the right bank with a gap of some-150 feet, and gracefully curving through the low trees. On the opposite side is a remarkable point, the Pedras de Bura do Jequitahy, horizontal strata of stone from which blocks and boulders have been washed into the stream.

As soon as the air became dusk we looked about for a nighting place; here the working hours are from sunrise to sunset. Boat-men will not travel in this part by night; even with the full moon, they cannot see the "Maretta," or ripple caused by snags below surface. Our men preferred the exposed left bank, which supplies wood; the right affords more shelter from the east wind, and from the storms which sweep up from that direction. In the river tongue, the latter is known as Banda da Bahia, the Bahian side, the western being the Banda de Pernambuco. These are old names, dating from the days when the captaincy of Pernambuco covered part of the present Minas Province.

This portion of the São Francisco, and, indeed, we may say the whole course, is more civilised, tamer, and less picturesque than the Lower Rio das Velhas; we passed hardly a league of land without sighting huts or improvements. Making fast at 5·30 P.M. to a sandy "praia," and climbing up the steep clay bank, we found a small tenement, surrounded by a dwarf field of manioc, poor bananas, and first-rate cotton, which seems to flourish everywhere. The maisonnette turned its back upon the

---

* Or Agatho, probably a P. N.

† M. Halfeld calls it. "Barreira dos Indios," a name given to a place further down stream.

‡ Or "Gequitahy," a considerable stream, 120 direct miles long, heading in the western counterslope of the chain that discharges eastwards into the Jequitinhonha. It is navigable for canoes, which ascend it three leagues in the dries, and twenty-eight during the rains.

west, the rainy quarter, and some trouble had been taken to build it. There was a kiln constructed in the river bank, a circle four feet in diameter by two deep, a clay floor pierced with holes, separating the fire below from the material to be fired : the latter operation appears very insufficiently effected both in pots and tiles. The Western Valley is bounded at a distance of about five miles by the Serra do Itacolumi ; the mists, however, robbed us of the view. On the opposite side was the Povoado do Olho d'Agua, a few thatched sheds, buried in orange trees and Jaboticabeiras.

To-day the stream has averaged some 1200 feet in breadth, and in places has widened to 1600 yards. The banks to which the flood swings are eaten below, and rise perpendicularly, whilst the opposite side assumes the natural angle. The height varies from 25 to 36 feet ; the material is a base of white or reddish sand, supporting hard "taúa," and the surface is rich humus, mixed with silt. The supply of wood will last for years, but the vegetation is uninteresting after the magnificent avenues of the Rio das Velhas. The surface is composed of swells and waves of ground, in whose hollows are Alagadiços, or stagnant waters. Now, also, begins the Ypoeira, which partly corresponds with the Igarapé,* or canoe path of the Amazons and the Lower São Francisco. When the bayou is considerable, it retains its water through the year, and is drained to the level of the dries by a Sangradouro. These little creeks carry a quantity of sand ; they are mostly disposed perpendicularly to the stream, and they assist in unwatering the waves of ground which are not reached by the inundations. In many places there are lumpy hills, forested or cleared, and on both sides the divides of the riverine valley are well marked with heights, which will disappear a few leagues down stream.

September 19.—We effected an unusually early start, but our men are paid " by the job." The right bank showed a mass of building material, argillaceous schistose sandstone in horizontal

---

* Igarapé is derived from yg, water—jara, lord (*i.e.* a canoe), and ipé, where (it goes). Of the Ypoeira feature I shall have more to say further down stream, when it becomes important. It is what Lieut. Herndon calls Caño on the Upper Amazon, a natural arm of the main river, opposed to the " Furo " (small mouth) and the " furado," an artificial (but sometimes a Nature-made) cutting. That traveller also remarks, "Igarapé is the Indian name for a creek or ditch, which is filled with ' back water ' from the river ; and the term Paranamiri(m)—literally, little river —is applied to a narrow arm of the main river running between the main bank and an island near to it."

slabs; on the other side was "As Lages," a clearing with bananas and oranges. Presently rose before us the Morro da Estrema, a terrapin-shaped buttress, disposed perpendicular to the stream, high above the floods, well wooded, and with good improvements below. The little village of the same name was at the bottom of a sack, formed by the river sweeping to a projection of the opposite left bank. It is built upon an inner ledge of rising ground, and a few poor tiled huts clustered about the little church, Nª Sª do Carmo.

At noon we halted for rest on the Pernam side, below the hamlet known as Serra da Povoação.[*] The hills of that name form a meridional line of scattered lumps running parallel with, and rarely three miles distant from, the stream. At the Serra or Serrote do Pé de Morro, they impinge upon the banks; the little crescent is called by the people Serra do Salitre, as it contains a saltpetre cave, and they declare it to be a north-eastern branch of the great Mata da Corda[†] range. Opposite it the Barra do Pacuhy[‡] forms the usual Corôa; a little below, on the left, we were shown a sand-bar, where a pleasure party of seven had come to grief some eight years ago. They were returning from a festival at Estrema, a little place of great debauchery; the "dugout" struck a snag, and all were drowned.

Passing the Riacho da Fome, an ill-omened name now not uncommon, we anchored before sundown at the mouth of a Sangradouro, called the Cachoeirinha, from an adjoining village.[§] The clay wall of the river is here some thirty-two feet high, and the streamlet draining a bayou is about a mile in length. The Mandim fish had awaked, and grunted like a gurnard, and his hunger in the afternoon suggested to the pilots that he foresaw rain. Presently a cold east wind arose, the clouds gathered in heaps, and the horizon gleamed lurid with the reflection of field fires, easily to be mistaken for electrical "weather lights." During the early night there were raw and violent gusts, and they presently induced a downfall whose steadiness promised persistency.

[*] Serrote da Povoação (M. Halfeld).
[†] Forest of the Cord, so called from its long, narrow line.
[‡] This stream runs almost parallel with the Jequitahy, and drains the Montes Claros de Formigas. It has no mines, but the lands are good for pasture and agriculture.

[§] The Pacú, according to Castelnau, is the genus Characinus of Artedi, and sub-genus Curimata of Cuvier. The carp-like body averages two to three palms in length, and is considered good eating; the Pacú vermelho being held to be the best.
[§] There is a Cachoeira hamlet on the right or opposite bank.

CHAP. XVI.] FROM GUAICUHY TO SÃO ROMÃO.

This day showed us a more than usual quantity of animal life. A Jacaré (cayman) stared at us from the bank, with the short round muzzle protruded in curiosity, and another lay dead upon the stones. Jacús (Penelopes) chattered on the tree-tops, and afforded fine practice, but the bush was too thick for bagging, although we worked like men for the pot. A large otter plunged close to us, and at times we heard their whistling cries, which the pilots compared with the quarrelling and scolding of old fishwives, and the frequent ejaculation of " diabo." There are two kinds, the Lôntra, or common species (Lutra brasiliensis), and the Lôntra grande, also known by its Tupy name, Ariranha. This animal is said to attain a length of six feet; the colour is a lighter brown than in the smaller variety, and a white ring encircles the neck. This species may have given rise to the Mãe d'Agua, or water fairy; it bites terribly, and dogs fear to attack it when it is making its escape over the rocks. The otter has an extensive range in the Brazil, it is frequent upon the streams of the sea-board, and if the " main d'œuvre " were cheaper, its skins should reach the markets of Europe. The people of the São Francisco destroy it because it is so injurious to fish. It lives in families, tunnels into the river bank, and drives a breathing shaft (suspiro) to the surface. The hunter stops both holes for a time, and then opens the entrance, the inmates rush out to take the air, and then they are killed *ad libitum*. Often, also, they are shot in the rivulets, and their bodies are found floating after some hours. The skins are of a comparatively high price, I bought none under 2$000.

September 20.—Ember Day again. In the morning the men looked like turkey buzzards during a heavy shower : they were so benumbed that we had some difficulty in avoiding the snags and a dangerous sunken rock, said to be of silex.* After two hours' work we passed on the left bank the Paracatú de Seis dedos, which M. Gerber has located on the right. The pilots praised it for good water (rio bonito), but none could explain how it came to have six fingers. Near its mouth was a hamlet and clearing on the finely-wooded banks, and the creaking of the water-wheel spoke of molasses and rum.

One league below that point we halted for breakfast on the left

---

* The pilots called it Pedra de fogo (of fire), or de Espingarda (of the gun).

VOL. II. R

bank of the great Paracatú River.* Its dexter jaw showed a point
or shoal which drove it to the other side, the centre was garnished
with dangerous chevaux de frise of embedded timber, and the
course, bending like a Turkish scimitar, was painted with the
red Páu Jahú. The sides, despite their height, are flooded in the
wet season, and the sandy ground, mixed with humus and clay,
slopes to bottoms where the trees show a water-mark of eight feet.
There was little undergrowth, and the surface was strewed with
dead leaves : it was cut in all directions by tracks and paths ; the
cattle fled from us, and the ticks soon caused us to beat a retreat.

Yesterday we had seen but a single bark creeping up the right
bank. To-day we find two ajôjos anchored at the mouth of the
Paracatú. The owner, a stout, healthy man, whose appearance
spoke well for the climate, was taking provisions to Capão
Redondo, a " Garimpo," or small diamond-digging up-stream.
In former days hundreds of arrobas of gold were sent from this
valley ; he declared the bank-diggings to be now exhausted, but
the bed to be still rich. M. Halfeld tells us that in his day the
active and energetic riverines supplied flesh and cereals to the Lower
São Francisco, even as far as Joazeiro, distant nearly 700 miles.
Our informant stated that the staple industry of the land was
stock-breeding, although agriculture still thrives, and the fine
Maçapé soil will produce any quantity of fruits, especially
mangos. He ended by predicting that we should not reach
São Romão that night, as he himself would probably not have
done. Of course we resolved to give him a practical *démenti*,
and we now thought little of discouraging reports which had
begun at Rio de Janeiro, and which will end there.

After receiving this " formidable tributary," the São Francisco
widens and shallows. At 11 A.M. we passed on the left hand side
a remarkable bluff, the Ribeira da Martinha,† which drives the
course nearly due east. Before reaching it the land was low and

---

* Dr. Couto and other old writers prefer
Piracatu (pyra-catú), or good fish river,
opposed to Parahyba (Pira-ayba), the bad
fish river. This important stream drains
2° 30′ of latitude by 3° of longitude. Its
northern branch, the Rio Preto, breaks
like most of the great western influents,
through the frontier chain of Goyaz, the
Serra de Tauatinga, which sets off the great
northern versant, the Serra dos Pyrenéos.
The mouth is about 1500 palms wide, the

normal breadth of the stream is 600 feet ;
twenty-eight falls and rapids encumber the
bed, and it is navigable, after a fashion,
nearly 260 miles, to the Porto do Buriti.

† Or Ribanceira (bluff) da Martinha
(P. N. of a Moradora, the proprietrix).
The up-stream end is the Barreira da
Martinha proper ; the centre, Ribanceira
de Amancio José ; and the eastern, or
down-stream, Ribanceira da Martinha.

thickly-wooded like an old river bed, possibly that of the Para-catú.* The Barreira is the butt-end of a ridge cut off by the stream: the material is compact argile of many colours, white and brown, pink and yellow, surfaced with thin humus; it stands up stiffly to a height of some eighty feet, and at the base it has fallen into the usual slope. After a total length of some 440 yards it thins out into "Cánga," and terminates in woodland. Below it the bank became sandy, and showed the usual huts and improve-ments which argue the approach to a place of some importance.

Beyond this Barreira the river is a mass of shoals, sand-banks, and sand-bars, whilst the stream varies from 0·87 to 1·28 miles per hour. The "remanso," or sluggish current, is dreaded by barquemen, and usually the General Trade forms a troublesome head wind. For some hours the low dark clouds, dissolved by the cold breath of the north, which in this section promises a continuance of wet weather,† had indulged us with a slow, steady rainfall: it began at 10 A.M., and lasted, with rare intervals, till 4 P.M. An ajôjo is certainly not a pleasant place during the "Cashew Showers:" on the other hand the heavy discharge from above silenced the gale.

At 1·15 P.M. we grounded in the narrow channel of hard gravel between the left bank and the Ilha do Jatobá: the men were obliged to take foot,‡ i.e., to tumble in, and to shove us off. Here the total width of the river, including the island, is some 1600 fathoms, and wonderful to relate, M. Halfeld proposes to block up the western channel with "stakes and fascines." The Jatobá is the normal type, an elongated lozenge with the side angles shaved off, and outlines of sand in all directions, except where the bank is highest. At this season it is double: up-stream there is a small, well-wooded clay formation, which a long flat sand-bank connects with a similar and larger feature to the north-east, and the latter boasts of a few inhabitants. Further down there is a "Pedra Preta," black blocks and green Arindá shrubs, as on the Lower Rio das Velhas, which drives the stream almost at a right angle to the west. The next turn is to the

---

* The pilots denied this, but their reason was that they had never seen the stream here.

† The cause is the cold wind setting in after a few days of hot sun and still, damp air. The people call these showers, which are normal in August, Chuvas de Cajú (of the Cashew nut), a term evidently derived from the Indians. They declare that the wet season does not begin till November; but this year they are manifestly out of their reckoning.

‡ Tomar pé, to find deep water by wading.

north, and presently after thirty-six miles we reached our destination for the night.

São Romão, or to give it the name in full, " Villa Risonha de
Santo Antonio da Manga e de São Romão," (the riant (?) town of
St. Anthony of the Cattle-ford, and of Saint Romanus)—takes its
nom de baptême from the holy martyr, St. Romanus, who pre-
sides over the 9th of August, and who is, I believe, generally
ignored by the English faithful. Two Paulista explorers, the
cousins Mathias Cardoso and Manoel Francisco de Toledo,
having killed an Ouvidor-judge, fled with their families and
slaves to the Sertão do São Francisco. The date of their
journey is not positively known, but it is supposed to have been
between 1698 and 1707. They were driven upon the islet opposite the town, and having beaten off the Indians they settled here
for a time, and then resumed their voyage, finally establishing
themselves at Morrinhos and Salgado. Between 1712-13 the
Bishop of Pernambuco, hearing that the Indians were of peculiar
savagery, sent the Padre Antonio Mendes to catechize them.
Before 1720 S. Romão was a Julgado belonging to the Comarca
of Sabará. The district was presently subjected to the arrondissement of Paracatú, a city then newly settled, and distant 200
miles—only. On August 16, 1804, the Bishop D. José Joaquim
da Cunha sent its first parish-priest (parocho), the Rev. Feli-
cianno José de Oliveira. A chapel was dedicated to Santa Anna
and São Luis at a place above the confluence of the Japoré with
the São Francisco ; this was removed to S. Romão on his own
day and the invocation became Santo Antonio. The settlers
throve ; São Romão, in 1804 a freguezia, rose in 1831 to the
honours of township.

I shall describe at some length this God-forgotten place, not
for what it is, but for what it will be. Many travellers have mentioned it,* and most of those who have visited it left with the
worst impressions. The last was a naturalist sent down the
river by Professor Agassiz ; he got into trouble by carrying
weapons. There is absolutely no reason why the settlement
should be so miserable, the people so barbarous. A good building-site is close at hand, the surrounding country is admirably

---

* St. Hil. (I. ii. 428) regretted " de   as "le poteau surmonté d'une sphère."
n'avoir pu visiter la Justice de S. Rumão ;"   The Monsenhor Pizarro had previously
and he defines the symbol of a " justice "   given a detailed account of it.

fitted for agriculture, and the town is well placed for the carrying trade. The day I hope is not distant when some wayfarer shall pass through São Romão and find my description of the São Romanenses utterly obsolete.

Near the town the stream, nearly 1300 yards broad, runs to the north, and hugs the left bank; it is broken by the island of São Romão, about four miles long by 400 paces broad, densely wooded, uninhabited, and still private property. At the "port" one canoe was drawn up, and about half a dozen were in the water; the only "ship-yard" is on the top of the bank. Staked to the side was a fine barca flying the Imperial flag. The crew, including the pilot, numbered seven, and the tonnage was 4000 to 5000 "Rapaduras,"—20,000 to 25,000 pounds.

We swarmed up the steep bank, some thirty feet high and buttery with rain; the lower part was yellow clay much mixed with silt and sand above. On the summit appeared a remarkable feature, a line of six enormous Gamelleira figs,* like those described upon the Tocantins River. At the point where the stream deflects a little to the east, a decayed stump shows that there was a seventh, and two of the giants are likely soon to be washed away. The pair to the south raise their majestic crowns of stiff and burnished ovoid leaves, and overhang the stream with an admirable umbrella of verdure. The trunks, instead of being as usual, low, thick columns, are bundles of compacted trees, five or six feet high, and of the horizontally projected branches, one, not the smallest, measured 100 feet. The birds had settled in colonies amongst the boughs, and but few epiphytes had sprung from the bark. In one of the two which front the landing-place time had dug a chamber used as a dwelling-place; the idea must have originated in Central Africa, where the bulbous Calabash acts alternately home and water-cistern.

Immediately beyond this ridge with its colossal growth, the land droops towards a bottom flooded during the rains, and thinly covered with bush; it must be a hot-bed of miasma during the retreat of the waters, and the sun must raise it well to the

---

* The Brazilians term the Gamelleira either Preta or Branca, chiefly from the colour of the bark. Koster (ii. 11) makes the latter useless, and the former distil, after incision, a gummy juice, which is taken internally for dropsy and cutaneous diseases. According to the System, the acrid milk of the white fig (Figueira branca, or Ficus doliaria) is an anthelmintic, but it adds that many other figs have the same properties.

level of the local face. The swamp is subtended by a rise corresponding with the lay of the ridge running parallel with the
stream, and facing the east. Here is the Rua do Alecrim, consisting of a dismantled hut on one side, faced by seven poor
tenements, of which one, by affecting a square box as an upper
story, ambitionizes the title of "Sobradinho." Beyond this
thoroughfare of flowery name, and lying side by side with it, was
the Rua do Fogo, a higher and drier site. Here we counted
fifty-four tenements, mostly with roofs of coarse tiles and mud
and wicker-work walls,* slightly washed with Tauatinga; the
large compounds are either railed or enclosed with pisé, coped
with thatch. The most pretentious show attempts at orna-
mentation, white scrolls of plaster on azure ground, doors striped
with blue, and windows with small lattices instead of the shutter
or the cotton cloth. Amongst them were three Vendas, the
main of whose occupation is to sell spirits; and the blacksmith
in his leathern apron, suggested the village Vulcan of Negroland.
The wealthier houses had wooden steps leading to the raised
floors, the poorer logs of wood above the level of the puddly
thoroughfare, by courtesy called a street. To the south some of
the tenements were propped up with stays and others were in
ruins; not a few had a closed room attached to the unwalled
tile-roof which the Tupys called Copiar or Gupiára, and sometimes Agua furtada. In this place the traveller is allowed to
swing his hammock and to cook his meals.

Going northwards we passed the Quartel, or barracks, hung
inside with carbines, and tenanted by eight soldiers, who on paper
appear as a battalion. These black-brown men in Kepis and holland blouses looked somewhat more surly, as in duty bound, than
the rest of the citizens; they eyed our patent leather waist belts
curiously, but they did not interfere with us. Beyond the Quartel
was the Lago da Cadèa, a tiled roof and an open scantling,
suggestively representing the future prison. João de Barro had
derisively built his domicile upon the cross beams, and upon not
a few of the wooden crosses profusely scattered about the settlement.

Beyond the northern end of the Rua do Fogo, and surrounded
by bush, was the old Rosario Church, definitively broken down.

---

* The citizens declare that they have no stone, when the river bed is a quarry.

Turning to the left we ascended the Rua Direita, an embryo thoroughfare which numbers twelve houses, including a farrier's. It rises gently from the river to a cemetery, denoted by a cross, from which half the instruments of the Passion had been abstracted. This village of the dead was fronted by a support of rough stones, while the rest was wholly unwalled; the surface was cumbered with timber, and littered with graves which lacked monuments. In the centre of High Street was the square of the new Rosario, a white-washed temple with three shutters, a very model of meanness.

To the west of this Rosario is the Rua da Boa Vista, the aristocratic quarter, numbering thirty houses; it commands a pretty view of the stream, the islet, the reaches above and below, and the low blue hills on the Bahian or Eastern side. I sent in a card to the Delegate, Sr. João Carlos Oliveira e Sá. He had probably never seen that civilised instrument, for he left us in the rain till a friend beckoned to us from the window to come in, and after eyeing the pasteboard much as a crow inspects a dubious marrow-bone, he returned it to me with a little weary sigh. Unwilling to accept defeat, I produced my Portaria or Imperial passport: he glanced over it and restored it in dead silence. My desire for information was likely to catch cold, when fortunately a decently dressed man walked in, and did not prove so chillingly uncommunicative. I told my tale all down the river, where men agreed in giving a good name to the Delegate; it is therefore only fair to suppose that he was exceptionally suffering under the influences of Saõ Romão.

Resuming our walk after this episode, to the south of the Boa Vista we found a second church, the Nª Sª da Abbadia; it boasted of the usual white-washed and two-windowed face, wearing a mutilated, noseless look. To the west or inland of this line are a few straggling huts, whose enclosures are hedged with the organ cactus. Here is the highest and healthiest ground, where the villa should be built; unfortunately it is too far from the business quarter, the river side. Therefore, as in our West African "convict stations," men will not move; they would rather see the floods walk into their windows. At times exceptional inundations put them all to flight. In 1838 the water rose in places five feet above the floors, and in 1843 the lowest street was nine feet under water.

The trees scattered about the town show the excellence of the soil. The Almecega or Mastich grows to the largest size. Nowhere in the Brazil have I seen finer tamarinds, the natural corrective of liver complaints. The Imbuzeiro (Spondias tuberosa) is a magnificent rounded growth; the juice of this myrobalan, tempered with milk and sugar, makes the favourite "imbusada" of Pernambuco and Bahia. There is an abundance of fruit, limes and oranges, papaws and bananas. In the higher levels, where the thorny mimosa and acacia flourish, cotton grows taller than the houses, and in the lower parts sugar-cane flourishes. Behind and above the town the vegetation is that of the campo, excellent for cattle-breeding. In the streets we saw a few small horses, the goats and poultry were tolerable, the pigs and sheep much wanted breeding. An idea of the popular apathy may be formed from the fact that whilst the river flowing before their doors produces the best of fish, and while salt may be brought from a few leagues, if indeed it cannot be washed from the ground, the townspeople eat the hard, dry, and graveolent "bacalhão," or codfish, brought in driblets from Newfoundland.

In 1822 Pizarro gave to S. Romão 200 houses and 1300 souls. Gardner, in 1840, reduced the number to "not above 1000 inhabitants." M. Halfeld (Relatorio, pp. 27—28) numbers 220 houses and 800 souls. The Almanak (1864) assigns to the municipality 8676 inhabitants, 723 voters, and 17 electors. According to my informants, in 1867 the houses, or rather hovels, amounted to 200, and the tenants to 450. When Saint Hilaire wrote,[*] the "village of S. Rumão" monopolised the carrying trade of salt between the river and Santa Lusia of Goyaz: it also exported a considerable number of hides. In those days it had its "richards," Major. Theophilo de Salles Peixoto, the late Lieutenant-Colonel Ernesto Natalista Amaral de Castro, the Capitão José Jacob da Silva Silveira, and others. A relic of the good times is the vicar, Padre Antonio Ferreira de Caires : hearing that he was a "curioso," [†] rich in local information, I called upon him; unfortunately he was at his Fazenda, and the Sacristan assured me that there was no such thing as a Livro do Tombo, or parish register.

---

[*] III. i. 216 and 359.

[†] St. Hil. (III. i. 104) remarks, "le mot curioso répond dans notre langue, à celui d'amateur ; mais il a un sens moins limité." In the Brazil, "afficcionado" is an amateur, and curioso also includes the non-professional expert.

About ten years ago the diamond-diggings at Santafé * and on the Paracatú River have caused a small exodus, hence partly the falling off of the census, and the exceptional number of old men, women and children. The fevers have greatly increased; we could read ague legibly written in the yellow skins, emaciated frames, and listless countenances of the people who suffer terribly during the retreat of the waters between May and July. The fomentors are, as usual, poor diet, excess in drink and debauchery, late hours and extreme filth, not of person, but of habitation. In this point they seem to have borrowed from the indigenes of the land, who bathed several times during the day, but allowed themselves to be littered out of their " carbets " (wigwams) by mountainous collections of offal.

The São Romanenses did not affect me pleasantly. I did not see a single white skin amongst them; they were a "regular ranch " of bodes † and cabras,‡ caboclos and negros. The lower orders—if there be any in this land of perfect equality, practical as well as theoretical—were in rags; the wealthier were dressed in European style, "boiled " shirts and velvet waistcoats, but their lank hair and flat faces recalled the original "Indian." They are devout, as the wooden crosses of squared scantling affixed to the walls show: scant of civility, they have barely energy enough to gather in groups at the doors and windows, the men to prospect, the women to giggle at the passing stranger. Some of the older blacks plied the primitive spinning wheel, but the hammock, despite the raw chilly weather, was in more general requisition.

São Romão, I have said, is well situated for trade. A good road, some sixty leagues long, runs up the valley of the Rio Preto, the northern branch of the Paracatú. A little beyond the settle-

---

* This place was described to me as a little village, with the rudiments of a church in the municipality of S. Romão.

† In the Brazil, " bode," or he-goat, is a slang term for a mulatto.

‡ St. Hil. (III. ii. 272) makes the Cabra (she-goat) a mixed breed between the Red Man and the Mulatto, and synonymous with the Peruvian "Chino." Here it is applied as a general term to those who are neither black nor white; addressed to a man, it is grossly insulting, but I have heard a boatman facetiously apply it to himself.

The wild men, I have said, gave the name macáco da terra ("country monkeys") to the African. Yet travellers have stated that they were fond of such monkey's meat, and all agree that their women had " un goût très-vif pour les nègres." Some have advised, by way of saving the "Red Man," to mix his blood with the black. This is indeed unanthropological. There is no need to preserve a savage and inferior race, when its lands are wanted by a higher development; and, in this case, the artificial would be worse than either of the natural races.

ment, called Os Arrependidos, it crosses the Serra of Goyaz, which here offers no difficulties. Thence it bends north to the old Villa dos Couros, now Villa Formosa da Imperatriz. Here there is communication with the great Tocantins tributary of the Amazons, viâ the Rio das Almas, the Corumbá,* and the Rio Paranan, which bears canoes.

At night-fall we returned to the Brig Eliza, lighted the fire, drew down the awning, and kept out as much of the drifting rain and cold shifting wind as possible. It was not easy to sleep for the Babel of sounds: here the dark hours are apparently the time

> When man must drink and woman must scold.

The Samba and Pagode seemed to rage in concert with the elements, the twanging of instruments and the harsh voices screaming a truly African chaunt, suggested an orgie at Unyanguruwwe. Evidently much reform is here wanted, and it will come in the form of a steamer.

---

* Men of information at Januaria, as well as São Romão, mentioned the Corumbá stream and village. I hope that they have not confounded it with another Corumbá, the great northern influent of the Southern Parnahyba, or Paranahyba. Usually, traders embark at the Villa das Flores, or the Paranan, or l'arana (St. Hil., Parannan), the eastern head water of the Tocantins. Castelnau (ii. 196) declares, "Le Parana peut être descendu en canot jusqu'au Pará." My informants described the river as very "bravo" above S. João da Palma, at the junction of the Araguaya, or great western fork, and some have spent six months in ascending it. Hence they say goods worth 0$700 at Pará on the seaboard, sell at the Villa das Flores for 5$000, and a bottle of wine bought for 0$500 fetches 4$000.

CHAPTER XVII.

FROM SÃO ROMÃO TO JANUARIA.

SECOND TRAVESSIA, 26½ LEAGUES.

STEAM-BOAT ISLANDS. — THE URACUIA RIVER. — THE VILLAGE AS PEDRAS DOS ANGICOS.—QUIXABA-TREES.—THE RIO PARDO.—APPROACH TO THE CITY OF JANUARIA.—VEGETATION AT VILLAGE OF Nª Sª DA CONCEIÇÃO DAS PEDRAS DE MARIA DA CRUZ. — REACH THE PORTO DO BREJO DO SALGADO.—THE PRESENT CITY OF JANUARIA.—ITS HISTORY AND PRESENT STATE : DANGER OF BEING SWEPT AWAY.—RECEPTION.—PETTY LARCENY. —CIVILITY OF SR. MANOEL CAETANO DE SOUZA SILVA.—THE PEQUIZEIRO.— MISSIONARIES AND MISSIONERS.—WALK TO THE BREJO DO SALGADO.—ITS ACTUAL STATE.—ROMANTIC LEGEND OF THE PEOPLE'S DESCENT.

> —— outro se engrossa
> De São Francisco, com que o mar se adoça.
> *Caramurú.*

Saturday, September 21, 1867.—The ceaseless drenching rain reduced the men to a manner of torpid hybernating state. After a start under difficulties we threaded the long line of shoals and islets. In some places as many as six sand-bars were in sight; all were of finely sifted material without the gravel of the Coróas in the Rio das Velhas. Passing the Roça do Porto Alegre and other clearings* we came to the first of many features which will last till we reach Remanso. It is a long, narrow bank of stiff sand, sharp fore and aft, and shaped like a river steamer in the United States; in places bushes formed the paddle-boxes, and strata the lines of caulking. We called them Steamboat Islands. The vegetation was generally of yellow-green, showing want of fat humus.

A head wind, driving misty blue clouds, drove us to the right

---

* M. Halfeld calls this pretty spot "Povaodo," or Village do Porto Alegre. His villages are mostly termed by the pilots "fazendas" here, meaning tracts cultivated by a number of settlers. The next was the Barra do Brandão, a long, low clearing on the right. The bank also shows improvement, but it must be extensively inundated.

bank. The mercury showed 71° F., but we trembled with cold; such is the effect of air in motion, which seems to despise a sun nearly overhead. Furling the awning we made a good "lick" to the entrance of the great Uracuia stream.* The right bank was wooded with a truly magnificent vegetation; it showed for the first time the Carahyba de flôr roxa, a tall tree with lilac-coloured blossoms, which will presently become common, and here we observed that every great western influent brings down with its waters a new growth. The mouth of the Uracuia is about 315 feet broad, and behind the woodland the low bank of yellow clay bears only thicket.

A white-washed house, now a novelty, appeared on the Bahian side, and presently we took the left of the "Ilha do Afundá," which is interpreted to mean that the water is deep; its material is a pure yellow and easily melted sand. The upper part of the islet is well clad with various growths. We then ran in mid stream past the second Afundá,† and, after eleven hours of hard and comfortless, dull and eventless work, we came to anchor at a praia on the left bank.

September 22.—The north wind which had raged all night blew itself out about dawn, and we set out with alacrity. The banks were dead flats, in places splendid with tall sugar-cane and tree-cotton, but generally showing second growth where magnificent forests had been. Passing on the left the little Acary tributary,‡ we found another high white bluff about a mile long,

---

* Also written Aracuia, which means, say the people, "fartura," or plenty, alluding to the fertility of its upper banks. It drains the southern slopes of the Chapadão (big plateau) do Uracuia, and is divided from the Paracatú Valley by the Serra do Rio Preto. Its area of drainage is latitudinally 2°, and longitudinally, 1° 30'. The stream, though broken by many rapids, is navigable for rafts and canoes as far as Campo Grande, 102 miles from the mouth.

† M. Halfeld calls this Ilha das Carahibas, and elsewhere spells the word Caraibas.

‡ According to the pilots, the true Acary is further down-stream.

The name of this fish (a loricaria of many species) is also written Acari and Acarehy. The Tupy name was Acará, with the terminations, -apuã, -assú, -tinga, and -peixuna. In this river we find the A. de Pedra, A. de Casca (or Cascudo), A. de

Lama, and A. de Espinho. It is the Juacaná which Marcgref saw at Pernambuco and the Cachimbo, or Cachimbáo, of Ilheos. A species is probably the Acará bandeira (Mesonanta insignis, Gunther) of which Mr. Bates (ii. 140) gives an illustration. It grunts like the Mandim, and the pilots say that when eating the mud and weeds from the canoe-bottom it rubs its bluff head against the wood, and produces the peculiar sound. They declare that it lives in holes along the banks; many deem it poisonous, and it is generally thrown away on account of the trouble of cooking it. Both the black and the white kinds have hard, spiny skins, with longitudinal lines of points, highly dangerous dorsal fins, and hooks above the caudal fins. Another well-known loricated and "grunting" fish is the "Cascudo," which abounds in the rivers of the interior. The people praise it, but I found the white meat soft, tasteless, and full of thorns.

and divided into two sections, the Barreira do Indio (do Honorio, M. Halfeld), and the Barreira Alta. Here we remarked the abundance of the Angico Preto acacia, on this part of the stream an ugly tufted tree; its timber is too dry for use, but the gum is given to consumptive patients; the bark abounds in tannin, and the ashes in potash.

About noon, entering S. Lat. 16°, we came to a new feature, "as Pedras" (dos Angicos), and we landed on the right bank to inspect it. Here a wall some forty-two feet projects from a shallow sag, fronts to west and drives the river to the north-west; it runs nearly a mile down-stream, and is found in a hollow, to the north-east of the little village. The outcrop is evidently the base of a bulge of ground observed on the east. The floor near the water was a hard bluish limestone, effervescing kindly under muriatic acid; above it was a stratum of laminated, friable clay-shale, capped by a bluer calcaire, with dislocations, broken blocks and horizontal bands, varying in thickness from three inches to three and a half feet. Water drops appeared upon the exposed slabs on the summit, which is always six feet above water, and it was revetted in parts with iron clay, whilst to one block is attached a small portion of quartz conglomerate. This is one of the many places which will supply admirable hydraulic cement.

The bank and the village showed a scatter of noble Quixabeira trees, huge bouquets of verdure, whose aromatic flowers and perfumed shade attracted hosts of bees.* The little chapel of São José, the brago of the place, is about nine feet above the floods, and yet boasts of a stone foundation. Walking up the sandy street, perpendicular to the stream and showing traces of pavement and bottom, we found the usual hollow parallel with the ridge and periodically under water. In the loose free soil, cotton, essentially a sun-plant, grows neglected, fifteen feet high, and the castor shrub twenty. In "Water-Street," whose houses and ranches were superior to those of São Romão, appeared three Vendas, with men sitting outside the counters, or using them as card-tables. Two shoe-makers and a dry-goods store seemed

---

* Also written Quichabeira, one of the Sapotaceæ, a tree which covers large tracts on the Rio de São Francisco, above and below the Great Rapids. It resembles the Zizyphus, produces an edible berry, and affords a fine shelter for cattle. The System mentions the Quijaba and the Catinga branca (here called Catinga do Porco), as leguminosæ abundant in stryphnum.

to be doing a thriving business. South of the village were three canoes selling fine water-melons. Under an old angico hung with dark moss stood the frame-work of a barca, very solidly built of cedar.* On the northern waterside washerwomen were plying their trade, whilst their bantlings swam about, or played on carts with wheels of one piece some eighteen inches high. Horses and mules were resting after being ferried across the river, and a little caravan appeared upon the opposite bank. This at once explained the prosperity and the civility of the place. The Delegado had at once sent to procure lodgings for us. It communicates with the Acary River, where, at the distance of ten leagues from the mouth, are diamond diggings. São José das Pedras dos Angicos now numbers ninety-five houses and a population of 500 souls; we left it convinced that it has good things in store for it.

Resuming our way in an exceedingly hot sun, we presently passed on the left the Barra do Acary,† which breaks its way through a sandy Corôa. Below the mouth are three "steamboat islets" of the same name, and the Ilha do Barro Alto, a wooded holm. Then came the mouth of the Rio Pardo‡ about 140 feet broad; here began the magnificent bosquets (capões) of cedar, vinhatico and balsamo (a myrospermum) found in every river and rill. Opposite this point we nighted. The air had become "muggy," damp and tropical, like Western India, and, for the first time after leaving Rio de Janeiro, we began to disuse the blanket. I need hardly say that we recalled to mind with regret the charming accidents of the Old Squaws' Stream; the clear limpid air rich in oxygen, the splendid forest scenery of the wild banks, the music of birds and beasts, even the song of the rapid and the fall, and the cheerfulness of nature in general.

September 23.—After an hour's paddling appeared the Barro Alto, a high bank of white clay on the right side where the bed is embayed. We landed a little below it, at the mouth of a Córrego known as the Braúna; it puts forth an under water ridge, extending from south-west to north-east, and ending in what the

---

* The smaller barcas here cost 200$000; those of moderate size, 500$000; and the largest (45 × 14 feet), 1:600$000.

† This Acary stream is not mentioned by M. Halfeld, nor is it in M. Gerber's map. The mouth is about 150 feet broad; the high left bank of yellow clay is garnished with grass and low trees, and on the opposite side the vegetation rolls almost to the water's edge.

‡ The Rio Pardo drains the Southern slopes of the Chapadão de Santa Maria. Its length is 1° 30', but it is navigable for canoes only twelve leagues.

crew called a "batida," a low sand-bank flaked with mud. Here we found the true diamantine "formação," the Cattivo, the Siri-cória, in fact all the symptoms, but not the gem. These evidences appear at scattered intervals; the people declare them to be apparently arbitrary, that is to say, the source from which they come has not been investigated.

Beyond this point the stream showed on the left heaps of stone; on the right, thrown out in a relief of bare or blurred line against the blue sky, rose the Serra do Brejo,* which from this point appeared a slope, a broken saddle-back and a lump swelling above the trees and sands. To starboard we passed the Riacho do Peixe, near whose mouth is the small Fazenda of a German settler, Doctor Otto Karl Wilhelm Wageman; further down is the Riacho dos Pandéiros, † whose winding course admits canoes for some five leagues; nearly opposite it the Riacho do Mangahy. The northern limit of the S. Romão municipality showed at the mouth a clump of magnificent trees, and, a little below, a large bed of Cascalho. In front rose the remarkable table-mountain known as the Itabirassába, corrupted to Piassaba; ‡ the word is translated "Monte de Fogo." We are evidently approaching an important place; the primitive vegetation disappears, nature begins to look drilled and disciplined, there are kilns, the huts are whitewashed and tiled, and the people offer fish for sale.

After a few features of no great importance,§ we reached a place which we had long seen in the shape of a lumpy line on the right bank, and we ascended a flight of steps cut out by the retiring waters. Here there is a bed of the finest white limestone some 10 feet thick. The place is known as Nª Sª da Conceição das Pedras de Maria da Cruz, and the first edition of the little chapel was built about 1725 by the Paulista Miguel Domingos, after the defeat at the Rio das Mortes in 1708. The mound that

---

* Brejo do Salgado, which we shall presently visit.

† The Pandeiro is a gipsy kind of instrument—a bow and calabash, derived from Africa. The wild men, as might be expected, greatly enjoyed its music; hence the name has been given to many places in the backwoods. Near this Pandeiros a man lately died said to be 107 years old.

‡ The head man declared, and with truth, that this saddle-back was the Serra do Brejo to the north-east of the former.

§ At noon we passed three islets near the Pernam. bank, and an hour afterwards we saw the Ilha das Pedras, a sandy and shrubby formation, with small clearings and barking dogs. On the bank opposite to it were ledges of Cánga; beyond it appeared, upon a base of light-coloured stiff clay, a wall of ferruginous argile, black and red, pudding'd with pebbles, varying in thickness from one to nine feet, and thinning off from north to south.

supports it had an old and remarkably good "ladrilho," or tile pavement, which commands a magnificent view. The river, broken by sand-bar and island, sweeps nobly from south-west to west, and at this point trends nearly due north. The wavy bank in front is clothed with immense trees, and about eight miles distant the western horizon is closed by the quaint-shaped Itabirassába, towering high above its chain.

The population of the little hamlet was scattered about in low huts of wattle and dab, thatched or tiled. Some of the women working the old pillow-lace were hardly clad with decency, and on the bank was a yellow girl with unveiled bosom, as if she had been in the Bight of Biafra. All, however, are more or less tinged, and here, as elsewhere, the brown buff simulates dress. Goats wandered about the bush, and seemed to enjoy the shiny, succulent leaves of the gigantic croton shrub, which grows to an abnormal size. This Jatropha Curcas* of many names supplied the physic nut for the Lisbon lamps, and thus for a time preserved certain of the Cape Verde Islands from starvation. It has an extensive range. I have seen it at altitudes between the seaboard and 3000 feet. The Guinea negroes administer with excellent effect the green seeds together with the pulp; the dose is, I believe, a quarter of the nut boiled in water, which is drunk.[†] Half-dram doses used to be given in the Brazil, but the "physic-nut" is now neglected, as it is a dangerous, and has proved a fatal, drastic. As the rains begin, everywhere sprouts a pretty pink flower somewhat like a primrose, solitary, and capping a thin and fragile stalk about one foot high; the people call it " Cebolla brava," or the poison-onion, and declare that cattle will not touch it.[‡] On the higher banks, where the floods do not extend, grows the Solanaceous Juá, still bearing the last year's blackened berries; the organ cactus; the Pitombeira (Sapindus edulis), a large tree with an edible fruit; the Pinguí, here called Imbárú, and the shady Aroeira de Minas, also known as Capicurú. The latter resembles the Melia Azadirachta of Hindostan, but the leaves are not bitter.

* Here it is popularly known as "pinheiro de purga," or "pinhão do Paraguay." The Tupy dictionaries give "Mandubi-guaçu" (the great ground-nut, or Arachis), a mixture of African and American terms. Labat has "Medicinier" or "pignon d'Inde," and when describing its effects, he offers sensible advice to tra-vellers—namely, not to eat unknown fruits which birds refuse.

† In Africa the unripe pulp, duly prepared, is, I believe, also taken as a strong medicine.

‡ In other parts cattle are, they say, poisoned by it.

As we advanced, the long reach, running nearly due north, bent to the north-east and showed in the far distance a whitewashed chapel and three large double-storied houses. On the left was the Ilha do Barro Alto, a long "steamboat island." We were obliged to round the large, flat sand-bars before we could make the Porto do Brejo do Salgado, the channel above the town not admitting even our raft. This is the most important place on the Upper Rio de São Francisco, and its only rival is Joazeiro, distant 190 leagues down-stream. The site is a dead flat on the left bank, distant four to five miles from the Serra do Brejo, a broken line to the north-west and north. A certain Maciel, of whom more presently, here built a chapel of brick and lime, the people assembled round it, and the Bishop of Pernambuco sent a curate, the Padre Custodio Vieira Leite. The principal settlement however was inland, at the base of the hills, and this povoação or hamlet on the river side took the name of Port of the Salt Swamp, abbreviated to "O Salgado," "the Salted," and this the people insist upon retaining. Of course the two settlements were rivals and enemies. In 1833 the Port became the "Villa da Januaria," christened after the sister of the reigning Emperor; in 1837 the honour was transferred to the Brejo inland; in 1846 it was re-transferred to the Port; in 1849 it again moved to the Brejo; and finally, in 1853, it settled upon the Stream.* The water side objects to the hill side that it is too far from the seat of trade; the hill side retorts that at least it is in no danger of seeing even its saints swept into the river. The municipality, which is large, and contains a considerable extent of uncultivated land, numbers five districts, namely, the City, the Brejo, Mocambo,† Morrinhos, São João da Missão, and Japoré, the latter distant about 20 leagues.

We had to fight hard against the strong current, which now shows signs of incipient flooding. We passed the tall sobrado of the Capitão José Eleutherio da Souza, fronted by a dozen stunted and wind-wrung palms, and a slope of Capim ussú (the big grass) stretching down to the stream. The herbage is of a metallic green, like young Paddy. It is not destroyed by the

---

* According to the Almanak, the parish was created by Royal resolution of January 2, 1811, and the Port was made the Chef-lieu by Provincial Law No. 288 of March 12, 1846.

† Much of the land in the Mocambo has no proprietor, and its admirable fertility suggests colonisation. Here the best land is worth at most 500 $ 000 per league, not, however, a square league, which would be nine geographical miles, but half a league each way. In these matters there is no regulation, and each man adopts his own system.

inundations, and cattle will eat it. The Port is formed at this season by two sand-bars fronting the left bank. It has been proposed to remove them, but the best authorities are agreed that they defend the side, to which a strong flood swings during the rains. The river is now upwards of 3000 feet in breadth, and the weight of water does far more damage than the superficial washing. It will not be easy to save the place; about twenty years ago half of the Rua do Commercio, the "water street," became the stream-bed. A few stakes have been planted to act as grains, and a stockade of tree trunks defends the sloping bank of sandy clay, perilously near whose edge runs a line of low, whitewashed, and red-tiled tenements. The principal danger is above the city, where a small channel admits a vast influx of flood water. Here it would be easy to throw up one of those levées with which we dyked the Indus near Hyderabad.*

We found in port a number of canoes and eight barcas made fast to the usual poles. The praia, as the bank is here called, at once recalled to mind the African market, and the monotonous chaunt of the negroes measuring beans did not diminish the resemblance to the scenes of distant Zanzibar. Women, now far more numerous than men, washed at the water-side, or carried their pots to and fro; the boys, more than half-nude, squatted on the sand-bars on tree trunks, or in their dug-outs, bobbing for daily bread. The dark boatmen, clad in the sleeveless waistcoat (Jalé or Camisola), and the cotton-kilt (Sayote) of the Guinea Coast, stroll about or lie stretched upon the slopes playing with the splendid and majestic Aráras,† which they have brought from down-stream, and whose plumes glittered in the sunshine. On the more level ground were planted seven shed huts of poles, mats, and hides. Here the merchant who disdains to hire a house exchanges his salt and cloth for provisions and supplies.

---

* In making these levées, it is well to dig a trench, and carefully to remove tree roots, and everything that can assist percolation. The dyke should have a base of 3 : 1.

† Ará, I have remarked, is a parroquet, or parrot; the augmentation ará-ará contracted to arárá in the large psittacus. It is regrettable that we have not adopted this pretty onomatopœia, instead of the grotesque half-bred Spanish macaw, and vulgarized the scientific "Arainæ."

The common wild varieties here are the Araruna (Ararauna) and the Arary, also termed Canindé, or Arara Azul. The former (Psittacus hyacinthinus), as its name denotes, a black, or rather a dark-purple bird, of smaller size than usual; it flies in pairs high, with loud screams, like the parroquet. The Arary (Psittacus Ararauna) is the well-known and magnificent bird, with a coat of the brightest blue, and a golden waistcoat. St. Hil. (I. ii. 376) notices the error of Marcgraf, who gave the name Ararauna, which means black or dark macaw, to the wrong bird.

When we had slipped into place I sent up my card and introductions to Lieutenant-Colonel Manoel Caetano de Souza Silva. Januaria showed her civilisation by crowding to inspect us with extreme avidity. A very drunken youth, with teeth chipped into feline shape—here fashionable—addressed Agostinho as "moleque"—small slave boy—very offensive to a big slave boy, and a "row" of the mildest nature ensued. Another stole an "Engineer's Pocket-book," and offered it for sale to a Portuguese, who at once returned it to us. The police authorities took no notice of the theft, perhaps because the robber was half silly with liquor, and consoled us with the intelligence that we might expect to be extensively plundered down-stream. This, however, was not the case ; Januaria was the only place where anything of the kind was attempted.

We were soon rescued from the situation by Sr. Manoel Caetano, who, accompanied by some friends, invited us to inspect the city. I greatly enjoyed the view from the bank summit. To the west the purpling hills were faint as clouds floating upon a sea of ruddy haze, the last effort of day. In front lay the River Valley, at least twelve miles broad, and suggesting a vast expanse of water during the floods. About two leagues distant rose the Morro do Chapéo, curiously shaped like a Phrygian cap ; it is an outlier of a long broken wall extending from north-east to south-west as far as we can see. This Serra dos Geraes de São Felipe is exceptionally rich, and supplies the river with lard, tobacco, and maize-flour. Its remarkable points are the Urubú peak, from this point a regular pyramid, the Serra das Figuras, the table-shaped Morro da Boa Vista, and the three round heads known as the Tres Irmãos.

The Nª Sª das Dôres is rather a chapel than a church, and at times, they say, fish have been caught in it; the building is fronted by a tall cross, enclosed in a dwarf square of short wall. At the other end of the settlement is a Nª Sª do Rosario blown down by the wind, and still unrepaired. The streets are floored with sand, and in places there are strips of trottoir, slabs of the fine blue limestone from the Pedras dos Angicos. Trees require a soil less lean; each house has its "compound," walled or staked round, but the largest growth is the papaw, and a palm here called the "Gariroba; " * it is a tall, dull-brown stick bear-

---

* Or Guariroba (Cocos oleracea, Mart.), a palm commonly found in the stunted growths of the Sertão.

ing small ragged fronds and a raceme of edible fruit about the
size of an egg. The thoroughfares are straight, but as usual too
narrow; their names are carefully inscribed upon the corners,
showing that the Camara does its duty, and the tenements are
numbered. The Praça das Dôres contains the jail, with barred
windows, where guards and sentinels loll and loaf, and near it
the humble Guildhall. A hospital is much wanted; we met in
the streets many cripples.

The total of houses may be upwards of 700, of which at least a
fifth are Vendas. In 1860, the famine-year of Bahia, the population numbered 6000 souls; five years ago it declined to 4000,
and now it may be 5000, slaves included. For some time past
the serviles have been traded off to Rio de Janeiro, and, only
lately, thirty head were sent down country. The city is supported by brokerage and the carrying trade. The Quattro-Maôs*
of the backwoods bring in a very little cotton, a quantity of sugar
and rum, excellent tobacco and provisions, especially hill-rice
and manioc, which flourish on the table-land beyond the riverine
valley. Fine canoes of the best Vinhatico and Tamboril,† forty
feet long here, cost 100$000, and are sent down-stream, where
large trunks are rare. The imports are chiefly viâ Joazeiro,
which the people place at a distance of 220 to 240, instead of 190
leagues; they are chiefly dry goods and salt. Those who have
not visited the inner Brazil will hardly imagine how necessary to
prosperity is this condiment. It must be given to all domestic
animals, cattle, mules, and pigs; they convert into "licks" every
place likely to supply the want, and even crunch bones to find it.
Without it they languish and die; in fact, here the desert may
be defined as a place where salt is not. A popular succedaneum
is oil and gunpowder, and even this is found better than nothing.
In 1852 the mule load of eight arrobas from Rio de Janeiro (200
leagues) ‡ viâ Diamantina, paid 45$000; it has now risen to
15$000 or 16$000 per arroba, nearly three times the price.
Consequently the capital sends only "notions" and "objects of
luxury." Bahia (186 leagues) adds hides and salt, pottery,
ammunition, and iron-ware; the price of conveyance varies from
12$000 to 14$000 per thirty-two pounds. Goyaz, like the

---

* Quadrumana : here the word is used
in the sense of Caipíra, country bumpkin.
  † A large leguminous tree. St. Hil.
(I. ii. 331) writes the word "tamburi,"
according to the pronunciation of the

Sertanejos.
  ‡ The distances are those given to me
by my friends at Januaria. They made
Diamantina 70 leagues distant, and Lençoës
76 to 80.

Geraes * lands on both sides of the river, supplies stock and provisions, " doces," cheese, and a little coffee and cotton ; some of them produce a small quantity of wheat. "Colossal fortunes," says the Almanak, " are rare," but there are men worth upwards of £4000, and money here breeds safely 24—36 per cent. per annum.

Our host was a distinguished "Liberal," who prefers politics to trade or farming ; he is made well known throughout the country by a greater generosity than is usual. He offered us the novelties of absinthe and cognac, he compelled us to sup with him, and he placed his house at our disposal. For liberty's sake I preferred the raft, also to escape from the screams of the children, which, throughout the Brazil, form the terribly persistent music of the home. The mothers, I presume, physically enjoy being noisy by proxy, the fathers do not object, and thus the musicians are never punished. Indeed you are considered a " brute " if you object to losing a night's rest by a performance, which could be settled in a second. The only place where the shriek of woman and the scream of babe are silent is, I believe, the Island of Madeira.

Sr. Manoel Caetano gave an invitation to visit him at his fazenda, where he intended to sleep, and promised to send animals at daybreak on the morrow, but apparently the light at Januaria dawns after 9 A.M. We, therefore, set out on foot under guidance of Sr. Candido José de Senna, ex-Professor of First Letters. The path led to the north across an inundated flat, which appears likely to disappear, and a line of mist showed the Corrégo Secco, that requires the levée. During the rains it is a flood, now it retains water-pits (poçoẽs) frequented by washerwomen. Ahead, and a little to the left, lay the table mountain, up which men have ridden ; at its foot is the fazenda of the Capitão Bertoldo José Pimenta, and near the summit, they say, is a natural well.

After walking a mile we made rising ground, and exchanged white sand for ruddy soil rich with humus. Here even the floods of 1792-93, which rose thirty-eight and a half feet above the mean level of the stream, did not extend. In 1843 there was another inundation, when a Surubim was caught in the church, followed

---

* In these parts the Geraes are generally named after their streams; *e.g.*, Geraes das Palmeiras, do Borachudo, &c.

by a third in 1855. In 1857 the citizens took refuge here, and
spent several days in picnics and jollity. It is called the Piquizeiro,* from the abundance in former times of the wild Caryocar
tree, and it will probably become the left bank of the Rio de São
Francisco. Evidently this is, even now, the fittest place for the
settlement, which a line of wooden rails would easily connect
with the Port; the air is cooler and healthier, water abounds,
there is plenty of building-ground, and the soil behind it is loose
and red, excellently adapted for cotton and sugar.

A scatter of huts is springing up around the Piquizeiro, where
a new cemetery has been laid out. Our host dug a leat to
supply the builders with water, and the place is strewed with
adobes and fine slabs of blue limestone. A tall cross of cedar
bears a little cross and the legend " Salus. P. R. G. C. 1867."
This was lately set up by Fr. Reginaldo Gonçalvez da Costa, a
vicar detached on a missionary campaign from his cure near
Montes Claros by the Bishop of Diamantina. He collected a
copper from the poor and a testoon from the rich. Some 6000
souls, mostly feminine, strewed the plain as he doled out the
Bread of Life, and the fireworks which ended the day are described as having the effect of a volcano in full blow. Januaria
had lately been visited by a convert, pervert or divert Spaniard,
in the pay of a certain Bible Distribution Society. When I was
there he had left to raise more grist for the mill at Rio de
Janeiro ; and he had bequeathed to a Portuguese clerk the work
of conversion, perversion, or diversion. The priests down-stream
were much scandalized by the distribution of "false Bibles," and
I could not but sympathize with them, knowing how easily in
these countries the local mind is unsettled by a small matter.
Surely it will be time to Protestantize the world when it shall
have been Christianized. Similarly the missioner † and the mis-

* According to Arruda, the "Acanta-
carix pinguis ;" the tree prefers the sandy
soils of the Taboleiros and Chapadas, where
the growth deserves all encouragement. Its
height is fifty feet, with proportional girth ;
the timber is good for boat-making ; and
the fruit, as large as an orange, supplies an
oily, farinaceous, and very nourishing pulp,
much enjoyed by the people of Ceará and
Piauhy (Koster, ii. pp. 486—7).

† The Saturday Review, when noticing
a book which I wrote after my return from
Dahome, remarked the use of the word

" Missioner." The Reviewer did not remember that of late years "Missioner" has
been adopted by the (Roman) Catholic, in
contradistinction to the Protestant "Missionary." Perhaps it would be more an-
thropological to call the former the phase of
faith at present adopted by Southern Europe,
opposed to the young Church which belongs
to Northern Europe, and to the Greek
Church, as old as the oldest which prevails
in semi-Oriental Eastern Europe. Similarly
we observe in El Islam that certain unimportant articles of belief — unimpor-

sionary, Jesuit and Church of England, have been let loose upon Abyssinia, whose church dates from the third century, and doubtless resembles the primitive form far more than those of Rome and London. A few massacres have been the direct, and an Abyssinian campaign the indirect, result of the merciful interference. Meanwhile, until quite of late years, the Galla accolents have been left in full enjoyment of their savage fetishism.

Revenons! After a walk of four miles we reached an admirable grove of mangos, perhaps the finest that I have beheld in the Brazil, lining the approach to our host's property, the Fazenda de Santo Antonio do Brejo do Salgado. It is on the right bank of the Salgado, or Salt Rivulet, which rises in a pretty plain, the Fazenda da Carahyba, and which feeds the São Francisco a little below the settlement, to which it gave a name. Here it breaks through the Boqueirão, a gap in the Serra do Brejo, where it acquires a cooling and salt-bitter flavour, which argues saltpetre. When floods in the main artery block up the mouth, it can be ascended by canoes, showing that the channel could be converted into a canal. The people avoid drinking the water, as it is highly laxative ; and after using it strangers must check the effects with an orangeade made of the sweet, fade and medicinal "laranja da terra." * In two years it has deposited on the wooden watercourse which turns the turbine, a coat of calcareous matter about three inches in thickness. Its lime and salts give a wondrous fertility to its little valley, the richest spot that we have yet seen on the Rio de São Francisco; and during the whole journey we shall see few that equal it.

Amongst the Mangos I detected by its circular crown of fronds an old friend in the other hemisphere, the Cocoa-nut, here called Coco da Praia. It was a fine tall and lusty specimen of the Cocos nucifera, hung with sixteen nuts. The tree is plentiful along the coast from Rio de Janeiro to Pará ; † except, however,

---

tant because neither the Koran nor Tradition has pronounced upon them — are adopted by one school of divinity because the rival school has preferred another view. "Ragban l'il Tasannun"—in hate against the Sunnis—is the Shiah reason for adopting some of its minor usages.

* The perfumed flower of this country orange is much admired by the humming-bird. I was here told that the fruit becomes bitter or tasteless, unless periodically refreshed by grafting, and they showed me orange-trees six years old, but still barren.

St. Hil. (III. ii. 409) says of Salgado, "Cette bourgade doit son nom à l'un de ses premiers habitans, et non, comme on pourrait le croire, à la qualité, un peu saumâtre, de ses eaux." This is, I believe, a mistake. Pizarro has explained the origin of the term correctly ; he remarks that the waters are stomachic, deobstruent, digestive, and capable of healing or diminishing goitres.

† It must be remembered that the Cocos nucifera was not found in the Brazil by the earliest explorers.

upon the river sides, it wanders but a short way inland, justifying the popular belief that it requires sea air. Here a bee-line to the Atlantic measures 350 miles, and we shall find it extending in patches all the way down-stream. The largest plantation is at the Lugar da Aldêa do Salitre, seven leagues south-west from Joazeiro ; the fruit is exported by Dr. Joaquim José Ribeiro de Magalhães, who preferred farming and road-making to being a Desembargador in the Relação of Maranham. Both these places have saline or saltpetrous waters. The Coco da Bahia, as it is also called, is found, however, in many spots where the ground, possibly an old sea-bed, supplies the want of sea air.

The host led us into his garden, and showed us, embedded in the soil at an angle of 45°, a semicircular fragment of "Cavitaria," the true white and black granite of Rio Bay, two feet broad, two and a half long, and three deep. The sides had been chipped, and the face had been used as a grindstone. An old Quattro-mão declared that the Geraes had whole hills of such rock, but no one believed him. It had probably been brought from down-stream, and about Joazeiro we found the formation common. The energetic Netherlanders, it will be remembered, built a Fort Maurice at the mouth of the São Francisco, and plundered Penedo ; it is more than probable that during their Thirty Years' War in the Brazil, they visited the upper stream. So M. Hal-feld remarks that the floods of 1792 laid bare in the river bank several tiles more than a foot long each way, and five inches thick. He believes them to date from the age of the "Hollandezes."

The plague of the garden is the "Cupim," and nothing but the plough will remove it from the rich fat soil. The coffee planted under the shade of the mangos or luxuriant jack-fruit trees, appeared to be subject to the caterpillar ; not so the leaves exposed to the sun. We saw a single tree dating from 1828, and were told that during its best days it had borne fifteen pounds per annum. The sugar-cane was remarkably fine, and once planted it lasts almost through a man's life. The arrowroot (a Maranta) grew well ; the Guandú pea was common, and there was a large grass whose dried root much resembled patchouli. The flowers were the perfumed "Bougarim," suggesting a white rose, lilies, gigantic snowy jasmines, and the "bonina," a kind of "pretty-by-night."

To the north-east we saw the solitary steeple of Nᵃ Sᵃ do Rosario gleaming against a green hill. South of it were the tiled roofs of the Barro Alto, a fine plantation, and behind them lay the "Boqueirão" estate, as well as gap, where the Church of Santo Antonio, built by Maciel, the Adelantado, lies in ruins. To the west-north-west peeped the summits of the Matriz do Amparo, the mother of Januaria city. And the background was the Serra do Brejo, pillared with cactus, capped with thin bush, and walled with banded grey cliffs of a stratification so regular as to resemble art, and stained here and there with a bright ferruginous red.

We then visited the sugar-house,* which had poor machinery, but an excellent article to work upon. Instead of troughs there were Jacás, cones of bamboo, each containing four bushels, and purging into pits below. Good mules were straying about the grounds; the natives cost 30$000, and those driven from the Province of Rio Grande do Sul, viâ Sorocaba and S. Paulo, a two years' journey, fetch 50$000 to 60$000. A "Jack" showed that breeding is here in vogue; further down the river asses become common. Flesh is not plentiful, and a cow of the small Raça curaleira, which gives good meat, commands 8$000 to 10$000. The "Curios" shown to us were broad-brimmed hats of the Imbé Vermelho, an Aroid used like the African "tie-tie;" its fibre takes a good colour; the leather clothing was soft as cloth; there were stout cottons, and woollen stripes and checks, worked by the women of Tamanduá, and stained with indigo, and a powerfully drastic cucurbitaceous plant known as the Bucha dos Paulistas.† We breakfasted at the usual bucolic hour, 9 A.M., preferred to Lisbon wine the "Minas wine," i.e., Restillo, and the peculiar cheese Requejão,‡ which here always accompanies coffee. We ended with Januaria-made cigars; the tobacco came from the hilly Geraes three leagues to the north-west of the city, and the

---

* The whitest sugar in Januaria came from Pitangui (120 leagues). It would easily have been crystallised, and moulded into loaves. I suggested the use of animal charcoal; but who will take the trouble to make it when clay is found ready made?

† Literally, the gun-wadding of the Paulistas. The specimen showed to us was a fibre containing dark oleaginous seeds. About one square inch of it is steeped in water over night, and drunk in the morning as an emetic, &c., by those who suffer from paralysis ("ar," or "stupor") induced by river fever. The System asserts that in S. Paulo it is known as the Purga de João Paës (Momordica operculata), and alludes to its various uses. We also heard of a smaller variety, said to be even more violent in its action, and the plant was described as resembling the Passion-flower. It is probably the Buchinha, or Luffa purgans, whose extract is used as the coloquintida.

‡ In making Requejão, the milk is curded, as if for cheese, and butter and cream are afterwards added. It lasts for two years, and is still soft.

leaf costs 8$000 per bushel; the cigars are retailed at a half-penny each, and they are better than many "Havannahs."

Finally, we mounted neat nags, and taking the western road, which goes to Mato Grosso, visited the venerable Arraial do Brejo do Salgado. It lies at the eastern foot of the Serra, which gives the air some similarity to the breath of a hothouse, and the curious limestone blocks were reeking with heat. The hamlet now consists of a sprinkling of houses round a square, whose centre is the Church of Nª Sª do Amparo, remarkable for nothing but red doors of solid timber, with tall bosses. Adjoining is a stone box with barred windows representing the jail, and a tall tiled roof, wanting the finish of walls, showed that it did not need enlargement. The people were yellow from eating fish and manioc.* Amongst them was a Polish Jew, Moses Mamlofsky, who did not speak in flattering terms of his new home; he had been in partnership with a German co-religionist, Samuel Warner, who called upon us at Januaria. The latter called himself a New Yorker; unfortunately he could not speak English; twenty years ago he settled in these parts, made money, and spent it.

The glory of the Brejo was the Conego Marinho, before mentioned as the historian of the movement of '42. He was equally distinguished as a liberal, an orator, and a statesman. We called upon several of the notables, who exhorted us strongly to visit the Lapa de Santa Anna, distant two leagues. Here the old conquerors found, or by vivid fancy thought that they found, stone crosses cut by the "Indians," statues of Saint Anthony, and so forth.† We heard, also, of another cave, in which a rocket could be fired without striking the ceiling; perhaps some more leisurely traveller may find it worth his while to inspect these places. At the Brejo we were told the romantic tale of its origin. When Manoel Pires Maciel, the Portuguese explorer, was descending the river, he attacked, on the Pandêiros influent, a powerful kinglet, who governed 120 miles of country between the mouths of the Urucuia and the Carunhanha streams. The redskins fled hurriedly, and the chief's wife hid her babe under a heap of leaves, as the

---

* They are not, however, a short-lived race. Our host's father, aged 81, rides like a man of 40, and the vicar, Padre Joaquim Martins Pereira, is still vigorous at the ripe age of 77.

† I will not positively assert that all was fancy. The Eastern Coast of South Ame-

rica, long before the age of Saint Columbus, was doubtless reached by Europeans and Africans, possibly by Christians, even as the western shores had Asiatics occasionally driven to them. I shall reserve the grounds of my conclusion for a future volume.

cayman is said to conceal its young. The Conquistadores' dogs found the pappoose, who was christened Catherina, brought up as a Christian, and finally married by her capturer. She bore to him two daughters, Anna, who settled with her husband João Ferreira Braga, upon the Acary River, and Theodora, who became the wife of Antonio Pereira Soares. The name of Maciel has been merged into that of many Portuguese houses, Bitancourt, Gomes, Morenas, Proenças, and Carneiros. Catherina's issue now forms a clan of 4000 souls, whose coal-black hair, brown skins, and sub-oblique eyes, sometimes "bridés," still bear traces of this Brazilian Pocahontas.

We returned to Januaria delighted with our visit, but justly anticipating some trouble in collecting a crew. The Guaicuhy men positively refused, despite liberal offers, to proceed; they were, doubtless, anxious to look after their wives. Sr. Manoel Caetano and his brother-in-law walked with me all about the city, and found that six of the barcas desired to start, but wanted hands. Many of the barqueiros had been carried off to the war, others had fled their homes, and some declined to leave the city, lest they might be enlisted in a strange land. Moreover, this is the season, as we were warned by the fiercely howling wind, which swept up the water from the Bahian shore, when the fields must be made ready. Finally, there is no actual poverty in this part of the world; the pauper has at least a cow, and a mare to ride, with unlimited power of begging or borrowing food from his neighbour; consequently, he will not work till compelled by approaching want. Those who did consent coquetted, demanding, at least, three days' delay, and one fellow, free, but black as my boot, could not start without his boiled shirt.

From Januaria to Joazeiro the hire of a barca is 1$000 per diem, and the barquemen are usually paid 14$000 a head, a poor sum, but the diet is some consideration. It was vain to offer 20$000, of course including tobacco, spirits, and rations. At last I closed with a pilot and a paddle-man, who demanded 35$000 and 30$000. My excellent friends had sent on board everything necessary for the long journey,* and we determined to

* The provisions bought at Januaria were :—

32 lbs. roll tobacco	. . .	6$000	Farinha	1$280
20 Rapaduras	. .	2$400	6 medidas of rice . . .	1$920
Demijohn of Restillo .	. .	1$800	5 lbs. meat	0$600
Lard	.	3$500	Quarta of beans . . .	2$000
			Total . . .	19$500

set out at once. There was an ugly frown upon the forehead of
the western sky, thunder growled, and lightning flashed in all
directions. The new crew shook their heads, and I began to fear
the loss of, at least, half the next day. However, they took heart
of grace, and we pushed off, to make fast a few minutes afterwards
near the ruins of the Rosario.

We shall miss the frank and ready hospitality of Januaria as we
advance, and going farther we shall fare worse in the little matter
of reception. The change will make us think more often of the
kind-hearted and obliging Lieutenant-Colonel Manoel Caetano de
Souza Silva; of his brother-in-law, Capitão Antonio Francisco
Teixeira Serrão; of the Promotor Publico, Luis de Souza Machado; of Gonçalo José de Pinho Leão, and others who took so
much interest in the passing strangers.

CHAPTER XVIII.

FROM JANUARIA TO CARUNHANHA.

THIRD TRAVESSIA, 30½ LEAGUES.

THE VILE WEATHER.—REMAINS OF THE RED-SKINS.—THE HAMLET AND LARGE
CHURCH OF Nᴬ Sᴬ DA CONCEIÇÃO DOS MORRINHOS.—DECAY AND DESOLA-
TION.—THE MANGA DO AMADOR SETTLEMENT.—THE SONG OF THE BIRDS.
—THE RIO VERDE, A SALT STREAM.—THE CARUNHANHA RIVER.—THE
MALHADA SETTLEMENT AND ITS RECEIVERSHIP.—LIEUT. LOUREIRA.—VISIT
THE VILLA OF CARUNHANHA.—DON RODRIGUES.—VILE NIGHT.

> Ergue-se sobre o mar alto penedo,
>   Que huma angra á raiz tem dos náos amparo,
> Onde das ramas no intrechado enredo,
>   Causa o verde prospecto hum gesto raro.
>
> *Caramurú,* 6, 18.

It was an abominable night. The storm, as often happens in
the Brazil, assumed the type of a Cyclone, passing round from
north viâ east to south, and about the small hours I thought
that the "Eliza's" awning would have been beaten down by wind
and rain. The new men, both now and afterwards, proved themselves real watermen; they talked much, but they worked more,
and better still, neither of them drank, nor had "sarnas." * The
pilot, José Joaquim de Santa Anna, officiates in a black coat; he
is silent and dignified, rarely consorting with the barquemen. Of
very different temper is Manuel Felipe Barboza, who rejoices in
the cognomen "das Moças," or "Barba de Veneno ;" he sings, he
roars, he improvises Amabœan verse ; he chaffs like a bargee,
and the fluency and virulence of his satire have made this
"repentista" † celebrated as "the Poison beard." Yet he has

---

* The "Indians," from time immemo-
rial, used to treat their "sarnas" by extracting, with a pointed thorn, the Acarus
(an Arachnid) which produced it. The
psoriasis is very common amongst the boatmen of the São Francisco, but they have
never adopted the wild system of healing it.

Some are loathsome objects, with blotched
and mottled skins, even after the sores
have become scars. As on the Lower Congo,
the disease is highly infectious, and very
difficult to cure ; in fact, many declare it
to be incurable.

† An improvisatore. I need hardly say

not ignored the main chance, and he expects to make money by investing capital in water-gugglets, straw-hats, and bricks (tijolos) of orange and other sweetmeats, which he will sell down-stream.

Thursday, Sept. 26, 1867.—The evil weather produced a start at 5 A.M. After passing some uninteresting spots * we were on the parallel of the Mocambo, which has been mentioned as one of the districts of Januaria. Beyond it,† on the left bank, rose the Morro do Angú, and its long sandy and partially cultivated island ; the heights are apparently an offset from the Serra do Brejo, a scrubby lump with scarped walls of grey and red-stained limestone. Presently the rain and thunder, coming from the north, drove us for refuge into a narrow channel formed by a "steamboat island," near the right bank. The hurricane proved a mere "peta" or feint, and after losing half-an-hour, we resumed the way and presently anchored on the Praia do Jacaré, opposite a small Arraial of the same name. We are now careful to take the windward or Bahian bank, and to avoid the vicinity of tall trees. To its north rose the Pico do Itacaramby,‡ a term which none could explain ; early in the day it had appeared to us like a tall blue pyramid. Here we found it to be the southern buttress of a line of scattered hills that trend to the north with easting. The low cone presented a curious appearance, the colour was somewhat darker than the slaty back-ground of lowering sky, and it seemed to vomit grey puffs of heavy mist, which formed conducting lines of electrical vapours girding the nimbus cloud.

Sept. 27.—The new moon brought with it for a time heavier

---

that the practice comes from Portugal, where the "justa," or trial of strength, is still popular amongst the peasantry. Here it met the "Indian" blood, which had also the habit of making impromptu chaunts.

* The Ilha da Boa Vista on the right ; the Ilha de Rodeador, fronted by houses, and the Vendinha islet, on the left.

† The Barra do Páu Preto, a small yellow stream from the right ; the Fazenda and large island of Amargoso ; and the Varginha, which showed a tiled house. After noon we passed the Ilha do Jatobá, a buttress lying to the left of the stream ; at the bottom of the sack, the Arraial do

Jatobá, fronted by canoes, and composed of mud and tiled huts, faced the river, which here must flood the banks.

‡ St. Hil. (I. ii. 24) mentions a Fazenda de Itacorambi, and derives it from "ita," a stone, and "and carambui," small and pretty ; certainly not applicable here. A better explanation is that given to him by a Spaniard of Paraguay, well versed in Guarani : "itaacábi," a mountain divided into two branches. Pizarro believes that this place was discovered in 1698 by the Paulist Captain Miguel Domingos. St. Hil. (I. ii. 303) attributes it to Fernando Dias Paes.

weather, and the air was wet and soppy. Presently the west bank showed a broad sandy ramp, the road to São João das Missões (or dos Indios), distant from the river three leagues, and the object of a great Patron (Romaria) on its Saint's-day. Here, removed eighteen leagues from their old home—the beautiful Brejo do Salgado, a savage paradise—are villaged the remains of three great tribes, the Chavantes, still powerful on the head waters of the Tocantins ; the Chacriábas (Xicriábas), and the Botucudos or "bung-lipped" races, an indefinite general name. Of old the Geraes hereabouts were held by the Acroás, vulgarly known as the Corôados or tonsured people, the Cherentes, and the Aricobís, who were dangerous till 1715. Now the nearest of the wild "Red-skins" are about Moquem,[*] in Goyaz, distant some 125 leagues.

After a succession of the usual features,[†] at 1·30 P.M. we saw Cascalho on the right bank, and washerwomen, the usual approach to a town. We ascended a natural ramp, and fell into a kind of street much broken up by the waters ; thence turning to the right we made the large square with its tall central cross, the beginnings of a second. Here the inundations have never extended. On the north there is a Casa da Camara, whose shutters are shut, and a jail whose grating is open. The twenty-one houses, including two ruins, are of the humblest, and down-stream are two parallel lines of thirteen to fourteen huts. The eastern side of the square is occupied by Nª Sª da Conceição dos Morrinhos, which gives a name to the place. It is a " delubrum miræ magnitudinis," which enjoys a wide reputation, and which makes the stranger inquire how it came here. It owes its origin to the piety of a certain Mathias Cardozo, before mentioned, who, with his sister Catherina do Prado, married in São Paulo to a Portuguese, settled in the wild, and for his services against the "Indians" obtained the rank of Mestre de Campo, a dignity to extend through three generations. He, and after him his son Januario, built, of course by the sweat of "Indian" brows, the fane, and the latter sent to Bahia for masons and carpenters.

---

* I have explained this word (and its verb moquear), which the author of the Caramurú defines as follows :

Chamão *Moquem* as carnes que se cobrem,
E á fogo lento sepultadas assão.

This is our "grushen." The term, however, is also applied to meat smoked or slightly exposed to the flame.

† The Ilha do Capão, where M. Halfeld places a village ; opposite it, on the left bank, the Fazenda da Barreira (H., As Barreiras). Then the sandy islet and Fazenda da Resaca (H., Resacca) made us take the left side.

The temple, facing a little north of west, rises from a platform of fine bricks, 1 span 4 fingers long, and 2 inches thick ; these and mortar compose the building.* It might easily have been made of stone, as massive calcareous blocks appear bald-headed above the ground. The façade has the usual pediment, protected by eaves with three rows of tiles, an attempt at a rose-light, and shuttered and railed windows above and below. Between the two latter is the gate, with massive doors, strengthened by large round-headed nails ; it is apparently never opened, and signs of fire appear near the floor,—bits of crosses, strings of beads, and decayed scapulars hang about it. The towers are massive, and capped with whitewashed pyramids like those of São Bento in Rio de Janeiro. The brickwork, however, is falling from above the windows, and poles planted against the front show that repairs are in prospect. On the northern and southern sides are fragments of cloisters, arches supported by six large square piers ; both end towards the east in rooms intended as sacristies. Outside the mortar is green with damp below, and stained red by the ochreous earth above. Inside, the northern cloister is heaped with sand and goat-dung ; opposite it the bulges of red clay dotting the floor betoken graves, a bier also lies under the arches, and a broken coffin is propped against the wall.

We had some trouble to procure the keys ; at last appeared the sacristan with the normal " tail." The interior was worse than the exterior ; the ceiling was partly stripped of its cedar boardings, the choir was ruinous — here decay generally commences, and the pulpits were likely to fall. The four side-chapels in the body of the church resembled portable oratories. A bold and well-built arch, revetted with fine wood and railings of turned Jacarandá, led to the high altar, which did not show any signs of gilding or whitewash. Below it a broken slab of slate from Malhada, bore inscribed :—

> AQVI IAS
> JANVARIO C
> ARDOZO DE
> ALMEIDA.

The date had been forgotten, and the sacristan could only tell

---

* Not "templo de pedra," as M. Halfeld has it.

us that at Morrinhos had lately died, aged 113 years, a man who said that the tomb was there when he was born.

We ascended the little hill at whose western foot the fane lies. The substance is blue limestone, in places banded with hard quartz, and capped with agglomerated sandstone; the soil stained with oxide of iron produces the red blots which marble these lumps. Formerly the Morrinho supplied saltpetre; it is now either exhausted or neglected. From the thorny summit we ascertain that the left bank is of similar formation, and even more subject to floods. Here we count four knobs of ground, the Morros da Lavagem, do Salitre, and so forth.

The smoky mists rising above a floating tree that left foam in its wake, formed a phantom-ship, which startled us by its resemblance to an expected steamer.* Descending on the right of a long island, the Manga do Amador, we saw the village of that name, advantageously situated upon the "Pernam side." It is the first unflooded settlement which we have seen on the high Saõ Francisco, and the superiority of site will tell in future years. Two barrancos or bluffs rise at least 100 feet above the brown tree-dotted bank, and are divided by a deep gulley, draining a bayou behind the village. The colour is a deep red earth—the finest of soils—extending down to the white clay of the water-side. I counted seventeen doors on the northern summit, and the settlement though young was not without ruins.

After a last hour's work we found anchorage ground near the Ilha do Carculo. Towards night we viewed the stars and planets like the faces of long-absent friends. The "wander-lights" † flashed through the darkness of the trees, the gull screamed at our intrusion, the bull-frog (Sapo boi), and the Cururu (Rana ventricosa), croaked like the wheel of a sugar-mill being set in motion; and again we heard the complaint of Whip-poor-Will, and of the Euryangú, which brought to mind the delightful wilds of the Rio das Velhas. The raw damp became mild and balmy, the lightning sank to the very horizon, and the north cleared to

---

* Chapter 25 will tell how we were disappointed.

† Vaga-lume, the firefly, also known as perilampo, and Cacafogo (Elater noctilucus).

These specimens will show that the Portuguese language has some of the prettiest and the ugliest of descriptive expressions.

a dull blue. In fact, Hope visited us once more, and very deceitfully.

Sept. 28.—We set out at the normal hour, 5 A.M., despite the heavy shower, and, after three hours' work, we landed on the right bank to inspect the mouth of the Rio Verde Grande. This stream flows from the northern slopes of Montes Claros; coming from the south it receives the Verde Pequeno, which drains the western Serra das Almas, a branch of the Grão Mogor range, and a counter versant of the great Rio Pardo that inosculates with the Jequitinhonha, the two uniting and bending to the north-west from the frontier of the Minas and Bahia Provinces. The stream is ascended by canoes, some thirty leagues from the embouchure.

At the mouth of the Rio Verde Grande is a broad "praia" which causes the stream to flow along the right bank of the São Francisco. Upon the water-side, which is caked with mud, we found, as might be expected, a finer diamantine "formação." The higher parts of the beach were occupied by a negro family, whose hut was in a little garden of beans and water-melons. Here the latter thrive upon sand, almost pure of humus, and where "corn" is short and wilted. They sold to us for three coppers five melancias, very cheap compared with what the fruit will be further down. Bees were busy among the flowers, the pink Crísta de gallo, like our "cockscomb," and the thorny-leaved, yellow-blossomed dog-rose-like Sarrão (Argemone mexicana) called "Cardo Santo," or holy thistle, from its real or supposed medicinal properties.[*] From this point we shall see its grey-green glaucous leafage all down the river. Another plant with white flower, pink stigma, long stamens, delicate leaves which curl up in the sun, and viscous stalk, will show its dull verdure in damp places near the settlements. The people call it Mustambé, and Mr. Davidson, after trial of the taint, declared it to be the "stink-plant" of the Mississippi valley. The Tiririca rush, so common on the streams of the Brazil, resembles papyrus, and towers over the Capim Amargoso (bitter grass), a large broad flag much loved by cattle. We saw but few animals on the banks, as the owners had begun to drive them inland. A few years ago one of

---

[*] Azara (i. 132) mentions its use in fevers; Prince Max. (i. 391) refers to it as a remedy for snake bites. The word suggests the Carduus Benedictus of the old world, concerning which we may ask, "Benedictus! why Benedictus?"

the breeders lost 300 head by the sudden flooding of the Green River.

The Rio Verde discharges through a bending avenue of fine timber a considerable stream about 150 feet broad. The water was of a dirty muddy green, "heavy," as the crew remarked, and sensibly salt, without, however, the taste of saltpetre. These saline influents on the Upper São Francisco were remarked by Dr. Couto; they attract swarms of fish, who enjoy them as beasts do the "salt-licks." From this part of the valley downwards we heard of many similar formations: the Riacho do Ramalho, ten leagues below Carunhanha; the Riacho dos Cocos, falling into the northern Rio das Egoas, and others. They are worth exploration; salinas or deep deposits of sea-salt would be better than gold-mines, and open a fount of wealth to some enterprising man. The water might be treated like saltpetre, made into lye, and left to sun in pans or troughs. At present salt must be here imported from the Villa da Barra do Rio Grande, and even from Joazeiro; consequently it costs per quarta * 8$000 to 12$000.

On the low right bank beyond the Green River there is yet more cultivation. We were charmed with the soft and amene scenery about the Fazenda das Meláncias, backed by the Serra da Malhada, a north-eastern offset from the Montes Altos, the further wall of the Rio Verde Pequeno. Presently we passed the mouth of the Japoré, a considerable stream draining the Geraes massed together in our maps as the Chapadão de Santa Maria; the confluence is known as the Barra do Prepecé, which the pilots could not explain. I believe it to be the name of some Indian chief. The next influent, also from the left, was the Riacho do Ypoeira. Mr. Halfeld translates this Indian term "lagôa," or "tanque d'agua," but it is temporary, whereas the "lagôa" is perennial. It is becoming a constant feature, found where the banks are flat, and have not, as above Januaria, waves of high ground perpendicular to the main artery, and dividing the tributaries from the east and from the west. We here miss these heights, which often extended to the water, rising above all inundations, and forming a natural dyke to contain the channel. The low lands subtending the river-ridge are periodically filled by the floods, which are thus

---

* The old "Broaca" of 24 pratos (each 4 lbs.) is no longer mentioned. The smaller measure is the medida of 4 lbs.; 20 medidas to 32 (at Joazeiro), or from 80 to 128 lbs., make the "quarta," which everywhere varies.

prevented from extending far, as on the upper stream. The swamps afterwards dry up or become "nateiros," *i.e.*, slimes. Wonderful draughts of fish, especially the Surubim and the Trahira, are taken from the Ypoeiras, and some parts of the valley are literally manured with fin. A boat-load is caught by dragging branches along the waters as they dry up, and the denizens may be caught with the hand.

We ran quickly past the Pontal da Barra do Rio Carunhanha,[*] a large western influent which drains the Serra da Tauatinga, and is a counter versant of the Paraná or Paranan, the eastern headwater of the Tocantins. It is navigable for some twenty leagues up to the Serra or dividing ridge, through which it breaks; then it becomes a succession of rocks and rapids. Large timber, especially cedar, is here felled; made into balsas or rafts, and floated down for sale. Seen from the south the low sole appears overgrown with trees, a view from the west discloses a river about 300 feet, curving through grand vegetation, and it has probably shifted south-westwards or up-stream. The left jaw is a mass of sand, disposed in waving lines; a little further down, and forming a dark line, is a deposit of fine purple slate in slabs or layers. It is not wholly neglected. I saw several pieces two inches thick and twenty feet long, and smaller sizes were cut into round and oblong tables; they were without stains or signs of pyrites.[†] The Carunhanha is the western frontier between the Provinces of Minas Geraes and Bahia, and at the Pontal or Ponto do Escuro a guard was stationed, and duties upon goods were levied. It was deserted on account of the malignant typhoid fevers, called "Carneiradas," which butcher men like sheep (carneiros). Since 1852 the receivership has been transferred to the right bank.

We made for the "Malhada," or to give it the full name, Nᵃ Sᵃ do Rosario da "Malhada," *i.e.*, a shady place where cattle gather during the hot hours. Here the São Francisco broadens to 2650 feet, and turns to the north-east; the Carunhanha pouring down the left channel strews the main stream with snags and branchy trunks, and forms a sand-bar, and a shoal which extends some way down. We were obliged to round the northern end of

---

[*] It is thus generally written; other forms are Carynhanha, Carinhenha, Carunhenha, and Caronhanha (preferred by Dr. Couto); it is supposed to be a corruption of Arinhanha, the large otter.

[†] M. Halfeld notices this quarry, and calls it Phyllado or argillaceous schist.

the latter, and to bring the "Brig Eliza's" head up to the south-east. The shore being exposed to the south-west wind, which came on heavily, yeasting the stream, I sent the craft to embay herself (ensaccar-se) leeward of the Corôa da Malhada, above the settlement. The single barca which was at anchor followed her example, but the canoes remained staked to the shore.

The Porto is a bank of sand and clay cut in steps by the ebb of the floods, grown with a few weeds, but bare of trees. A few horses and mules lingered over a scanty meal, and boys were fishing and bathing near a sandspit, where the water is too shallow for the dreaded Piranha. The settlement faces a little to the north of west, the houses on the bank are of mud and tile, one only being whitewashed; the long ends, of which the greater part is occupied by the door, fronts the stream, and the rails of the compounds are used as "horses" for drying clothes. The settlement consists of a water-street and two parallel thoroughfares, with a central square. Here is the Rosario Church, a ground-floor fronted by a deep sheltering porch; before it stands a rude black cross, bearing amongst the instruments of the Passion a very rude cock, and planted round with the Barba de Barata, or "cockroach's beard."[*]

The houses show a water-mark three feet high. Above the Malhada is the Sangradouro de Santa Cruz, which every year for about a week in January or February, permits the floods nearly to surround the settlement. After that the flow is surrounded by stagnant water, in places so deep that a boat-pole does not touch bottom. Of course this evil might easily be remedied, but who will undertake the cure which is "everybody's business?" To the east the land becomes sandy, and produces good cotton and sugar, the castor shrub, and the ever-green Joazeiro-tree,[+] a gigantic shady umbrella for man and beast. The level here begins to rise above all floods, towards the Serra de Yúyú, or Iúyúy,[‡] distant six leagues. It is a segment

[*] This is the Poinciana pulcherrima, a brilliant leguminous shrub, supposed to have been brought from Asia. According to the System it is rich in "stryphno," the astringent principle.

[+] "Zizyphus Joazeiro" (Açeifafa Joazeiro), a species of Jujube tree; an ally to the hawthorn (Prof. Agassiz). According to the System, its acrid, bitter, and astringent bark promotes emetism. Here, as in the Sertão of Ceará, it preserves during the dry season its foliage, which is eaten by cattle.

[‡] This is a Tupy word, which no one could explain. The range is also called Serra da Malhada.

of an arch extending from east to south-east, and opposing its
concavity to the river ; there is apparently a projecting elbow or
a buttress which forms an apex fronting to the west. It is said
to be calcareous, and to abound in saltpetre. The western bank
of the São Francisco is a vast level ; the nearest range is distant
about fifteen leagues. This Serra do Ramalho, more generally
known as "A Serra," is also calcareous and off-sets from the
great dividing ridge between Bahia and Goyaz.

I had a letter for Lieut. Silverio Gonçalvez de Araujo Loureiro, Administrator of the duties payable to the Provincial
Treasury of Minas Geraes (Administrada da Cobrança do The-
souro Provincial da Provincia de Minas). We called upon him
at his house in "Water Street," and sat there talking over
coffee. He hails from Ouro Preto ; and having spent twelve
months in this vile hole, where of his escort, a sergeant and four
men, all but one are dead or absent, he purposes to leave it as
soon as possible.

Lieut. Loureiro gave me a printed paper, dated October 19,
1860, and showing that the several "Recebedorias" collected a
total of 600*l.* to 800*l.* per annum.* Here imports and exports are
both taxed, and only salt going up-stream does not pay. Three
per cent. are taken on cotton, minor articles of provisions, worked
tobacco (including Pixuá, a kind of Cavendish prepared for
chewing), clothes, pottery bowls, canoes and woods for furniture ;
hammocks, whips, saddles, and so forth. Coffee is rated at three-
and-a-half per cent, and six per cent. is recovered from grain,
raw provisions, including poultry, which is the best thing in this
place ; hides, ipecacuanha, quinine, and precious stones, the
diamond only excepted. The horse, valued at 5*l.*, is taxed
3$160 ; the native mule (8*l.*), 4$960 ; the São Paulo mule,
5$000, and black cattle, 0$600. These animals are driven to
Bahia by a vile road which their hoofs made ; it crosses difficult
Serras without bridges or any "benefits," and the distance is 130
leagues.

A white man walked in whilst we were sitting with Lieut.
Loureiro, and astonished us by his civilised aspect, amongst all
this Gente de Côr ; he was introduced as Dr. (M.D.) João Lopes
Rodrigues, who had graduated at Rio de Janeiro, and had settled

* In 1852–54 M. Halfeld makes the   34,500*l.* ; balance in favour of the latter,
exports reach a value of 21,200*l.* ; imports,   13,300*l.*

at Carunhanha. No one had the indecency to ask him the reason why; he complained of the Preguiça do Sertão—the idleness of the wild country—and of stimulus being totally wanted, except when a stranger happens to pass. I have heard the same in Dublin society; possibly Dr. Rodrigues, like a certain Abyssinian traveller, found "making up his mind" a severe and protracted process. He had suffered from the climate of the River Valley, always cold-damp or hot-damp, so different from the dry air and sweet waters of the sandy table-lands on both sides of, and generally at short distances from, the river. He had none of the pretentious manner and address usually adopted by the Bahiano, who holds himself the cream of Brazilian cream, and he readily accepted a passage in the raft to his home, about two miles down stream.

The denizens of the Malhada have a fever-stricken look, and their lips are bistre-coloured as their faces. Yet within the houses we heard singing and clapping of hands, after the fashion of Guinea; and, as we embarked, a little crowd of women collected to prospect us. The dress was a skirt of light chintz or calico, a chemise or rather a shirt, generally a shawl, and above and below comb or kerchief, and slippers.

We dropped down the still fierce stream, here treacherous and much dreaded. The strong up-draught often keeps craft in port for fourteen days; they load heavily, and the waves are likely to damage the cargo. The weather looked especially ugly, but our companion consoled us by declaring that we were fast outstripping the rains. Here showers had begun to fall only five days ago, and were called "chuvas de enramar," of branching. The wet season will not set in till November, when the Vento Geral will shift to the south, the normal quarter. We escaped swamping with some difficulty, and presently reached the head of the Ilha da Carunhanha, which splits the stream into two channels of about equal depth.[*] The course of the river is here to the northeast, and the western arm is apparently widening; formerly children could swim across it. The islet is about two miles long, sandy, but of admirable fertility. It grows fine cotton, and as upon the São Francisco, lower down, manioc planted during the Vasante Geral (March and April), produces a large root

---

[*] M. Halfeld remarks that the right channel is low and full of shallows. The eastern bank is only half the height of the western.

fit for farinha before the flood-time in November and December.
Here is a good site for a bridge to connect Minas with
Goyaz.

At the landing-place are large blocks of Pissarra or Saibro do
Rio, a felspathic clay, yellow-tinged with iron; this bank is 55
to 75 feet high, or 5 to 25 feet high above the annual rise.   It
is, however, much cut up by a surface drain, now an Esbar-
rancado, but a Córrego during the rains, dividing it into waves of
high and low ground, and loudly calling for a levée. São José
da Carunhanha is a larger place than it appears from up-stream;
there are some 450 houses,* none double-storied, and mostly
flanked by the Gupiára or Agua furtada.   Though noble timber
is here, the wood-work is mostly sticks.   Young cocoa-nut trees
grow well in the court-yards, and the produce of the adult in
this saline nitrous soil is 200 nuts per annum.

In the north of the town we found an enormous square, the
Largo do Socavem; † it has a cross and symptoms of a chapel.
Beyond the settlement a Sangradouro with a sandy bed, based
on hard reddish clay, breaks the bank with a gap some fifty
yards broad, and the floods form a back-water which does
not extend far.   The best houses are in the southern square,
where fewer people squat on logs before their doors; there is
a Camara and a prison; in the latter our Januaria man
found a friend who had been resident for nearly four years,
after kniving a brother boatman in a drunken quarrel.   The
Matriz of São José da Carunhanha suggests nothing but an
old termitarium, yet it has a bell which sounded for us the
Angelus.

It now becomes difficult to collect local information.   The
great Province of Bahia is behind most of her rivals in popular
topographical works, and those which she possesses are too cumbrous and discursive for the traveller, whilst Minas Geraes has
her Almanaks, and São Paulo has two handbooks.   Carunhanha,
dismembered from the Villa da Barra, rose to township thirty
years ago, and is now capital of the Comarca of Urubú in the Bahia
Province.   Its municipality formerly extended to the Rio das
Egoas, the western branch of the Paracatú; here, however, a
villa has lately been established under the name of Nª Sª da

---

* In 1852 there were 265.
† It is the name of a town in Portugal,

but none of the Carunhanha people knew
what it meant.

Gloria do Rio das Egoas. This municipio still numbers about 10,000 souls, of whom 1000 to 1200 are in the town ; slaves are rare, and few fazendeiros have more than 40 to 50 head. The post arrives three times a month,* and each side of the river has a fair-weather road to Januaria—distant thirty leagues.† The principal imports are from Joazeiro, and include salt and dry and wet goods. There are no rich men, and the chief people breed cattle for export. They also send "sola"-leather, hides— here worth each 1$250, and at least double below Joazeiro— a little sugar and dried fish. The land would produce rice and cotton in abundance. Hereabouts also the Geraes grow a medicinal root known all down the river as Calemba or Calunga.‡

Dr. Rodrigues led us to his house in the square, and offered us the luxuries of sofa and rocking chair, wax candles, and a map of the war—moreover he gave me his photograph. I sent an introductory letter to the Delegate of Police, Capitão Theotonio de Sousa Lima. That young man did not even return a message ; possibly he, a Liberal, had seen us walking with the doctor, a Conservative. Again the stranger was tempted to exclaim, "Confound their politics !" Unfortunately for us, the Juiz de Direito of the Comarca, Dr. Antonio Luis Affonso de Carvalho, was on leave at Bahia ; all spoke well of this distinguished "Curioso."

We reached the raft in time to prepare for a night of devilry let loose. A cold wind from the north rushed through the hot air, and precipitated a deluge in embryo. Then the gale chopped round to the south, and produced another and a yet fiercer downfall. A treacherous lull and all began again, the wind howling and screaming from the east. The thunder roared and the lightning flashed from all directions ; the stream rose in wavelets,

---

* The 5th, the 15th, and the 25th are the days appointed, and this tri-monthly delivery is the rule of the river. Of course punctuality is not to be expected.

† The reaches will now become straight, and the land routes, which everywhere connect, are but little shorter than the water lines.

‡ Probably the word is taken from the African Colombo or Calomba (Cocculus palmatus), which gives the radix Colombo.

It is mentioned by St. Hil. (III. i. 164-5). The System (p. 93) calls it herva amargosa (Simaba ferruginea or Pichrodendron Calunga). The bark of the root and trunk of this Rutacea, which is much valued as a simple, has an unpleasant, bitter, acrid, and astringent taste ; it is stomachic and anti-febrile. I heard it everywhere spoken of, but no specimen was procurable under a couple of days' delay.

which washed over the " Eliza " and shook her by the bumping
of the tender-canoe. At last, just before day-break, the crisis
took place, and we snatched a few minutes of such sleep as hot
heads and cold feet, and dogs persistently baying at the weather,
would permit.

CHAPTER XIX.

FROM CARUNHANHA TO SENHOR BOM JESUS DA LAPA.

FOURTH TRAVESSIA, 24½ LEAGUES.

A GOOSELESS MICHAELMAS. — THE LUGAR DA CACHOEIRA. — THE PARATÉCA STREAM, AND THE DISPUTED " RIO RAMALHO."— DIAMANTINE DEPOSITS. — THE ALLIGATOR NOW KILLED OUT. — THE CONDE DA PONTE. — THE ASSASSIN GUIMARÃES. — THE MOUNTAIN OF THE HOLY CAVE DESCRIBED. — THE VILLAGE. — THE HOLY CAVE. — THE STOUT-HEARTED VICAR, REV. FRANCISCO DE FREITAS SOUEIRO. — THE " UNIFORMITARIAN " ENVIES THE " CATASTROPHIST."

> . . . lapa que esconde alto mysterio.
>
> *Caramurú,* 7, 8.

MICHAELMAS DAY found us gooseless, worn out and cross ; the song was hushed, and silenced was the voice of chaff. After a couple of dull leagues we reached the Lugar da Cachoeira, famed for pottery. The clay is made into neat talhas (jars) and quartinhas (gugglets), ornamented with red tauá, placed upon the naturally-yellow groundwork before burning. What is here bought for two coppers fetches six at Joazeiro ; our men made a small purchase, and the prospect revived their spirits. The Cachoeira took its name from a ridge of rock forming a diagonal rapid across the stream. A sand-bar has now been thrown up, and we passed over the place ignoring its old break. On the left bank, which rises above the floods, and which is drained by two "bleeders," there are a few huts. Further down is the Fazenda dos Angicos,[*] where the yellow variety of Acacia is common.

We halted at noon on the left bank near the Fazenda do Espirito Santo ; it has a large grove of Joazeiro or Jujube trees, whose bark is sold for tanning. The straight reaches, some twenty

---

[*] M. Halfeld, I have remarked, calls villages (povoações) what the pilots speak of as fazendas. The words are here nearly synonymous ; the fazenda is a breeding or agricultural establishment, often containing a little chapel and a dozen huts belonging to as many proprietors in partnership.

miles long, and the narrowness of the stream, 1460 feet, greatly
increase its flow, which averaged three knots an hour. The morning rain had diminished to a spitting, but a strong wind came up
from the south and played about the west. Here the people do
not shout

Honor be to Mudjekeewis,

who is also of the Pau-puk-Keewis family. These signs and
symptoms induced the men to caulk the port canoe, which had
scraped bottom till the cracks formed a leak. At 3 P.M. we had
to repeat the operation on a large praia to the right, opposite
the Fazenda das Pedras. Here we found bits of pure saltpetre,
and a trunk of Braúna tree almost lignite, while the diamantine
" formação " appeared under water, between wind and water, and
above water. About 5 P.M. we " knocked off work " at a long
beach near the mouth of the Paratéca;* called a river, it is a
mere streamlet, a fillet of water now coursing down the right
bank, and even during the floods it admits canoes for only two
leagues. A barque and sundry dug-outs were being repaired by a
dark carpenter, who told us five lies in three minutes, and who
apparently would have ridden twenty leagues to unburden himself. He pointed to the " finest place on the São Francisco,"
the Barra da Ypoeira on the Pernambuco, as the boatmen still
call the now Bahian bank. It was the usual high bluff, red above,
white below, with sand up-stream and bush down-stream. The
neat huts upon the level ground reminded me by their small size
and " natty " look of the pretty one-street villages on the Old
Calabar and the Gaboon Rivers.

Sept. 30.—During the night rain fell again. At dawn, low,
mist-laden clouds lay heavy where Carunhanha was, and lighter
vapours coursed from south-east to north-west, but far behind us.
Presently the climate became that of Malabar, and before 8 A.M.
the pilot actually removed his black coat. About noon a strong
southerly breeze swept through the well-washed atmosphere.
There was nimbus to the south as well as to the north, but we
were not molested, and the weather was peculiarly comfortable
and good for work. It was a " dies notanda," this our first fine
day upon the Rio de São Francisco.

---

* This stream also shows signs of diamonds in its sandbanks.

We set out at 5 A.M., and, after passing the usual features,[*] we landed at 7.30 below the Sangradouro da Volta de Cima to inspect the large R. Ramalho, which in Mr. Keith Johnston here enters from the west. Nothing appeared but a mere ditch, a Riacho.[†] Most men agreed that the Rio do Ramalho is a branch of the Rio do Corrente, further down stream. Hence, possibly, the confusion in our maps, which give a Rio Corrente entering the imaginary Ramalho, and to the north a Rio Correntes, which is the true Rio do Corrente. The beach again afforded good sign of diamonds, including the cattivo, the crystal, and the cánga-stone. Barboza, " Barba de Veneno," picked up a wax foot, some votive offering that had remained here en route to the Bom Jesus. He forgot to leave it at the shrine, and thus all our little accidents and evils were chaffingly attributed to him.

We passed in succession the Barra do Riacho das Rãas, from the right, and the Pitubinha and Pituba, formerly Fazendas. The Rio das Rãas, also on the east, is a mere rivulet, whose waters are said to be fetid. The opposite side showed a regular and tabular bank of soft greensward, adorned with tall trees. At the Ilha da Corôa Grande, a sand-bar and clump of vegetation, there was a shallow and a tide-rip. We took the right channel, and both abound in snags. Of this part M. Halfeld says, " there are many caymans (Jacarés), of ashen-brown colour, and one with a yellow throat, called the Ururáu, which is the Crocodile (!) [‡] Frightful numbers appeared, and my boats were surrounded by more than thirty." He also mentions Capivaras, which similarly have "made tracks."

About the "Frogs' River" we sighted a long blue range perpendicular to the stream, and extending far inland.[§] At its mouth was the Ilha da Batalha, a memento of some forgotten struggle with the wild men. At 3.30 we passed the Ilha da Boa Vista, a sand-bar in mid-stream. On the left bank was the

---

[*] A green islet on the Pernam side leads to " As Barreiras," a red bluff, wavy in outline with projections and bays ; the central depression is the only part subject to the "tip over" during floods. Then appeared the Ilha da Volta de Cima, where the stream bends to the east-north-east ; it is a strip of yellow-green vegetation, with its ruddy bluff a league long, and its Sangradouro.

[†] Some authorities told me that a little

Ramalho Rivulet exists near Pitubinha. M. Halfeld shows a drain, but does not name it.

[‡] This is probably the Jacaré de papo amarello (yellow-throated cayman), which is supposed to be more dangerous than the common Crocodilus sclerops. I do not know whether there be, as has been suspected, any specific difference between the two.

[§] Mr. Keith Johnston's map places along the stream a range which does not exist.

Fazenda of the same name. Here in old colonial times began the enormous property of a Portuguese, known only as the Conde da Ponte; the family has long left the river. The Fazenda da Boa Vista afterwards belonged to the "Assassin" (Antonio José) Guimarães, who sixteen years ago murdered his brother, the Commandant Superior José Guimarães. He was afterwards killed in Goyaz, it is said by a party of mule-troopers. A canoe was fastened to the bank, and we counted twenty huts, faced by a tall thin wooden cross. The men indolently stretched under the trees, replied gruffly to the extempore songs and bawling chorus of my crew. Here they are contented with a curral or fenced enclosure for their animals when driven from pasture, with railed-off plots of manioc and corn, melons running over the sand, and in rare places with a few stems of arboreous cotton. The furniture of the tiled hut is a giráo or cot, a sleeping-hide, a few benches, riding apparatus, wooden bowls and cooking pots, whilst the gun and the line never allow them to see the face of hunger. These are humble comforts, but they far exceed those attainable by the dwellers about the Great Rapids. The wigwam was as well furnished, even to the wooden ferule for thrashing the women, which hung to the ceiling.

Near the Fazenda da Volta de Baixo, on the right bank, we heard the dash of falling water, and at 5.30 P.M. we landed for the night upon one of the three "Ilhas do Campo largo." The clear dry minute sand crunched with a peculiar chipping sound, like snow under foot-friction; and here again diamantine deposits lay in lines parallel with the water. We are now in about the latitude of the Serra das Almas, whose eastern horn, the Serra de Sincorá, is one of the richest diamond districts in the Brazil. And it is evident from the state of the sand that it has floated from afar.

Oct. 1.—During the night the water fell, and we had some delay in pushing off. Observing the cirrus and cirro-cumulus high in air, the pilot quoted a proverb similar to our own.[*] The channel between the sand-bar was very foul with timber. On the right was the head of the Ypoeira or bayou, which spreads out into a little lake about its central course, and returns to the

Céo pedrento,
Ou chuva ou vento,
Ou mudança de tempo.

A stone-paved sky, rain or wind high, or change to dry.

main artery above the " Lapa." Below it is the Ilha do Medo—of Fear—another reminiscence of the dark and bloody days. As we bent to the right, or north-east, the Serrote da Lapa rose tall and abrupt over the vegetation based on the river sand. Above them was a slight central depression, and a yellow gash noted the position of the mysterious cave. Below it ran diagonally to the stream a thick avenue of Jacaré * and other trees, showing where the bayou re-enters the parent stream.

As we advanced northwards, the Serrote viewed from the west changed its form to that of a headless sphynx, or a crouching lion, the popular comparison. And now we could distinguish the peculiarities of a scene, whose novelty has raised it to sanctity. It is the mere skeleton of a mountain, disposed with a north-east to south-west trend, and lying lone upon a dead level. It is remarkable for perpendicular lines bristling against the air, with ribs which resemble finials or pinnacles. The sides, fretted and jagged like the flying buttresses of a Gothic temple, are cut up into salient angles, and are sharp-pointed by weathering. It has cleavage rather than stratification ; deep black cracks, at altitudes varying from ten to thirty feet, run horizontally, forming gigantic courses of masonry. On the north-eastern side these courses are slightly dislocated, dipping towards a bushy depression in the centre. The south-western end is a vertical precipice, with a long broad yellow stripe, where the stone had been removed. The colour of the mass generally is grey-slate, breaking blue, with fine crystals of the whitest calcaire.†

A few tiled roofs, and one white-washed house, rising in their line at the hill base above the trees and shrubs, directed us to the Port. We landed on the right jaw of the bayou, which during the floods becomes a harbour of refuge. A tall bank, much water-washed, led to a plain grown with grass, shrubs, and tall trees ; one of the latter, an acacia, with golden blossoms, emitted a heavy cloying scent. Deep pits, cut for adobes, showed the nature of the ground, sand and clay, with scatters of lime-stone. Hence cultivation here flourishes ; the people plant garlic and onions, melons and water-melons, pumpkins—especially the

---

* So called from its rough scaly bark ; the word is possibly a contraction of Jacaré ihuá (or igá, a canoe ?), which supplied the "Indians" with dug-outs twenty to thirty feet long.

† Colonel Accioli calls it a granitic formation ; it is, however, all limestone.

Girimú—haricots, and the castor bean, Quiábos or Hibiscus; rice, and a little maize, sweet potatos, and excellent cotton. We also passed a well-railed field, whose freshly-cut grass preserved the aroma of hay.

Presently we entered the settlement, which is detestably situated; even the African avoids the vicinity of great rocks. Here eighteen houses, disposed in arch-shape, front towards an unfinished church, which stands at the base of the great stone pile. They are all of the ground-floor order, built upon foundations of rough limestone; and one is solidly made, with attempts at pilasters. The total of the tenements may number 200, and, as all are inhabited, the population cannot be less than 1000 souls.* We found fresh meat, and bought tipioca cakes, whilst every vendor applied to us for medicine. We can hardly wonder that they suffer from psoriasis, cutaneous eruptions, terrible fevers, and inflammations of the spleen (opilacoẽs). Besides the limestone reverberator, they have the full benefit of a large Ypoeira swamp. Thus the stone raises the temperature of the air, and the heavier marsh poison rushes in to supply its place.

At the crescent a party of pilgrims were mounting their animals, and were being dismissed with a "Bom Jesus da Lapa guide ye!" We walked to the south-west, noticing in the occidental face of the buttress several ogival entrances, doubtless natural. In the higher levels, wherever the rock had been degraded to soil, trees displayed the filmy light-green foliage of spring; the most conspicuous were the Joazeiro, the Angico, and the delicate Pitombeira myrtle. The stone was clad with lichens and air-plants grey as itself. At the south-western end is the tallest bluff, which contains the grotto. Here a huge column, horizontally fractured in three places, and separated from the main wall by a perpendicular fissure, threatens to fall. At the cliff-foot is the Ypoeira channel, and here large fragments of limestone, cut into curious shapes by the water, block up the ledge which once allowed a path.

Six rough steps of blue limestone lead up to the Lapa, which faces west. A stout wooden door, with ponderous lock, and above it two shuttered windows, with "rose-light" and drain

* M. Halfeld says 128 houses and 250 souls, a very unusual proportion, except where absenteeism is the rule.

pipe, are flanked by thin pilasters of the burnt brick and lime composing the entrance. Inside, ten steps of brick, placed edgeways, and dangerously narrow for crippled devotees, admit to the body of the Holy Grotto. I looked in vain for aught to justify the vivid imagination of Rocha Pitta, which saw here an entrance large enough for a city, a stone bell * made by Nature's hand, marvellous columns of stalactite, and a high altar with collateral shrines ready for human use.

The Cavern, a very vulgar feature, bends to the right, and extends forty paces in depth, widening from ten to twenty paces at the far end. The floor is of tamped earth, which, being like all the Serra, miraculous, is collected by coloured people to be used as medicine. It is the sovereignest thing for a headache. Near the entrance the ceiling is flat, water-worn, and smoke-blackened; over the shrine it is somewhat arched. Down the length of the blue limestone runs a light-yellow band, forming truncated stalactites. In the vicinity of the steps there is a stalagmite resembling a Hindoo "lingam." The narrower end, and both sides of the grotto, are supported with masonry. On the left of one advancing towards the altar, wooden steps lead to a box covered with red silk, and lace fringed with cloth. The awning of this pulpit is a projecting ledge of stone. Further on is a shallow recess in which some hermit has been buried. Opposite it, at the broadest part of the tunnel, projects the varanda or balcony, a natural opening in the wall. Here, upon a bench, lounged a few idlers, chiefly negroes, enjoying the fresh draught from the green-avenued bayou below. The atmosphere reminded me of Yambú, yet the thermometer showed only 85° (F.) †

The high altar is at the further and broader end of the Cave. It is approached by a raised platform of dislocated wooden oblongs, showing old graves. The shrine is fronted by a tall central arch, between two of smaller size, all three lined with painted wood, and hung with ex-votos. That to the right opens upon a narrow passage behind the adytum; the ascent is bad, the boarded floor threatens to fall, and there is an odour of death—perhaps the calcaire may be of that kind vulgarly called

---

* Meaning, I presume, a thin plate of stone, which could be used like a gong. The only bells now are two small articles, hung to the usual wooden gallows, and protected by a small tiled roof.

† M. Halfeld found it 95° (F.), nearly blood-heat. The bats of which he complains have disappeared, leaving no sign, and the dead are no longer buried within the cave.

" stinkstone." The left archway is the mouth of a recess heaped up with heads and faces, arms and legs of beeswax, and other offerings which commemorate the sanative powers of the spot.

In the highest part, under the central arch, and protected by a wooden tunnel ceiling, stands the Senhor Bom Jesus da Lapa. The little crucifix is, to judge by the ghastly style of the colouring, modern. A polite devotee assured me that it had been found here, and that, despite many attempts, no one had ever been able to remove it.* Upon the ledge at its feet are statuettes and two candles burning. On the altar below there are more images, and six lights, whilst a massive and expensive silver lamp, bought at Bahia, hangs from the ceiling outside. Beyond the railings of painted wood stand portable chapels of Nossa Senhoras, each about ten feet high, sentinelling the shrine. Also, most important of all, a strong box of iron, labelled in the largest letters, " Papel—Cobre," catches the first glance.

This place of pilgrimage has the highest possible reputation ; devotees flock to it from all directions, and from great distances, even from Piauhy. Sometimes there may be a crowd of 400 visitors.† The average daily receipts, I was told, amount to 20$000, and on Sundays to 50$000. The "esmolas" are paid to a certain Lieut.-Colonel Francisco Teixeira, who is the Procurador of the shrine. My crew when exhorted to visitation, lest they should call their employer " herége," pleaded " who prays, pays." They went, however, and the pilot gave fourteen vintens, the rest two. I left something at the foot of the crucifix ; the old Sacristan did not readily find it, and he hurriedly sent a message, asking the amount of my alms.

We left the fane very little impressed, except by the damp heat. Our next step was to the Porto, on the right bank of Ypoeira. This is the seat of trade. We found a few houses, half-a-dozen sheds, one barca and five canoes. The principal industry is making saltpetre, which is here found in quantities at the south-eastern side of the Serrote. It is a constituent of all these calcareous soils, the effect of atmospheric air decomposing the limestone. The process of extracting it is a mere

---

* Thus, at Cairo and in other Moslem cities, tombs are seen let into the walls of the domiciles. This is where the bierbearers have been unable to contend with an obstinate corpse, which insists upon choosing its own sepulture, and becomes so heavy that none can carry it.

† From Januaria the best road is on the eastern bank of the river.

lixiviation; the chocolate-coloured earth, mixed with stone, is thrown into a bangue or strainer. This is generally a square pyramid of boarding, with the base upwards, equally useful for extracting saltpetre or soap-lye. The poorer people use a hide, supported by four uprights, and both act like jelly-bags. When exhausted with hot water, the nitrous particles find their way, duly filtered, through a tube leading to a "Coche" or trough, often a bit of old canoe. The "decoada," as it is now called, is a thin greenish liquid, which must be boiled in a "tacho," or metal pan, like that used for sugar. This "tacho" is sometimes mounted upon an ant hill. It is purified by repeating the process, and it appears in regular six-sided columns of yellow-white colour. The price is here six coppers; on the Upper Rio das Velhas it sells for 10$000 per arroba. In the Sertão saltpetre is used medicinally for nitre. My specimens were unfortunately lost, and I cannot decide whether the material is or is not the nitrate of soda like that of Chile, which though usefully employed in composts and nitric acid works, attracts so much damp, that it is of little value for making gunpowder.*

We introduced ourselves to the Vicar, the Rev. Francisco de Freitas Soueiro, a native of Lamego, near Douro. He spoke with great reserve about the miracles of the place, and declared that the image must be some 100 years old. The Lapa Sanctuary, however, dates from 1704, and was founded by a Lisbonese, the Padre Francisco de Mendonça (alias da Soledade), a man of considerable property. He set up the figures of Nª Sª da Bom Jesus, and Nª Sᵗ da Soledade, and the Archbishop D. Sebastio Monteiro da Vide,† after sending to it a Visitant, made the Lapa a chapel, and the Padre its priest.

By no means so reticent or so sensible was the Padre Baldoino of the Villa da Barra, who was calling upon the Vicar. He gravely assured me that all the Serrote was blessed by Heaven, and consequently that it must contain gold and diamonds. The crucifix, he said, was at least 367 years old—about the date when the Brazil was discovered—and was worshipped by the wild people before it was found by Christians. His red face became

---

* Contraband gunpowder has, however, often been made with saltpetre brought from Minas, even in the days when the former was a royal monopoly; the latter in 1816 sold for 4$600 per arroba at Rio de Janeiro. A lately-made analysis of the brown Bahian saltpetre gives a fair account of it.

† This ecclesiastic issued the "Constitutions" of Bahia in 1707.

redder when I asked him if another would not do as well. He
declared, with various inuendoes, that the efficiency resides in
that particular figure; that it was the work of a miracle; that it
was formed by a miracle, and that by a miracle it remains. Sub-
stitute for it anything else, and all virtue departs from the Lapa.
I afterwards heard that this reverend was once a person of fair
attainments, but that his devotion to Bacchus had dislodged part
of his intellect.

The Vicar had lately recovered from an abscess in the leg,
which, despite Lanman and Kemp's Salsaparilla, had nearly
killed him. When we spoke of ascending the Serrote, he concealed his ailments, and offered to guide us. He proved himself
a good man, and actually climbed up in his slippers. At the
base of the hill began a thin grove of Xique-Xique, here a kind
of Cansanção or Jatropha urens. This is a tall shrub, with
patches of sharp and venomous thorns radiating from common
centres. It extends to the summit in clumps, and is much feared
by the people. Another unpleasant growth is a small Bromelia,
with cruel serrations. In the lower part I found sundry young
shells of a pink-lipped Achatina (No. 2), which here grows to
a large size. John Mawe (i. Chap. 12) records his astonishment at seeing the eggs laid by this "new variety of helix."
The air was perfumed with the odour of peppermint from a bright
blue floweret, which seemed to have no name. We ascended the
wooded central depression on the western side, behind the main
bluff, and a steep rough path had been worn by the fuel-seeker.
In the shade the thermometer was 94° (F.) The small red ant
stung viciously, and huge iguanas eyed us as if the lazy things
disdained to run away. We found adhering to the lime a hard
red sandstone, with black spots like syenite, and silex with a
conchoidal fracture, which had the tint and the compactness of
Rosso Antico.*

Reaching the summit of the gulley, we started flights of urubús,
which had whitened the pinnacle tops. Here there is no soil
except where the rock is resolved into its original elements. The
jagged surface is like the waves of a cross sea, and in places it
looks as though rain-drops had splashed upon a soft substance.

* This appeared to be a sign of igneous
action; our glasses could detect no signs of
shell in the limestone; and the glazed iron
stone and conglomerate scattered about the
base suggested exposure to heat.

A rude triangulation from below had given 150 feet, a total of about 180 above the stream.* Between the thorns we enjoyed a noble view of the São Francisco, whose inundations extend in places three leagues across. The broad band which glittered in the sun with silver and gold winds in majestic sweeps round the Island of Bom Jesus, the well-cultivated " Canabrava," and the " Itaberáva," or Shining Stone.† On the north is a blue knob, the Brejo de São Gonçalo, beyond the Rio do Corrente, and to the north-east a long purple line, the Serra do Bom Jardim, and the two low domes, near Urubú. Nearer is the Fazenda of Itaberáva, where only the stream-edge is flooded; its green pastures are rich in horses and black cattle. And at our feet lies the village, with its three small streets branching from their nucleus, the square.

In this grand lump of limestone there is sign of convulsion or catastrophe. The growth or upheaval must have been so gradual, that the long horizontal lines are still hardly broken. It is greatly to be desired that some catastrophist, writing upon " geological dynamics," would state precisely the ground upon which he believes that the ancient oscillations, dislocations, and inversions of strata are not wholly explicable by existing phe-nomena, with the Hindu ages and the Tropical and glacial epochs behind them. And when the Uniformitarians shall have won the day—and I presume that the believers in continuity, in the " orderly mechanism " of slow and long-continued movements broken by periodical paroxysms, will win it, seeing how much they have already won ‡—it is to be hoped that they will do better than the Cosmos, which includes under vulcanism, or vulcanacity, " crust-motion," together with earthquakes and volcanoes. Archeus has been proposed for the honour of naming that slow growth which belongs to the earth as to other inanimate things ; so has Ennosigæus. We want something which does not hail quite so far back.

---

* M. Halfeld gives 240 palms (= 172 feet).

† Itaberáva or Itaberába " pedra que luz," is, according to Rocha Pitta, the name of the whole Lapa. The Fazenda formerly belonged to the Conde da Ponte.

‡ In the beginning of the present cen-tury, M. Boubée and others explained the appearance of aerolites, erratic blocks, and similar " problems," by supposing that the earth had dashed to pieces some minor star or planet. This is but a modification of that semi-barbarism which sees in the world-plan disorder and destruction, the work of offended deities. Buckle (i. 800) complains, with feeling, that many men of science are still fettered in geology by the hypothesis of catastrophes ; in chemistry, by the hypothesis of vital forces.

CHAPTER XX.

FROM SENHOR BOM JESUS DA LAPA TO THE ARRAIAL DO
BOM JARDIM.

FIFTH TRAVESSIA, 26½ LEAGUES.

THE RIO DO CORRENTE.—THE SETTLEMENT "SITIO DO MATO."—THE "BULL'S
EYE" AND STORM.—VISIT TO THE VILLA DE URUBÚ.—URUBÚ WILL NOT
BE A CAPITAL.—WE RESUME WORK.—COMPLETE CHANGE OF CLIMATE AND
ASPECT OF COUNTRY.—THE SETTLEMENT "ESTREMA."—REACH BOM JAR-
DIM.—ITS RIVULET AND FINE DIAMANTATION.—TRUE ITACOLUMITE.—BOM
JARDIM A GOOD SITE FOR A CITY.

> Os tres reinos aqui que a opulencia,
> E bases são da humana subsistencia,
> Em Minas e animaes e vegetantes,
> Tão uberrimos são e tão patentes,
> Que não resolve a subida subtileza
> Por onde mais pendeo a natureza.
> *Frei F. de S. Carlos Assumpçao*, Canto 6.

WE bade adieu to the good Vicar and resumed our journey,
although it was already late. Presently a bad storm followed the
sultry "Mormaço," or stillness of the atmosphere, and came rushing up from the south. The lightning, seen through the rain,
appeared a white fire, whereas it was remarkably pink in the dry
air. Dripping with wet, and anything but merry, we made fast,
at nightfall, to the Sitio do Mato, a well-cultivated island; we
fed and we "turned in," to "bless the man who invented sleep."
Mixed with the sounds of mankind, the cry of the night heron
resembled that of the ounce, and the fish splashed a treble to the
grim bass of the falling banks.

Wednesday, October 2, 1867.—Cirrus again and "mackerel's
back" prepared us for more bad weather. We set out, however,
at 4·45 A.M., and ran down the island which had sheltered us;
it thinned out and showed an even richer cultivation than above.
At the bottom of a high bank on the left, came in the Rio do

Corrente,* so called from the swift currents which sweep round the salient angle. We crossed the mouth, some 500 feet broad, of this great stream, which here runs from west to east; its right jaw projects in a long sand-bar, and a dark avenue in its left cheek shows the line of an affluent, the Riacho da Barra.

Below the port, which is flooded, the bank rises 35 feet, driving the main stream to the north-east. The high ground is divided into two waves, and, in the hollow between them, is the manga or kraal for cattle, communicating with the ajôjo raft, which passes them over for Sincorá and the Bahian Chapada. Above rises the village Sitio do Mato, running nearly north and south, a line of mud huts and three whitewashed tenements. We landed below it upon Tauá, a stiff white clay, underlying a steep, sandy ramp. Opposite was *the* flash house—roof-corners adorned with pigeons of white plaster and so forth—belonging to a cattle breeder, Theodoro Antonio de Oliveira. He turned his back to us, as we were walking past him, and, of course, he was a " cabra " or a " bode," probably the latter. Further to the north is a tiled shed covering, a portable chapel and a cross, with its sudarium ; behind it lies the railed cemetery, and a heap of adobes, intended for a mortuary sacellum, whose beginnings were washed away in 1860.

Inland, the bush extends up to the settlement, and the out-lying lands are said to be good for cotton and castor. West-ward, and not in sight, rises a range known as " A Ribeira ; " † between it and the village are many lakelets and ypoeiras, which do not recommend the " Sitio do Mato " for a future capital. The village proper is to the south ; here the floods enter between the waves of ground, and extend to the habitations behind the " manga." The small industries are cotton-spinning and making soap-lye ; we shall now find the " bangue " everywhere ; the animals are barking curs, and pigs, and poultry, especially turkeys. When we wanted to buy fish, the fisherman refused to sell, saying that he had a large family ; and under a shady Joazeiro tree we found, in excellent repair, the good old

* This great influent drains the meridional spine that separates Bahia and Goyaz. Boats navigate it, despite snags, as far as the Porto de Santa Maria, 28 leagues from the mouth ; the banks are said to be grandly forested, and, in places, to be cultivated. One of its many tributaries is the northern Rio das Eguas, and this again has a considerable influent, the Rio Acanhuão.

† The right bank showed a long blue range which the people called de Sant' Inofre (Onofre or Onofrio).

"tronco,"* or village stocks, which have but lately disappeared from rural England. Here they are two long boards, planted upright, and pierced with ten holes, accommodation for five men, "in log," as the Africans say. At times it is used as a pillory, but the offence must be very grave.

Pushing off from the Sitio do Mato, we found the water so deep that the pole would not touch bottom. The effect of the Corrente River is a great sack to the left, and then to the right. The eastern shore is only nine feet high, and the interior is still lower; during the rains boats cut across country to the Villa de Urubú, despising the risk of submerged trees, and the annoyance of insects. On the side is the Fazenda da Bandeira, and below it, a section of the eastern shore, the large island of Santo Antonio,† from which another cross-cut, setting off north-east to Urubú, joins the other. An ostrich appeared, pacing along the shore, but the people have not yet learned to kill it for its feathers.‡

At 1·30, as we were going north with easting, opened up a full prospect of what we had dimly sighted for five hours, and which prepared us for a change of country and climate. On the left bank appeared a "neat's-tongue," projecting in regularly shaped treeless mounds of brown-red hue. This is a spur of the Serra Branca, which, according to M. Halfeld, is a calcareous range; the specimens shown to me were sandstone grit revetted with quartz.§ Behind the Serra begins the plateau known as the Alto do Paranan, rising almost imperceptibly towards the heights which feed that stream. Along the southern side of this neat's-foot begins the highway to Goyaz city,‖ which is here said

* Trunco in St. Hil. (I. ii. 42 and III. ii. 101), who describes it minutely, but makes it like the "Tornilho," a military punishment, and refers to the neck being placed in the pillory. The invention is probably due to the Arabs, whose "Ma-kantarah" has extended to the Zanzibar coast in East Africa.

† Mr. Keith Johnston places on the right bank, about half way between the Lapa and Urubú, the town of "Santo Antonio," which is a mere fazenda or Sitio fronting its large island.

‡ The Welsh colony in Patagonia are buying, I am told, Ema feathers for three-pence per pound, and expect to sell them in England for thirty shillings. In these civilized days, when no head requires to wear the colours which Nature gave it, surely the grey plume of the American bird may, by bleaching and dyeing, learn to pass off as an African.

§ The citizens of Urubú declare that from this Serra an old Minas negro, who was prospecting for gold, brought rounded steely grains, which in the cupel proved refractory. The discoverer died, and the discovery was lost on the road to Bahia. Platinum, of which the people have seen little and heard much, is naturally suspected.

‖ The country lying to the west of the city, is one of the few which the Brazil still offers to the explorer, as opposed to the traveller.

to be distant 150 leagues. The road is described as being safe, and abounding in game and water; the sole inconvenience is a desert tract, 30 to 40 leagues broad, where provisions must be carried. On the right bank was the second distance, a straight blue wall, the Serra do Boqueirão, three leagues beyond Urubú; and the third, still further east, consisted of a saddle-back, a ridge and two lumpy heads, parts of the Geraes attached to the Boqueirão.

Shortly afterwards, the left side, red above and white below showed the Povoado do Mangal, and its Rosario church, with falling front. Beyond its island the stream bent to the north-east, and already, behind a large central holm of vivid red, we descried the white dottings of a town. But now the effects of "mackerel's back" declared themselves. Boulder clouds surged up from west and south, hiding the hills with hangings of rain sheet. To the east appeared the ominous "Olho de Boi," or section of Iris that promised a "temporal." We made, with might and main, the windward bank, where at 4·15 P.M., the roaring gale compelled us to anchor, and to bush the Eliza.* We passed a night of scanty comfort. The guinea-fowl clucked in the village till dawn, and there was another nuisance. Hitherto, we had slept near Corôas or Praias to avoid insects, which are very properly termed "immundicities." Here the weather compelled us to roost under a ridge, with a fall inland, a mere cattle-trail, and a rich breeding-ground for a small and almost minute mosquito, whose sting was like a needle-prick. As a rule the river has been wonderfully free from insects as from snags; this part, however, is an exception. When we least wanted a calm, the gale fell dead, and when light was worthless, the stars hung like lustres from the cloudless sky. The pilots declared that we had escaped from the rains to fall into the power of the wind; it will be seen that they were right. Our course was against the sun, which will presently bring up with him wet weather, but the heavy showers, now falling behind us, must increase the evaporation, and open a way for the cool dry Trade.

Oct. 3, 1867.—At earliest dawn began angry puffs from the red

---

* To prevent the waves washing over these shallow rafts, the pilots have the sensible practice of cutting off the heads of young trees or leafy branches, which, fastened alongside or to the bows, act as screens.

eastern sky, which was striped with cirri of a dull vermilion, and was mottled with clouds, standing out hard and solid as if cut in dark grey paper. This appearance will soon become familiar, and cause many an impatient sigh. The stream turns nearly due east, so every capfull was a head-gale. On the left bank rose the Povoação de Pernambuco, a hamlet of dingy huts nestling below the Ponto do Morro, the south-eastern buttress of the Serra Branca. Here the stream is broken into two arms by the rich and fertile Ilha do Urubú, a mass of grass, bush, and trees, one league long, and shaped like a leg of mutton, with the knuckle-bone down stream. The left channel is the broader, the deeper, and the straighter; we took the right, upon which the town is built, and at once grounded upon a sand-reef. Both sides are low and liable to floods; on the right, at a "port," denoted only by women with water-pots, is the mouth of the Sangradouro, which, during rains, admits canoes to the Sitio de Santo Antonio.

Presently we landed to inspect the town of Urubú, the "Gallinazo," the turkey buzzard. The riverine plain is here low and caked with mud, soon trodden to impalpable silt. A bush of the "Araticum" Annona—here the people mention three varieties of the shrub—shows the limits of the floods. Beyond it begins the vegetation of a dry and sterile land. I saw, for the first time, the "Favelleiro," that arboreous Jatropha, with sinuate leaves, described by Gardner. It varies between the size of a blackberry and an apple tree, and the stiff, quaint look at once attracts the eye. The leaves, resembling those of the oak, but dangerous to touch on account of their cruel, poisonous thorns, are, as often happens amongst "Campo" plants, only terminal, not axile, and planted in tufts at the end of fat twigs. The leaves are used to narcotize water, and to catch birds; the fruit, described as resembling that of the castor plant, supplies oil for the table. The rhubarb-coloured gum, with a faint perfume, is compared with gum arabic, and the wood is made into spoons. The Aloe family musters strong, especially the "fedente babosa," which Liliacea can only be rendered "fetid slabber chops;" the leaf-juice, mixed with oil, and called "Azeite de babosa," is used to correct baldness. A flock of dirty-white sheep, whose fleeces were torn to rags by the thorns, wandered about, seeking what they could devour.

A walk of 200 yards leads to the town, which is the usual long, shallow line, fronting to the north-west. The items are chapels, adobe houses, palm-frond huts, railed compounds and rude gardens, in which the cocoa-nut, with its rounded tuft, rose conspicuous. The main street, the Rua de São Gonçalo, runs along the whole length, and is raised above flood level. Two houses displayed the civilization of glass windows, amongst shutters, lattices, and squares of calico ; of those twain one was a Casa Nobre.* I sent in my letter to the Juiz de Direito, Dr. Joaquim Rodrigues Seixas, who asked us in, gave us coffee, and gallantly exposed himself to a well-furnished fire of questions.

The judge complained that he had lost his memory by living in such a hole, and I can readily believe him. The climate, as so often happens in dry places, unpleasantly close to damp situations, is dangerous. Fevers, or rather "chills," are mild, yielding easily to native practice, tartar emetic and quinine ;† they generally, however, end in spleen diseases. About August, pleurisies are dangerous when treated with the popular simples, fatal when exposed to scientific practice of "lancers and leechers," copious blood-letting, tartar emetic, heavy doses of nitre, and ptisane of a certain emollient Hibiscus,‡ the only harmless part of the "cure."

Santo Antonio de Urubú was formerly known as the Urubú de Cima, the upper turkey buzzard, opposed to the nether turkey buzzard (Urubú de Baixo), a pleasant name now changed to Propiá or Propriá, on the Lower Rio de São Francisco. According to the citizens, this place began the diamantine discoveries, which presently spread to the Chapada Diamantina, then in the district of the Villa do Livramento do Rio das Contas. It may be remarked, however, that in 1755, gems were discovered at Jacobina, on the eastern flank of the Bahian Chapada, and that the Prime Minister, Pombal, forbade the working of the vast buried treasures, for fear of injuring agriculture. The effect of these days of ignorance endured till 1837.

---

* It belonged to Sr. Gualteiro José Guimarães, a merchant who at the time of our visit was pilgrimaging to the Lapa.

† Sulphate of quinine is much used in the Brazil, and with little prudence by the people ; thus while it relieves one disease, it often brings on another. Homœopathy has done much good by preaching against the abuse, and by substituting pilules for doses of six to ten grains.

‡ Cozimento de Althéa, which Moraes translates Malvaisco (Hibiscus). The System (60) also gives Alcéa, and describes the use of the Sida althæifolia.

The judge congratulated himself upon the fact that, under his
jurisdiction, there had been only four murders in four years.
The municipality contains only 3051 voters; in 1852—54, M.
Halfeld gave the district 731 fires, and 7204 of all sexes and
ages.   The town cannot contain more than 300 houses, and
when full 1600 to 1700 inhabitants.   They live and die in the
greatest ignorance.   I was astonished at the absence of all
progress in these western outstations of the great Bahian Province,
whose chief city was once the metropolis of the country, and
whose seaboard is now one of the most prosperous and populous
portions of the Empire.   Everything that we see denotes
poverty, meanness, and neglect; a Fazenda in the interior of
São Paulo or of Minas is equal to a town here; and whilst the
majestic São Francisco flows before these hovels, and there are
excellent lines for routes both to the seaward and to the interior,
the people have wholly ignored their communications.   This is
at once the cause and the effect of their semi-barbarism; they sit,
calling upon that Hercules, the Imperial Government, but they
will not put shoulder to the wheel.

Urubú will not be a capital.   The port is bad, the lands are
deeply flooded every year, and the Serra do Boqueirão is too far
to be utilized.   I heard, however, of olhos or water pools, which
possibly exist in it, and these metamorphic formations may be
found to be rich in minerals.   All vaunted the fertility of the
inner country to the east and to the south-east; they declared
that four shrubs give three pounds of uncleaned cotton, formerly
an item of export to Bahia.   The so-called "Irish" potato is
small but very good, and onions grow from their own and not from
imported seed.   In addition to the usual list, the soil produces
cucumbers, ground-nuts (Arachis hypogæa, here known as Man-
dubi, Mundubi, or Manobi), and oriental Sesamum (Gergelim or
Jerxelim).   Oranges and limes* grow, and the tamarind,
though stunted, produces an abundance of fruit, which the
Africans know how to prepare, while the Bahians do not.   I
also heard of soils in which the "Mandioca brava," the poisonous
manioc, spontaneously becomes "Aipim" or "Macaxeira,"
the sweet kind.   The Judge and Juge de Paix, Dr. Claro Fran-

---

* The sweet lime (Citrus Limonium) is   limetto (C. Limetta) is simply limão or
known as limão doce; the sour lime or   lima.

cisco Negrão, also assured me that they had seen three colts got by stallions out of she mules, adding that the offspring was a most unsightly animal.

The principal " curio " shown to us was a bit of compact uncrystallized alum from Mocahubas,* a town fourteen leagues to the south-east. It is said to appear like stalactites in the caves which riddle the Serra do Machichi, and, as we were floating down stream to the north-west, the pilot pointed out a white mark which he declared to be the mine in a range right behind us. The people ignore the easy art of purifying their " pedra hume." The ruddy resin of the Angico Acacia, which here forms their forests, was vaunted as a pectoral and an expectorant ; and the yellow gum of the Jatobá, light as amber, serves to caulk boats. The chief of the small industries is weaving hats, for which the Aricuri palm† supplies material; here they are worth 0 $ 200, and they sell down stream for 0 $ 500.

We walked up the Rua da Palha, which runs parallel with and inland of the São Gonçalo ; two lines of very humble houses led to the large square behind the Matriz of Santo Antonio. This fane is built of brick, mixed with boulders (rolados) from the opposite Ponto do Morro, and with iron-stone from the river banks ; as yet the belfries are wanting. There is a Casa de Camara, a detached jail and a vicar-general's house, but no such things as parish registers or public documents. Here the dry, sandy, and silty plain is covered with the Quipá, a dwarf cactus, about eight inches high, with fine, hair-like, but sharp thorns, radiating from white spots. Its flat plates contrast curiously with the tall " organ," the five-sided chandeliers (C. candelabriformis), the short, thick cylinder (C. brevicaulis), and the serpent cactuses around it. My friends showed me upon the Quipá what appeared a white web, but after crushing it, the fingers were stained with a rosy-pink juice. This is the indigenous cochineal-insect, and it extends throughout the dry riverine regions. It is looked upon, like most unknown things, as a magnificent mine of wealth, but years must pass before it can be made useful in commerce. I told the bystanders about Tenerife, which had imported from

---

* In Mr. Keith Johnston's map, " Macaúba." ·

† Commonly spelt " Ouricury," also written Aliculi, Aracui, and Arari (Cocus schizophylla). According to the System, the juice is used in Bahia for curing ophthalmia.

Mexico the large succulent nopal, and the fat insect. They manfully supported their fellow-country growth, the Quipá,* which was juiceless as a shoe-sole, declaring that during the rains it swells to thrice its present size. Here, as elsewhere in the Brazil, men hold the "esprit du mieux ennemi du bien;" to advice they are untameable as flies; their minds must grow, like those of infants or "Indians," by example rather than by precept, and though intelligent and imitative, they always require improvements to be subjected to the faithful eyes.

Our friends "convoyed us a bit," gave us oranges and limes, and saw us off at 11 A.M. The north-east wind, cold in the burning sun, blew in strong blasts (refegas), frequently repeated till 3 P.M., and hindered progress. And now we noticed that a complete change of soil and formation, climate and physiognomy, had taken place—the frontier being Urubú, and its portal of hills. The limestone country, with its great productive powers, and the rich Maçape clay, have passed into sandstone, and the wooded banks have altered to a "Carrascal," or low bush. This ground in places produces the small maize, but agriculture and breeding flourish only in the "Geraes," or inner lands. The river, which before could spread far over its wide, flat valley, is now narrowed, and the overflow is checked by bounding ridges, through which the larger tributaries must twist; the eastern wall will last with breaks till near the Great Rapids, the western till the Villa da Barra. There is no general name for the range, each place christens its own section; that to the right is usually spoken of as "A Serra," while near Urubú the opposite wall is the Serra Branca; it then becomes Serra de Santa Catharina, the O Furado (or Serra Furada), the pierced, and so forth. The effect of these containing walls is to form a funnel, up which the Trade, now to be our deadly enemy, blows violently; the greatly increased evaporation is carried up due south, hence the lands on the higher stream are drenched, where here all is bone dry.

These Serras are disposed in straight and in slightly waving lines, which viewed from the stream appear to be great lunes and crescents, approaching and diverging. The regularity of their shape, the flatness of the summit-line, and the steps and benches, which run in straight course along them, suggest that they were

---

* The fig of this cactus is eaten, but it is full of seeds.

formed under water, and that presently they rose to be river branches. As the bed, whose general course is from south to north, winds between them, the ridge of one side is often confounded with that of the other. From the plains connecting their feet with the stream-banks, rise detached and mound-like knobs, here single, there in groups, now perpendicular to, then running diagonally from, the bounding Serras ; in places they form bluffs, striking the bed at right angles. The material of all the heights is sandstone, in places revetted with quartz, and containing, according to the people, gold ; we often see the strata exposed in the precipitous flanks facing the river. Further down we shall find iron in the lowland lumps. The surface of these formations is a poor, shrubby growth, chiefly thorny, and here the giant Cactus, the Acacia, and the Mimosa are kings.

About 3 P.M. we touched at Estrema on the right bank, which, though high, is swept by great floods ; here was a whitewashed house, a few huts, and various "timber," post and rail, and snake-fence. We had been told that the owner had a goat for sale ; he was absent, and we were disappointed. At sunset we made fast to a corôa, opposite a little hamlet, the Riacho das Canôas. The crew was living upon a bit of dried "bacalháo, or salted cod, whilst the fish leaped and splashed in all directions ; they had no bait. Ashamed for them, I made the youth, Agostinho, arm a hook with a bit of meat, and in a few minutes we had enough for a day's food. The worst was the Curuvína ;* the Matrincham† is not bad, and a kind of Pirá‡ bit freely.

Oct. 4.—Sunset and sunrise had both been red, nothing could be more delightful than the dawn, but we felt that, as in Hindostan, the noon and the afternoon would make us do unlimited penance. The gusts and raffales which blew at times during the night, fell into a fitful slumber, which, however, did not in any way delude the watchful suspicions of the pilot. Here the river itself offers prime conditions to the breeze ; it will broaden to a

---

* Gardner writes Curvinha, M. Halfeld has Caruvina. The fish is about two feet long when full-grown, scaly, with pale, soft meat, anything but delicious. The head contains a white bone, which is pounded and administered for various diseases.

† Gardner writes the word "Matrixám ;" it is one of the Salmonidæ, smaller than the Dourado, and very common in the upper waters. Yellow and scaly, it grows to the length of three to four palms, and is a favourite food with fishermen.

‡ Also called Tamanduá ; it is a long-headed fish, with light-blue tinge, about two feet in length, and tolerably good eating. One variety is the Pira de Couro, another the Pira-pitanga (M. Halfeld, Pri-petinga). There is also a sea-fish of this name.

mile and a half, and split into channels, often of equal depth, and both filled with stranded trees and snags. The river islets greatly increase in size ; we shall presently pass one about a mile in breadth, and five miles long. These formations are mostly of sand, covered with thin humus, 'green with grass, in places cultivated, and bearing tall trees, amongst which the Grão de Gallo is conspicuous.

After a few minor features,* and a prudent halt at an " espera " on the Bahian side, we sighted at the bottom of a "big bend," the Arraial do Bom Jardim. Tiled huts appeared on the right bank, a wave of higher ground offsetting from the Serra ; they lay some five miles behind or to the east. This range was patched with green, suggesting that it is better watered than the hills about Urubú, and the nearer surface appeared as if the bush had been burnt, or that a cloud was fitfully shading it in patches. Dark streams and sheets, apparently spread by an eruption, invade one another, alternate and strive for mastery ; at last, puzzled, I ascended a hill side, and found the gloom to be produced by a matted aromatic shrub, with leafless twigs of umber-brown, and growing between stones, set off by the light of golden yellow grass.

The left bank opposite Bom Jardim is a lower level, a mass of tangled forest, cut by many an ypoeira, and nothing but an embanked causeway could render it passable. The bend is fronted by the western containing ridge, Serra Furada, a tall and regular line, running north and south ; here it is some seven miles distant from the stream, but below only about a league. On the waterside appeared the hamlet, Passagem (do Itahy or Bom Jardim), with its ruinous chapel, Nᵃ Sᵃ do Bom Successo. Where piles can be fished out of the stream, no one thinks of planting them under their floors, and of thus securing ventilation and escape from the floods.

We landed at the Riacho de Santo "Inofre,"† above the set-

---

* After one hour we passed the large green Ilha do Saco, and on the left bank, when the thalweg is to the right, rises the Fazenda (H., Povoado) do Saco do Militão. A rugged line in front, apparently on the Pernambucan, really on the Bahian side, presently shows peaks and distances, and in the pure clear air it seemed to be at no distance. Another hour brought us to the

Ilha do Gado Bravo (H., Ilha do Barreiro), some two miles long. We took the normal line, the western channel, and facing to the north-east, we were compelled to anchor by a head wind, which, meeting a current like a mill-race, produced an angry tide up.

† In Mr. Keith Johnston the "R. S. Onofrio" is marked with dots, and made to

tlement. It rises to the south-east, draining, with the aid of its affluent, the Boqueirão, the north-western face of the Serras das Almas de Sincorá and dos Lençoës;[*] the eastern slopes forming the great Paraguassú. Small canoes ascend it for some leagues, during the floods, to the Vargem de N^a S^a da Guia. During the hot season it is nearly dry, but leats and courses would readily create reservoirs in the lower levels. The mouth of the green avenue is about forty feet broad, the left jaw is a sandpit, the right is a stony platform, composed of ferruginous "cánga" and pebbly conglomerate, pasted with hydrate of iron. In time it will become a steamer-pier; the stream swings to it, always allowing a deep-water approach; it is flooded for a few days during the year, but a levée higher up would, if necessary, obviate the inconvenience. At present it is used only as a ground for washing linen. The shallow pits and pot-holes supplied the finest sign of " diamantation; " the people, who leave it unworked, declare it to be brought down by an eastern influent, the Riacho do Pé da Serra, where they still dig gold.

Below the mouth of the stream lies the little arraial. The water froths against pure pottery-clay of dull, dead white, worn into holes by the tongues of cattle ; in the upper levels it is mixed with sand. The settlement consists mainly of a single line, whose railed backyards extend to the river-brink. There are scatters of houses inland of this line, including a ranch for travellers. The total may be forty, whereas in 1852—4, there were 300 inhabitants under 103 roofs. The people live by breeding cattle, by agriculture, and by fishing. We bought a three days' provision of the fine Caçunete [†] for ten coppers (0 $ 400). Behind the village lies a sandy plain, about 100 paces broad, with thin pasture, and showing symptoms of flood. Beyond it the ground, thickly bushed over, rises high above all inundations, and here will be the site of the settlement. At present it is occupied by the vicarless church, N^a S^a da Guia, whose windowless front had been freshly whitewashed. Like the hamlet it faces to the west with southing. A heap of torrent-rolled

come from the western versants, which send to the Atlantic the Rio das Contas.

[*] In a map lately published by the concessionees of the Paraguassú Valley line, the " Paramirim " is the main western drain corresponding with the Paraguassú to the last. The details in the text were supplied to me by the people of Bom Jardim, and therefore are open to doubt.

[†] A fish with few spines, highly prized, and supposed to be a kind of Surubim.

stones (pedras de enxurrada) lay at the wall-foot, and at once showed the origin of the diamonds and the gold. There were large pieces of laminated quartzose sandstone, in fact, true Itacolumite. Mostly it was reddish, like a half-burnt brick, exceedingly compact, and streaked and dotted with finely disseminated mica; other specimens were purely white, and their coarser texture showed the grain distinctly. The formation is found upon the hilly Geraes, three to six leagues to the north-east of the river's right bank, the strata are often too thick and solid for use; it supplies, however, the country-side with the slabs for flooring massive ovens, and it equalled in size those "Pedras de Furno," which I had seen near Camillinho of Diamantina.

We were much prepossessed by the general appearance and the capabilities of the land; even the phlegmatic German exclaims, " É esta uma das mais agradaveis paragens à beira do Rio de São Francisco." The people appeared comparatively healthy after the wretched pallid faces of Urubú, and even the horses seemed better bred. The prospect is charming, and this must always form a great consideration, when estimating the future value of a place. The channel is narrow, compact, and unencumbered with shoals, while the current is not too rapid; sweeping to north-east, and frequently to north-west, it throws its main current against the bend, whilst the general wind, being easterly, and blowing over a high and dry country, the evils which might arise from ypoeiras, bayous, lakes, or lakelets in the low riverine valley are corrected. Building-room is endless, material abounds, and in the vicinity are hills which will allow change of climate.

Bom Jardim, a name of good augury, is the only site yet seen which deserves to become a city, or which can pretend to be the capital of the long-expected province or territory. In some points, especially as regards river-navigation, it is better than, in others it is inferior to, its rival down-stream, Chique-Chique. The position is central, about equi-distant from Januaria on the south, and from Joazeiro to the north. It is nearly due west of São Salvador, metropolis of the opulent commercial province of Bahia; it is nearly due east of Palma, one of the most important cities, in agricultural and cattle-breeding Goyaz, where the navigable Paranan or south-eastern branch unites with the Rio

Maranhão to form the grand Tocantins. It thus connects with the Atlantic by two roads, more and less direct. The water-way is down the Rio de São Francisco. The land route is *viâ* the line of the Paraguassú River, which passes by Cachoeira city, the head of Bahian steam navigation. I will say nothing about the steam tramway, which proposes to run along the southern valley of that stream, as the ground is absolutely unknown to me beyond Cachoeira. A glance at the map, however, will show that this has the advantage of a riverine plain, whereas both the Pernambucan and the Bahian Anglo-Brazilian main-trunks are distinctly "cross-country." Meanwhile it has been strongly advocated by Mr. John Morgan, of Bahia, who has had the advantage of a thirty-five years' residence; and the works have, I am told, commenced under every advantage.

Finally Bom Jardim connects by land and water with that Brazilian Mediterranean, the Amazons; and we may safely predict for it high destinies, of which it is at present naïvely unconscious.

CHAPTER XXI.

FROM THE ARRAIAL DO BOM JARDIM TO THE VILLA DA BARRA (DO RIO GRANDE).

SIXTH TRAVESSIA, 29½ LEAGUES.

THE CARNAHUBA, OR WAX-PALM.—VINTENS OFFERED TO SANTO ANTONIO.—
FIRST SIGHT OF THE ARASSUÁ RANGE.—THE GULL-FAIR.—BIG CRANES.—
THE TOCA, OR CAVE OF SAINT ANTHONY.—THE THORNS.—THE VILLAGES
OF THE PARÁ. — THE LEATHER-COAT BIRD AND THE CHAMELEON. —
APPROACH TO THE VILLA DA BARRA DO RIO GRANDE, A PROPOSED
CAPITAL. — THE RIO GRANDE AN IMPORTANT INFLUENT. — THE VILLA
DESCRIBED.

> Onde a natureza
> Bella e virgem se mostra aos olhos do homem
> Qual moça Indiana, que as ingenuas graças
> Em formoza nudez sem arte ostenta.
> *(Poesias B. J. da Silva Guimarães.)*

As the wind fell we put off, and presently landed on the right bank, below Cachoeirinha. At this point a short projection of stone makes the water dash and murmur, but in no way injures the thalweg. We broke through the tangled bush and found a sandy plain between the stream and a knob of thickety sandstone hill, distant about 100 paces. The surface sloped away from the river-ridge to a hollow paved with flakes of mud; it must be a water-course during the rains. All the ledge was cut by paths leading to the various settlements, cattle grazed the thin grass, and the sheep besides being fat, were woolly and not hairy.

Amongst the Angicos and the Myrtaceæ, one of which was mistaken by the " Menino" for a Jaboticabeira, now alas ! no more, we observed a white-blossomed bush, much resembling in perfume and physiognomy, English " May." And here we saw for the first time in situ the beautiful wax-palm known in the Brazil as Carnahúba (Carnauba), and Carnaíba (Corifa cerifera, Arrud. Copernicia cerifera, Mart.), the Caroudaï of Spanish America. Its habitat is the riverine land upon the streams of the Pernam-

buco, Parahyba do Norte, Ceará and Piauhy; during the last
few years it has been introduced into gardens near the coast.

The Carnahúba, when first appearing, is a mere bunch of fronds
projecting above the ground. As it advances the trunk is clad
in a complete armour of spikes. The fronds, as they fall off,
leave their dull brown petioles in whorls or spirals winding round
with or against the sun. When not higher than a man the
youngster's pith or heart yields, when crushed in water, a fecula
somewhat like tapioca, white as manioc, and useful in times of
drought or famine. At a more adult age it puts forth a thin
shaft, smooth, clean, and grey, like dove-coloured silk, which contrasts strangely with the six feet of corrugated chevaux de frise—
the magnified thistle—which protects its base. After the fifth
year it assumes its full beauty, the cruelly-thorned leaves distinctly fan-shaped, and with long rays rising from a spindle which
attains a maximum of thirty-five feet, are peculiarly picturesque
In old specimens the trunk is raised, after the fashion of palms,
upon a lumpy cone of fibres or aërial rootlets, a foot high. Some
eccentric individuals have narrowings and bulgings of the bole,
others encourage creepers to form in masses upon the frond-
petioles below, and suggest the idea of a tucked petticoat. The
vitality of the tree is great, it resists the severest droughts, and I
have seen instances when the trunk lay upon the ground and the
upturned head was still alive, fighting to the last. It grows to a
great age; people mostly decline to mention the number of its
years.

The Carnahúba is justly considered, both for man and beast,
the most valuable palm of the Sertão. Its gum is edible and the
roots are used as salsaparilla. The mid-rib is rafted down the
streams for fences, the fibre is worked into strong thread and
cordage. The leaves are good food for cattle,* they form excellent thatching, and the fibre is made into "straw-hats," ropes,
and cords, for nets and seines. The fruit is in large drooping
clusters of berries, which in places strew the ground. When
green the nut resembles a small olive; it ripens to a brilliant
black, and attains the size of a pigeon's egg. The pulp, boiled to
remove its astringency, becomes soft like cooked maize; it is considered good and wholesome, especially when eaten with milk, and

---

* I have read of, but have not seen this: the part usually given to cattle is the
miolho, or pith of the young tree.

animals readily fatten upon it. The ripe berry is usually eaten raw.

The most notable property of this palm, according to Koster, was discovered in 1797, by the Portuguese naturalist, Dr. Manoel Arruda da Camara;[*] the latter communicated it to Frei Jozé Marianno da Conceição Vellozo, who published an account of it in the "Palladio Portuguez." The leaves of the young tree, when two feet long by about the same breadth, are cut and dried in the shade. They then discharge from the surface pale grey-yellow dusty scales, which, melted over the fire, become a brown wax. Cereous matter is also procured by boiling the unripe berries,[†] and chiefly by scraping the central spike, which prolongs the tree. The wax occurs mixed with heterogeneous substances, bark or fibre, and it loses considerably by sifting. The material is tasteless and soft to the touch; the smell has been compared with that of newly made hay. Its chief fault is its brittleness; this, however, is remedied by mixing with three-parts of vegetable one-part of animal wax, or 1-8th to 1-10th of tallow. Carnahúba candles are made upon the seaboard; but I saw only one "dip" upon the Rio de São Francisco, where, a little lower down, the palm is found in forests. The colour was that of rhubarb, yellow or brown sugar, and the light was not to be compared with the worst "Paraffine."[‡]

Another league placed us at the head of the Ilha da Pedra Grande, the largest yet seen, and where the river contained a greater breadth of land than of water. We took the right-hand channel, although the left is marked in the plan; perhaps the crew did not wish to land at the cave of Santo Antonio in a rock lump (Morro da Imagem de Santo Antonio), near a remarkable

---

[*] He published at Rio de Janeiro in 1810 two brochures, which were analysed by Koster. Appendix, vol. ii.

[†] This also is from books. I do not believe that the fruit is used to extract wax.

[‡] Koster tells us (quoting vol. xxxi. p. 14, Trans. Philos. Soc. 1811) that the Count of Valveas (the minister Pombal, Count of Veiras) sent from Rio de Janeiro to Lord Grenville a specimen of the "carnaubaa" wax as an article of export, produced between N. lat. 3°—7°. The brown-yellow colour of the dust was attacked with weak nitric acid, and exposure to air on glass plates. After three weeks it became a pale yellow, with a surface almost white. The same change was effected by reducing it to thin plates, and dipping them into an aqueous solution of oxymuriatic acid. Made into candles, with properly proportioned wick, it burned uniformly and with perfect combustion. It was found to differ from other species of vegetable wax, such as that of the Myrica cerifera, lac and white lac. The latest authority upon the subject of this palm "Notice sur le palmier Carnauba," was published at Paris, 1867, by Sir M. A. de Macedo, 1 vol. 8vo.

buttress, the Morro do Pichaím.   They contented themselves with throwing a vintem into the water, reminding me of my Beloch escort and their slender gift to the holy but angry Shaykh, who lies upon the banks of the Pangani River.   We cheated the mosquitos by anchoring upon a sand-bar below the Fazenda do Barro Alto, and were regaled with the music of song and drum, which extended into the smallest hours.

October 5, 1867.—Appeared in the yellow of dawn a pretty site, the Limoeiro Fazenda, backed by the Serrote do Limoeiro, an assemblage of sandstone heaps and hills, here and there tied and compacted with ribs and ridges; its containing wall vanished to the north-west.   From the Fazenda Grande further down, a man put off, bringing for sale a neat new saddle, like the Egyptian donkey pad, and priced at 8$000.   At "the Carahybas" a boat-load of the last night's revellers greeted us with shots, and we returned shouts.   The hierarchy of the river formerly was established with a certain rigour which, however, is fast disappearing before the " levelling tendencies of the age."   The canoe was expected to halt and compliment the raft, by trumpeting or blowing the conch; the raft showed the same deference to the barque, and the saluted craft passed proudly on without deigning reply.

Shortly before noon, as we passed the islands do Meleiró and do Sabonete, the wind fell to a dead calm; all Nature seemed to take a siesta, the air was cloudless, and the long level in front showed a silver plate of water narrowing near the horizon to a thread.   Behind lay a charming prospect, strata of golden sand supporting emerald bush, a warm ruddy buttress flying from the back-ridge of sandstone, a mound of purple distance, and a far perspective of sky-blue peaks.   About noon we opened the Riacho das Canôas;* this is the half-way house for the pilots of Joazeiro, as is the Villa da Barra for those of Januaria, and thus the boats overlap.

The stream, now bending east, showed a brown saddleback, apparently on the left bank, and quite close; it was the Morro do Pará, on the right shore, and distant.   At its foot

---

* Mr. Keith Johnston gives the "R. Canoas," making it head near the Rio do Corrente to the south-west.   It is a brook of little importance.   At the mouth is "Passagem," a small well-situated settle-ment on a wave of ground; it lives apparently upon a ferry-raft used by passengers and animals, bound to the Bahian Diamantine range and to the provincial capital.

seemed to nestle the Penedo da Tóca, above yellow with dry tufty grass, and below dark, with water-glazed sandstone. The far distance was bounded by a broken blue range, on the Bahian side by a tall ridge with a pyramid peering above it, a central saddle-back connected by a low wall, with a lion couchant on the left. This is our first view of the " Serra de Arassuá."

As we approached the Penedo buttress, the sudden curve made the stream run swiftly, and form, near the left side, an eddy and a boil, which the pilots called a " Remanso." A sand bank to the right showed a kind of gull-fair. The larus and the sterna, essentially wandering and restless birds, may have been trooping preparatory to a jaunt during the approaching rains. Amongst them the rosy Spoon-bill (Platalia Ajaja) gathered in patches forming a flower bed ; and the Guara, or red Ibis (Ibis rubra, or Tantalus rubra),* with still brighter plume, reminded me of fla-mingo-companies. Amid the variety of gloomy divers and snowy herons, large and small, stood aloft the Jaburú (Jabirú),† here also called Tuyuyu (or Touyouyou, Mycteria americana, Linn.), about four feet tall, with a bare jet-black head capping its purely white plume. It haunts the banks and sand-bars, where it passes the time in fishing ; ‡ hence the people do not eat it, declaring that it tastes of fin as much as of feather. We shall often see it all down-stream, especially in the morning, when it wings its way in regular triangles, flying low enough to be shot down ; and amongst the chatter and the screams of the smaller birds its loud hoarse voice sounded " like the chaunting of a friar." Mr. David-son compared it with the sand-hill crane of Florida.§ I could not but remember the " adjutant-bird " of old.

We paddled to the left bank, were swept down-stream by the eddy, and poled up to the landing place, at the base of the rock. A rough cross to the east directed us to the " Tóca de Santo

---

* This ibis was of importance to the "Indians," who used its fine plumes in their full dress. There are several kinds, the white and the green (Tantalus Cayan-nensis), which the Tupys called Garaúna, blue or dark Ibis, and which was corrupted to " Carao."

† Mr. Bates (i. 282) mentions the Ja-burú-Moleque (Mycteria americana), a powerful bird of the stork family 4½ feet in height.

‡ Prince Max (iii. 146) heard that it was a bird of prey which devours other volatiles. This the pilots deny. Lieut. Herndon found the Tuyuyú grey on the Amazons ; the pair " which he succeeded in getting to the United States were white." He also mentions a " large white crane, called Jaburú " (p. 229).

§ Other common species are the white Courica (Ciconia americana). A Tanta-lus albicollis, with white and black plume, and loud harsh voice, is mentioned by Pisen and Marcgraf. The Garça real (Ar-dea pileata, Lath.) has a black head and a yellow-white coat.

Antonio—holy caves are now becoming banal. This tunnel, seven paces long and six feet broad, opens to the south a mouth eight feet high. The ceiling is pierced with a natural shaft; the floor is of dry caked mud, and the highest water-mark is ten feet high at the entrance. We found inside a flight of bats, whose perfume was the reverse of pleasant, and a taper of the usual brown bees'-wax, curled up like the match of a matchlock, was stuck up against the wall. The formation is a hard, red, laminated Itacolumite, with dots and particles of mica; the dip is nearly vertical.

Seen from the stream in front, this "penedo" appears a sharp roof-ridge of stone, somewhat like a cocked hat, tapering to the north-west. Externally the profile has a strike nearly north and south, and cleavage lines dipping 45°, split by other fissures nearly at right angles. We failed to ascend the eastern wall, which was worse than precipitous. Where it thins out, however, the slope is easy. The summit, 100 feet above the plain, bristled with slabs serrated and set almost on edge. The Itacolumite was striped with broad bands of white quartz, and the junction may be the birth-place of the diamond. The stone would readily have cooked a beefsteak, yet it sheltered the goatsucker, which rose in pairs, flitted past as if thrown from the hand, wheeled suddenly above ground, and hid itself nestling a few yards from our feet. On these rocks also the coney had his refuge. The brown Mocó (Caira rupestris)[*] peeped out of its home, stared curiously from side to side, and, scenting danger, sprang back with the action of the rabbit. The riverines hunt this animal, and declare the flesh to be excellent eating. It is a congener of the tame variety which, preserving its voice, changed its coat during the process of domestication, and deceived the world by calling itself Guinea pig and Cochon d'Inde. I was simple enough to ask, when in Guinea, whether it was at home there.

Santo Antonio has not been so successful with the thorns as was St. Peter with the frogs. We scraped unpleasant acquaintance with the Macambíra, a Bromelia whose thorns, shaped like a bantam's spurs, are sharp as awls. The gregarious Quipá Cactus

---

[*] It was called Kerodon by M. Fréd. Cuvier, and is mentioned by every traveller in the Brazilian interior, from Koster to the present day. In the Sierras of Peru Lieut. Herndon (chap. 4) seems to have found a dish of stewed Guinea pigs very good.

did its little best to sting. The ugliest customer of the nettles by far was the Urticacea which the men here called Cansanção bravo (Loasa rupestris), a poison nettle. The tall stem was garnished with short sharp bristles which seemed automatous, finding their way through the air. Worse than any Dolichos, they penetrated the skin in dozens, caused a violent itching, and raised an eruption, which disappeared only after suppurating. The only non-spinous tree that grew upon these rocks was a stunted and silvery Cecropia. Thus the ancient "Indians" found growing together the two shrubs, large nettle and the sloth-tree, which supplied fibre-thread for their thick, heavy, and enduring cloths.

The rock top gave a fine view of the glorious river-plain below. The stream, dotted and patched with islets, made a long sack from south to east and north. The Morros do Pará and da Torrinha, on the right and left, seemed planted to keep it in place. To the north-east the Arassuá range displayed its huge folds and slopes, and far to the south-east giant ramps stretched between earth and cloud. Between the blocks was a dead level which, according to some informants, extends as far as the northern breakwater of the great Paraguassú valley.* The riverine plain is populous and well cultivated. It showed the usual features, hut clumps, bright green clearings, dark green woods, and yellow grass, which four several prairie-fires canopied with a long purple awning of smoke.

Once more shooting across the eddy, we reached the elbow upon whose right bank stands the Povoação do Pará; where "Barboza of the Girls" struck up the "riling" ditty—

Não me querem bem, não me querem máu
Pará e longe, não vou lá.

The mouth of the Para-mirim, or, as the pilot called it, Paranamirim,† opened with a line of green to the south of the settlement, and formed a sand-patch upon which cattle basked. The houses of red clay, and ashen grey thatch, set off by a few domes of fresh-foliaged trees, ran in lines at the south-western foot of the umbre-

---

* The road passes by a town, known as N. S. do Bom Caminho: despite which, many informants complained that it was in a desperately bad state.

† This is the Paramirim which in Mr.

John Morgan's map encloses the Riacho of Bom Jardim. According to the people it is of no importance, and certainly the mouth does not argue a long course.

coloured hill. The next feature was the Morro da Torrinha, a stony ridge beginning at the water-side and forming a double tongue, the more distant lump being the higher. At the point were tall trees, and above rose brown bush. This is the Fazenda laid out by the Commendador Antonio Mariani, and the ten huts and houses to the water front are so disposed that the people can fly from the floods to the knob-top. Passing sundry islands, all more or less inhabited, we anchored at night-fall near a low sand-bar below the Ilha do Timbó. Our visit disturbed hundreds of water-fowl, and again at night we heard a concert of drums and voices. There is no want of "jollity" here. Yesterday, however, a blind white had begged alms with the true drone and whine of the professional "asker"—an event rare enough to be chronicled.

Oct. 6.—At night, the Vento Geral gave way to the westerly land breeze, and the sensation was of unusual cold. When we awoke the river had risen some eighteen inches, floating away one of our paddles, and placing us at some distance from the sand-bar.* These "repignetes," as the barquemen call them, are swellings and subsidings preliminary to the flood of the year; according to the pilots they occur three or four times in succession. The morning was pleasant, but it showed distinct signs of wind. As the sun, between 6 A.M. and noon, warms the earth and water, the cold breeze comes up with puffs, and blows hard till about 2 P.M., when the equilibrium of the atmosphere is restored. Then by slow degrees succeeds a calm, which often lasts till evening. Near Remanso we shall have alternately one day of wind and another of rain.

Setting out at dawn, we presently sighted, from a distance of four to five miles, the Serra do Brejo, or western containing-ridge, trending to the west, and bending north; it is faced by the Assarauá, rising like a gigantic insulation, and capped by a high white cloud, like a second storey of island in the light blue sky. The near banks were flat, grassy ledges, producing an abundance of the hard, gnarled, and dark-barked Jurema Acacia.† The

---

* We had, I have said, an anchor with us, and this proved of no little use. Generally rafts, and even barques, are made fast to upright poles, and many an accident has taken place from their breaking loose. The men work hard, especially if they wish to reach a town in time for some fête, a watch at night is never set, and the craft would be amongst the rapids before the sleepers would awake.

† This Acacia was first noticed about Malhada and Carunhanha, where it is sup-

trees were tasselled at the branch ends with nests two or three feet long, bags of dry and thorny twigs, opening with a narrow entrance at the upper end, and comfortably lined with soft grass. Probably, like the clay hut of João de Barro, it receives an annual "annexe." Here the tenant is called Casaca de Couro,[*] or "leather-coat."

We had to battle with the winds and the wavelets, which rose as by magic; and off the "Angical" Fazenda the enemy had the best of the affair, and kept us at halt for three hours. This is a large breeding estate in a sack on the right bank, which is sandy and produces fine Cocoa-nuts, Carnahúbas, and Quixabeiras. From a point a little below this, canoes during the floods make a short cut across country to the Ypoeira of Chique-Chique. Approaching the Ilha do Camaleão,[†] of the Chameleon, we saw ahead, the white houses of the settlement, attached to a huge pile, projecting over the green left bank. The northern approach to the Villa da Barra do Rio Grande is by the narrow "Corredeira," or channel, formed near the western side by the long thin island-ship, the Ilha do Laranjal; to the east is the main line of the São Francisco, a mass of sand-bars and beaches. The course is then across the mouth of the Rio Grande, which here runs to the east with northing, and discharges into the São Francisco. Its right jaw pushes out from inundated thickets a clay point thinly covered with bush, and in the centre there is a shrubby island. The current at the confluence, where 1200 feet of breadth rush to meet 6100 feet, strikes heavily upon the Pontal, or projection which faces to the south-east, and separates the two rivers. The material, fortunately for the town, is a perpendicular bank

---

posed to give the finest charcoal. It will become more abundant as we approach the Great Rapids. The people speak of two qualities (species?), the Jurema (alias Gerema or Geremma, Acacia Jurema), and the Jurema Pesta. The ample growth of Acacias and Mimosas Angico, barbatuirão and Ingá, combined with the saline soil of this part of the valley, prove how well-fitted it is for camel-breeding.

[*] It may be the same as the Gibão de Couro (leather-jacket), a gobemouche (Musicapa rupestris): I did not see the bird. Prince Max. (iii. 95) described a similar nest of the Anabatis rufifrons, or Sylvia rufifrons, with an opening at the lower end: he found the bird in the upper storey, and below it a kind of bush-mouse (Rat des Catingas, Mus pyrrhorhinus).

[†] The author of the Caramurú asserts (vii. 58) that the Camaleão feeds upon wind. In the Brazil, however, the Chameleon is a lizard (Lacerta iguana), which changes a little the colour of the skin, but which cannot be compared with the true chamæleones. This animal in the wilder parts of the Sertão is considered to be more delicate than the chicken; but the people are not particular, they devour the ounce, the cayman, the wild cat, the Siriema-bird, and other strong meats.

of hard clay, strengthened with hydrate of iron, at this season six feet above the water; it extends some leagues down the left side of the Rio de São Francisco. From the mildest of heights we can see the low-lying valley of the Rio Grande winding up from the south-west, where there is a break in the blue curtain which subtends the plain. It is a flat Delta of dense vegetation, at least twenty miles across in a bee-line. These confluence towns run a double risk, from the main artery and from the influent; the heavy downfalls of rain are often local deluges, and thus one stream may do damage when the other is peacefully inclined. During the last night the Rio Grande rose several inches, when the São Francisco fell: the people declared that they never saw this happen so early, and began to predict that water would be wanted when most required.

The town runs from west to east, along the northern bank of the Rio Grande,* beginning about a mile up-stream, and extending to the "Pontal." It has a mean look, the houses are low and small, with roughly railed courts on the water-side, where the floods prevent building, and sundry are un-finished, mere tiled roofs without walls. Here and there, on the higher levels, is a platform of rough stone and lime brought from Porto Alegre, six leagues down stream; it supports a whitewashed back wall or a tenement half-white, half-yellow, set off with pea-green shutters. The Porto,† a dirty landing-place of sand and clay, is the common sewer; in the mornings it becomes a fish-market, during the day seines hang on poles to dry, children pelt the dogs, the asses, here a "feature," and the long-legged pigs, ducks and fowls, wander about in com-

---

* This great influent has been surveyed by M. Halfeld, who devotes to it three charts. Well deserving the name, it drains the eastern side of the northern dividing range of Goyaz. The mouth is in (approximately) south lat. 12° 10′, and west long. (Rio) 1° 3′. It is navigable for 45 leagues to the Villa do Campo Largo, where it is still 350 wide; its depth is about 4 metres, the current 0·77 per second, and the discharge about 190 cubic metres, or nearly double the Seine at Paris. Beyond this point there are difficulties, but small "dug-outs" go 20 leagues further to Limoeiro. The Rio Preto, its great north-western fork, gives a navigable line of 32 leagues *rid* Santa Rita to Formosa; but this is a troublesome journey. From these lands are exported rice, farinha, maize, legumes, rapadura, and other provisions: some salt is also made at the Barra do Boqueirão, 16 to 18 leagues from the em-bouchure of the Rio Grande. The Rio Preto is the stream whose waters Lieut. Moraes would throw over the mountains into the Parnaguá Lake. I have alluded to this wondrous project in Chap. 26.

† I do not understand what M. Halfeld means by "este porto parece ser artificial." It is rare to find anything more wretchedly natural.

pany with half-tamed cranes, white and ash-coloured, and the women wash in company. Water for household use must be brought from up-stream: here it is dark, foamy, and tainted. A number of canoes and barcas ride at anchor attached to their poles, and a favourite conveyance appears to be the "balsa," or raft of "Burity" fronds. The long bundles are lashed together in five or six places, and are kept in position by cross-pieces; they rise about one foot above water, and, being elastic, they are less likely to be injured by shoals and rapids. They carry down the river huge "pipas," or hides full of grain, and similar "trem:" at their destination they are broken up to make posts and rails, which are tolerably durable.

This is a great "festa," the peculiar day of the Padroeiro, or patron saint, "S. Francisco das Chagas." As we approached the town, we saw the F. F. in accurate black, riding small horses and smaller mules, along the unclean Praia to join in the office. The rest of the crowd was in hats of sorts, chimney-pot, Burity-straw, or felt, and in brown or white cotton clothes. There was the usual grotesque old negro, wearing a caped cloak of the thickest blue broadcloth, in an atmosphere of 98° (F.). The women are all in church till the ceremony ends, and the men cluster at the door like a swarm of bees. Presently the "function" ended with a discharge of fireworks—it was still broad daylight—which seemed to administer much spiritual comfort. A procession issued to perambulate the streets, and the dignitaries, by their red and white "opas," or short cloaks, caused no little sensation. Girls dressed in the brightest coloured stuffs, and small youths in the lightest of clothing, and very little of it, charged wildly about the place, dodging round the corners to "catch another sight." I visited in the evening the little chapel of Bom Jesus, which has stumps where towers should be—a man in uniform without epaulettes. The illumination was not brilliant, but it showed me that the feminine element predominated: the principal duty seemed to be to kneel down before a table, and to kiss the Saint's very diminutive feet—the principalest to deposit a few coppers upon an adjoining table. The night showed not a few of the scenes which one expects to see at a commercial port on festal occasions.

M. Halfeld speaks with enthusiasm of the townspeople.* I found them civil and courteous, as indeed is the rule of the Brazil, but the Bahiano did not shine after the Paulista, or the Mineiro. My letter of introduction to the Lieut.-Col. Joaquim Francisco Guerreiro was not followed by any results; on the other hand, the Lieut.-Col. Carlos Mariani, the grandson of a Corsican who had emigrated to the Brazil, "in the days of the Genoese Republic," came at once to see me, led me to his house, and showed me all his curiosities. He had octahedral pieces of magnetic iron (ferragem), which is found scattered about the fazendolas (little estates), and on the Vareda do Curral das Egoás, beyond the western containing-ridge. His rock crystals came from an eastern Serrote; whilst the Tauatinga Range and Natividade in the Tocantins Valley supplied red sandstone with attachments of quartz, showing at the junction regular lines of free gold, and diffused traces of copper. He informed me that a wandering German had lately been robbed of some opals, which are supposed to be found near the Villa de São Domingos, en route to Cuyabá in Mato Grosso.

I spent the main of my time wandering about the town, and trying to detect its latent merits. Beginning at the east and walking round by the north, we find that the site is a great Varzéa, or river plain, raised 18 to 20 feet only above the low level of the stream. The land immediately behind the town is flooded six feet, and even more; to the north there is a large swamp-bed, which has its own drain to the east. Many of the houses in this direction show a water mark of 3 to 4 feet in height; and some have sunk twenty-four to thirty-six inches into their sopped and sandy foundations. It is probable, however, that this may be accounted for by the deposit of the inundation; the Mississippi, in some places, leaves annually a coat of mud and sand two or three feet thick. On the north-west is a whitewashed cemetery, and beyond it another of clay. In this part also is the Tezosinho (little rise) da Conceição, a "Retiro," where the townspeople huddle together when their houses are under water; it is the resisting bluff which prevents the plain being swept

---

* "The noble and loyal character of the inhabitants of the Villa da Barra, especially the higher classes, evinces, in all their acts, civil and religious, cordiality, the most gentlemanly politeness, and in social life an extreme delicacy of manners which rivals the most civilised Courts."

away. At the west end we find the origin of all these evils.
Here is the tip-over, the "transbordamento," where the waters
of the Rio Grande enter, form an Ypoeira, and, with the assistance of the swamp, convert the site into an island.* The
bayou-head can hardly be embanked, it is too broad and the
soil is too loose and silty to form a levée. Lime being expensive,
clay is used in its stead, and the deep holes dug for this material
form, under a sun that burns at 6·30 A.M., another fomenter of
marsh disease. The only remedy is to remove to a better place,
but the question is where to find it.

The town is in the usual long narrow form, with silty and
sandy thoroughfares, all bearing names, none boasting pavements. Behind, or north of Water Street, is the Rua do
Santissimo; behind it the Rua do Rosario has at the west end
a Praça, a huge cross, and a two-windowed ground-floor chapel;
still northwards is the Rua do Amparo, a wild suburb, and
beyond it the "Retiro." These long lines are connected as
usual by Travessas or cross streets. There are a few sobrados
and meio-sobrados, fronted by the usual bits of brick-edge trottoir,
and proudly displaying glass windows. Most of the houses are
small, with large projecting eaves under-boarded; many, even
in the highest parts, appear half interred. There are a few shops
of dry goods, and a photographic establishment, which sells
cartes de visite at the rate of 8$000 per dozen; a butchery sup-
plies tolerable meat, and a host of Vendas sell spirits and rapadura, onions and garlic.

The nucleus of the settlement is about the Largo da Matriz.
The people determined to show their spirit by building to São
Francisco a church of the grandest description. Such things
begin vigorously in the Brazil. The Provincial Government
gave £400, which alms and contributions raised to £2400.
Bahia was applied to for a plan and an architect; the person
chosen was a German, Herr Heinrich Jahn, who brought with
him his family. The first stone was laid on Oct. 4, 1859. The
building is, or rather will be, 100 feet long by fifty broad,
double towered and with a clerestory. The material is brick

---

* Since the little deluge of 1792 the
town has often been threatened with destruction, especially in 1802, 1812, and
1838. In 1857 the Villa da Barra escaped
better than Januaria; the latter, as well
as Urubú, was not so fiercely visited as the
former in 1865.

and lime upon a foundation of ashlar. The front has the usual three entrances and five windows, and the graded pediment has introduced a little change into the popular monotony of façade. In the interior party walls set off two sacristies, which seriously diminish the space. At present all is scaffolded with Carnahúba palms, and the works are stopped by lack of funds. The whole affair is out of place and size, and the Villa da Barra looks like an annexe to its Matriz.

On the south-east of the church square is a detached Casa da Camara, with a bell and six windows above, and a grating which shows the jail below. At times the floods have rendered it necessary to save the archives in canoes. The prisoners appeared, like the rest of the people, "jolly," and here they need never sing with the starling, "I can't get out." The military force, paid by the Province, consists of one sergeant and ten men, whose duties seem principally to sound the bugle. The sentinel at the door leans against the wall; he has neither collar nor shoes, his only weapon is a bayonet, and he much reminded me of the items which composed a certain corps on the Gold Coast, now disbanded. The last of the public buildings is the Hospital de São Pedro. The Government assisted with funds a Brotherhood, which subscribed 1$000 each per mensem, and continued to do so for a short time. The house still remains, but the inmates are at most two, and the good work may be said virtually to have been dropped.

The Villa da Barra dates from 1753—4. Its municipality contains 10,000 to 12,000 souls. There is only one freguezia— São Francisco das Chagas. In 1852—54, the houses in the town numbered 660 and the population 4000; neither had increased in 1867. Its connection with the seaboard is very imperfect. The road to the city of Lençoës (sixty leagues, each of 3000 braças), was a mere "picada" in 1855—a line of river fords, muds, and mountains barely passable, but passable. The best road to Bahia is through the old town of Jacobina (seventy-five leagues), a long leg to the east. It is described to run over a plain with three "jornadas" or stages of twelve to fourteen leagues each, waterless during the dries; the mule troopers, however, accomplish each one in the twenty-four hours; then comes the Serra do Tombador, leading to the town, a stony ladeira or ascent, for which, however, the mules are unshod, and lastly from Jacobina to Cachoeira City all is comparatively level.

The people of the Villa da Barra breed cattle and a few mules; their chief occupation, however, is the carrying trade,[*] and, like the West African seaports, they act as brokers between strangers and the people of the interior. We are now on the outskirts of the great salt formations, which, however, does not prevent the condiment being imported from the coast viâ Joazeiro. The saline matter is deposited by water chiefly in the vicinity of streams, and rock salt (sal gemma) has not yet been found. We visited further down several places where salt had been "planted," that is to say, mixed with the soil, with the view of making it spread and, as it were, breed. The " Salineiros " collect and make it between the months of July and October. It is treated like saltpetre, strained in bangués (coffers or hides), evaporated over the fire, and allowed to crystallize. Sometimes it is exposed in "coches " or huge troughs to solar action only, and this simple operation would pay better if done on a grand scale. What it chiefly requires is purification, and the separation of the other salts, magnesia, for instance, which are equally disagreeable and deleterious. Some of it is white and fine like sea salt; often, however, it is bitter and brown (amargoso e trigueiro), fit only for beasts. Finally it is packed for exportation in hide-bags called Surroés (Surons).[†]

The Villa da Barra do Rio Grande has a high and unmerited reputation. I soon found how it had risen to fame. The Mineiros wish to see Januaria the capital of the new Province. The Bahianos prefer Carunhanha on the Villa da Barra, and the cause of the latter has been ably espoused by the ex-Minister and Senator, João Mauricio Vanderley, the Baron of Cotigipe.

---

[*] The following list of my purchases will show the prices then current at the Villa da Barra :—

1 Garafão (4 bottles) of country rum	.	.	.	0$500
2 lbs. salt	.	.	.	.	.	.	.	0$130
10 lbs. beef	.	.	.	.	.	.	.	1$000
16 lbs. lard	.	.	.	.	.	.	.	3$000
10 lbs. rice	.	.	.	.	.	.	.	1$600
1 string (resta) onions	.	.	.	.	0$100
½ quarter of farinha	.	.	.	.	.	0$800

Total . . . 7$130

[†] The measure varies everywhere; here the Surroč is of 24 pratos, say 50 lbs.

This influential Conservative is a " son " of the place, and has a filial regard for its prosperity. My conviction is that the Villa is one of the worst sites that I have yet seen, and that it is fitted only to be a port or outpost for Bom Jardim or Chique-Chique.

CHAPTER XXII.

FROM THE VILLA DA BARRA (DO RIO GRANDE) TO THE VILLA OF PILÃO ARCADO.

Seventh Travessia, 29 Leagues.

"Ce beau pays peut se passer de l'univers entier."—*Voltaire.*

We did not pass a pleasant night. The air during the early hours was still and sultry (82° F.). Then the cold land-wind set in. At first a long monotonous song made the hours unpleasant; afterwards came the lively splashing of Piranha, the "devil-fish," and the muffled growl of the stream, which seemed to be mischievously inclined.

Below the Villa da Barra the São Francisco broadens, the containing ridges retreat, and the riverine valley is a dead flat. The heat greatly increases, although the channel trends between north and north-east, the direction of the Vento Geral. Boats must sometimes remain embayed for days near the low ragged sand-bars, and the crews congratulate themselves on having a dozen clear working hours. Accidents are so common that there is hardly a boatman who hereabouts has not been wrecked at least once. Suddenly, in the clearest atmosphere, the breeze dashes down upon the wide surface, the waves rise, and the canoe or raft is swamped. The greatest care is given to observing the premonitory symptoms, especially the "redemoinhos," columns of sand sixty to seventy feet high, which career whirling over the plain.

Boats creep along the windward or sheltered bank and make ready for the refuge-place before the "vendaval" or squall bursts.

We could not set off before 9·30 A.M. The wind began early. The first league showed on the right a blind channel, the "Ypoeira funda," which, during the floods, gives direct communication with Chique-Chique. A little beyond Cajaseira * of the Capitão José Vicente is another bayou also impassable at this season, converging to its southern neighbour. These should be carefully examined. The narrow opening, made practicable to admit craft at all times, would greatly assist navigation to Chique-Chique, and relieve the latter place of its main difficulty, direct approach. The channel, it is said, would be easily managed. We shall presently skirt it when riding inland from Chique-Chique. On the other hand it must be taken into consideration whether such opening will not throw the thalweg to the right and greatly increase the amount of flooding. Fazendas and fazendolas were scattered in all directions over both banks. We landed on the right side to examine a place reported to contain limestone. It proved to be a mere "barreiro." After we had passed sundry cultivated spots, † and Carnahúba groves standing up like huge palings, the furious wind compelled us to anchor at the head of a little sand-bar, the Ilha do Mocambo do Vento. This "Maroon village of the wind," an ill-omened and appropriate name, is considered one of the worst places. The channel bends to the east and the south-east. The bed is unusually broad, and the stream flows in the teeth of the Trade. Upon the Corôa we found diamantine crystals, and a scatter of acary, the armour-plated fish, had been thrown away from the seines. The spiny outer skin had been mummified, and the attitude was still that of the death-throe.

Tuesday, Oct. 8, 1867.—The wind, after a fierce struggle through the night, made a feint of falling. It rose, however, with the sun and filmed over the Corôa with a gauze of sand which reminded me of the Arabian wilds. Even at 1 P.M., when we set out, advance was difficult. The left bank was dotted with small detached hills, and between Areia branca and Hycatú‡ we entered

---

* Probably from Acaya or Acajá (Spondias venulosa, in Tupy ybamétara), a Burseracea resembling the Imbu or Imbuzeiro.

† Sambahiba, a little village with red milho hanging to dry upon a Varão, or three-poled gallows. The Arraial do Porto Alegre, near a fine wooded rise of Carnahúba and Caatinga on the left, and so forth.

‡ The "good water."

a land of "lençoẽs," or shrouds, as they were called by the old
Portuguese explorers.   Sheets and heaps of the whitest sand, the
degradation of Itacolumite, glittered in the sun, like the patches
that lie about Diamantina.   Here and there it was dotted with
black points, dark green tufty shrubs, which at times the mirage
converted into tall forests.   In parts the substance becomes yellow,
and resembles even more the low dunes lining an ocean shore.
The underlying rock is probably limestone, and the formation
will extend for many leagues down stream, especially on the left.
Nothing could be more picturesque than this bit of the Sahára,
especially when backed by a gloomy pall in the northern sky—
here a sign of wind, not of rain—and when fronted by the steely
stream, damascened by the golden glories of the setting sun.

The main channel runs far to the north-west of Chique-Chique,
and there was not water enough to float us over the direct line,
about two miles long, passing to the south of the Ilhas do Gado*
and do Miradouro.   We were therefore compelled to skirt the
whole western shore of the latter, which in length is at least seven
miles, with four miles of extreme breadth.   At its north-eastern
extremity the navigable channel, a continuation of the blind
Ypoeira, doubles back to the south-west in order to reach Chique-
Chique.   It is at least eight miles long, not including the numerous windings.   This is the "Barra da Picada," so called from a
small place at its mouth.   There is yet another passage, at times
practicable, between the main stream and the Ypoeira, the "Barra
da Esperança," which passes between the smaller Ilha do Gado
and the Miradouro.   This portion of the São Francisco is exceedingly complicated, and the network of channels can hardly be
understood without a map.

The great artery widens to upwards of a mile, and is marked
by snowy sand-heaps, set in the darkest verdure, opposite the
mouth of the Barra da Picada.   This channel begins with a
breadth of 500 feet between terra firmâ and the Miradouro, which,
at its north-eastern extremity, fines off to a swampy point, the
Ponta da Ilha.   It presently narrows to 200 and 150 feet, and
where it joins the Ypoeira to the north-east of Chique-Chique it

---

* This is the lesser "Ilha do Gado," to
the west of the great Miradouro Island.
The larger "Ilha do Gado" is the space
included between the blind Ypoeira and
the main channel : it is south of the Miradouro, and it is insulated only during the
floods.

widens out to 700 yards. At first it makes a long "horseshoe-bend" to the west. After that its course is direct. The depth will admit boats at all seasons, and the breadth is hardly sufficient to allow waves to form. Its tranquillity, especially enjoyable after the roughness of the great river, reminded me of those West African lagoons which subtend the shores of the boisterous sea, and which aid so much the loading of slave ships. The low banks on both sides, the dense bush, at times broken by a bare talus, and the little patches of spinach-green fields with their rough fencing, vividly brought to mind the features of Dahoman Whydah.

We passed a few tiled huts on the proper right bank of this quiet channel, and the whitewashed chapel and hamlet of Santa Anna do Miradouro* on the eastern margin of its islet. We then entered the broadening mouth of the Ypoeira—at this season a backwater, and found a safe anchorage where the gusty north-wind can do little damage. At the port were a number of canoes belonging to fishermen and melon vendors. A barca had been stranded, and another was being caulked upon the beach. Above us rose the town, which was not less "jolly" than its neighbours. Drum and song, dance, laughter, and shouts of applause, prolonged till dawn, showed that, despite the absence of festival, the "folía" was not wanting.

The next day opened so badly with the wind-clouds that I determined to rest the crew, and to indulge myself in a short visit to the nearest diamond washings. We began by inspecting Chique-Chique. The "porto" along the eastern bank is formed by a natural pier, a dwarf cliff, at this season some four feet above water. The material is a silicate of white-grey lime, in places granulated with iron stone and puddinged with large and finely disseminated quartz. Containing silica and a considerable proportion of clay,† it will make the best hydraulic cement. This and the Lapa are the principal lime quarries. Chique-Chique annually sends up and down stream, between the Villa da Barra and Joazeiro, 1500 to 2000 alqueires. On the beach were canoes full of the finest water-melons. Horses were being groomed by the usual process of dashing water upon them from a large cala-

---

* In Mr. Keith Johnston's map "S. Anna de Miradouro" is made a small town upon the eastern bank of the São Francisco.

† Ordinary hydraulic limestone contains 15 per cent. of argile, the good, 16 ; and in those which make the best cement the proportion rises to 25 and even 30 per cent.

bash. Lads in naturalibus were preparing to bathe, and washer-women and carpenters plied their trades. Spoonbills (platyrhyn-chus) stalked amongst the dug-outs, which had brought for sale an abundance of fish. The birds were not improved by civilization, and their delicate pink plumage had turned grey with mud.

Upon the bank-top we found a large space open to the stream, with a central cross supported by a heap of stones. At the bottom, facing to the west-north-west, is the Chapel of N. S. Bom Jesus do Bomfim.* It is a poor, mean pile of brick and lime upon a stone foundation. The usual preposterous front was four windows, and no belfry-towers. The interior, anciently a burial ground, displays a blue and gold high altar, with frescoed ceiling, and two side-chapels where swallows had nested. The walls show a single miracle paper, dated 1804, and the congregation consisted of three old women, two in uniform black, the third girt with the white cord of St. Francis. The town extends on both sides and behind the church, thus forming a truncated cross. The tenements near the creek show a water-mark two feet high. They can easily be raised upon platforms. The floods do not extend to the higher parts, and the people boast with justice that their "assento," or site, is the best upon the river. The heavier rains begin to fall in October, and continue with breaks till May. The inundation lasts five months, from November to April. Already there is a freshet of six palms, and the indirectness of the water-course here makes a rise of one foot to four or five in the true São Francisco.

There is some excitement in visiting and describing these places, now the most wretched of "rancheries," but destined to become the centres of mighty States. Chique-Chique runs nearly north and south; as usual the long straight streets are parallel with the creek, and here they are almost sufficiently broad. Pave-ment is as yet unknown, but scatters of ironstone upon the hard ground render dust and mud equally impossible. A triangular "square," south-east of the church, surrounds a detached Camara-cum-jail, and the iron-bars of the latter are fixed into wooden frames. Farther to the east there is a neat, whitewashed ceme-tery, with incipient catacombs. One Casa Nobre, with a balcony of quaintly painted wooden railings, and a few half-sobrados, have been built. The rest are ground-floor tenements, each with its

---

* N⁴ S⁴ do Bomfim (M. Halfeld).

large compound and little "hanging-garden" of geranium, basil, and lavender (alfazema),* of onions and choice vegetables; the latter is mostly a trough or a bit of canoe, raised on poles beyond the reach of ants and pigs. The tenements may number 180, but many of them are opened only on fête-days, when 1500 souls find lodgings.

The country behind the town is a field of various Cactaceæ, which form contrasts. The dwarf of the family is the Quipá, with its large crimson fig, so much enjoyed by the parrot (Psittacus cactorum) that the beak is stained red. Another pigmy is a bulb nearly a foot in diameter (Melocactus, or Echinocactus), ribbed like a melon, and guarded at the angles by terrible thorns; upon the top is an inflorescence, like a Turkish fez, and the people know it as the friar's head (cabeça de frade). Horses learn to like the soft spongy substance, which the plant takes so much care to preserve; it keeps them in condition, and they fetch a higher price than those who refuse it. The people declare that riding animals and black cattle learn to open the armed exterior by striking it with the hoof. There is the common flat Opuntia and the "Xique-Xique,"† which is planted in hedges, and gives its name to the settlement. According to M. Halfeld, this is a kind of cactus which, roasted and peeled, has the taste of a batata or sweet potato. The almost general word is differently used in the several places. Here it is applied especially to a tall "Organ-Cactus," which is almost a tree; the angles vary with its years, in youth it is many-sided, and it ends life almost cylindrical. The shape also varies; here it runs serpent-like along the ground, there it stands stiffly upright. One kind has a fleshy white flower resembling wax-work; another (C. mamillaris) is patched with white fleece, as though it had robbed a sheep, and almost conceals its dark-red blossom. We shall meet with other forms further down the São Francisco.

I was surprised to see about a place so rich in Cactus, goats so small and stunted, whilst here were the finest sheep of the Brazil, and mutton is justly preferred to beef. Hardly any pasture, except thorns, was upon the ground, yet a perfect assimilation of

---

* The women are fond of these perfumed herbs, and ornament their hair with the flowers.

† Gardner writes Shuke-Shuke, as the word is pronounced. I have preferred the form "Chique-Chique" for the settlement, "Xique-Xique" for the plant; but the distinction is not recognised by the people.

food, as in the Somali country and on the Western Prairies of
North America, kept the animals in the highest condition. The
lambs wore a thick fleece, which disappears from the adult ; some
are white, others are brown, all are thin-tailed, and not a few are
bearded. No trouble is taken to breed them ; the owners, however, have sense to pen at night the flocks, numbering thirty to
forty.* The usual price is 2 $ 000, or 3 $ 000 when the animal is
very fat. Horses, small but hardy, and with signs of blood, cost
60 $ 000 ; good riding mules, which make Jacobina (sixty short
leagues,) in four days, rise to 80 $ 000 or 100 $ 000. The cattle
is neat and sleek, apparently untroubled with ticks or " bernes."
Besides stock-raising the country supplies, every year, 1000 to
2000 alqueires of salt to the Upper São Francisco ; manioc planted
after and taken up before the rains, gives good farinha ; maize
and excellent tobacco are brought from the Assuruá Range.
The people boast that their land is one of the richest, if not the
richest near the river ; it produces gold and diamonds, fish and
salt, and the wax-palm grows in vast forests.

After some trouble about conveyance I hired for 3 $ 000 each,
a horse and a mule, with the owner as guide. Cyriaco Ferreira
was a tall, thin, old black, with a preposterous masticatory apparatus, and a scanty, scowling brow. He consulted me shame-
lessly in the presence of his wife concerning a certain " Gallica ; "
here even white men talk about it before their families as if it
were a " cold in the head." The frequent mutilations which
now begin to meet the eye doubtless proceed from the use, or
rather the abuse of mercurials, to which are added the ignorance
and the recklessness of the patient, who, even when the facial
bones are attacked, will drink spirits and take snuff.

Our negro had been a good man and true as a slave ; a false
idea of charity had emancipated him, and with freedom appeared
all the evils of his race. Fawning as a spaniel to those who knew
his origin, he was surly as a mastiff to us ; recalcitrant as a mule,
he would loiter when we wished to advance ; he " trod upon our
toes" at all opportunities, and with the real servile style he proceeded to give his orders. Travellers who have even a constitutional aversion to a " row," are forced into it at times. When it
is thrust upon one the only way is " to go into it," tooth and

---

* The Brazilian variety, called on the Amazons " sheep of five quarters," was not
seen here.

nail. This was done ; a few rough words, and clearing decks for
action, soon brought back the old slave, but at times it yielded to
a passionate outburst of the new freedman.

Riding down the Rua das Flores, we struck out into the open
country, towards a long blue rock with a table-top summit, south-
east of the settlement. This Serra do Pintor will be conspicuous
for several days down stream ; it appears a frustum of a cone, a
second distance rising above a long sloping ridge. Cotton of
smaller than usual size grew in the suburbs ; and the district
beyond it, the Praia Grande, was clay strewed with iron pyrites,
which unless neutralized by underlying lime, must produce injurious sulphuric acid. Our path lay along the left bank of the
great Ypoeira Funda, which bulging out forms a lake round a
wooded central islet. Higher up the bed, it sends to the south-
east a canal or navigable arm, which we shall presently sight.
The Fazenda da Prainha was built upon the most unfertile soil,
which produced only dwarf thorns ; attached, however, to the
ranch was a large stockade of palm-trunks, and wandering about
the fold were the fattest of sheep. Few people were on the road,
all were armed, and most of them were talking about a late murder
in three acts—a drinking bout, a stab, and a shot. An old proprietor rode by with two immensely long pistols projecting far
above his holsters, and the attendant slave followed with a gun
slung over his shoulders. A typical sight was a woman on
foot and a man on horseback carrying the baby. The tropeiros
mostly drove horses ; here, however, we are getting into the
country of the pack-bullock. These men boast that they travel
all day, not only till noon, like the muleteers of the Southern
Provinces, and that thus they cover an umber of leagues. But
almost all were mounted upon pads supported by two broacas,*
which carried their salt and grain ; moreover the leagues are
short, and it is easy to walk over two in an hour and a half.

My companion could not travel without wanting to drink water,
which greatly amused the Brazilians. For this purpose we
halted at the Fazenda de Suassica, one of the many breeding
establishments—tiled huts, ranches, and large folds—scattered
at short distances. Two youths, the sons of a neighbouring proprietor, who with half a dozen whitey-brown lookers on were

* These square saddle-bags, with the hair outside, are now generally known as "Surroẽs
de Couro."

playing dominoes in a clay room hung with hammocks, came to the door and asked us to dismount. When coffee was finished came the usual query, "Pois, que trouxerão de negocio?" The inevitable reply puzzled every brain; they must have thought that they had entertained unawares "diabos"—government men on no angelic errand—but they preserved their courtesy to the last, and held our stirrups when we remounted.

Beyond Suassíca the land became a deep sand of ruddy colour, and presently passed, as the house-walls showed, into a blood-red clay; it was scattered with lime, and it is doubtless exceedingly fertile. The Favelleiro (arboreous Jatropha,) stunted near Chique-Chique, is here a tall and goodly tree. The thorny Mimosas and Acacias are hung with golden and silver blossoms, and the charming Imbuzeiro perfumes the air. Here the growth is low, drooping its flower-laden branches almost to the ground, and forming a shady bower, like the wild figs on the banks of the Lower Congo River. Many trees have the smooth barks and straight spindles of the Myrtaceæ, especially the Páo branco, which supplies the hardest wood; they contrast curiously with the gnarled Imburana* (Bursera leptophloeos, Mart.), whose bole is hung with burnished yellow rags, the peeling off of the cuticle that exposes the green-blue cutis beneath. This tree yields a greenish-yellow gum or balsam, resembling turpentine, and the scent is a favourite with the wild bee, as is proved by the many places cut away by the hatchet to reach the combs.

These strips of forest support, chiefly on the outskirts, a variety of birds. Plovers course across the opens, large green paroquets rise screaming from the boughs, and Aráras of the usual two species, the red and the black, appear to us for the first time in a wild state. The "Encontro branco," or large blue and white winged pigeon of Diamantina, here called "Pomba Verdadeira," is a visitant from the hills; it apparently prefers Itacolumite formations. The "Alma de Gato," a large, light-brown Coprophagus (?), seeks lizards and such small cheer. On the topmost twigs, especially of the shrubs, balances itself a snow-white bird with black wing-feathers, probably a Muscicapa; we see it now for the first time. High in air wheels the Urubú

* St. Hil. (I. ii. 105) explains Imburana by the Guarani "ibirañae," meaning baril, sébille, tirvir. But the termination "-rana" in the Lingua Geral, equivalent to the Portuguese "bravo" or "bravio," means poisonous.

Caçador, or hunting vulture, with crimson head and silver-lined wings.

We rode slowly through this interesting tract of wood, and presently we came upon a bit of African scenery; hedges of Cactus fencing a large field, whose " black jacks " were about three years old. This is the Fazenda do Saco dos Bois, with the little chapel of Nª Sª do Amparo, and a scatter of huts, inhabited by the proprietors in partnership. We were civilly received by a man who was lying stratus in umbrâ, under a thickly leaved and now blossoming Juá.* The site is high ground, never inundated, although within a few yards of the " Canal," the south-eastern arm of the Ypoeira, which we passed near the Fazenda da Prainha. The back water was then flowing up it towards the Assuruá lakelet, which it floods during the rains, and drains during the dries; it was covered with water-fowl, but the fluid was so muddy and impure that our beasts refused to touch it. The civil agriculturist, peasant I cannot call him, advised us to lose no time ; the hills, formerly blue walls, now looked near, and we could distinguish slips of white rock and patches of sun-burnt grass. But distances are deceptive in this unsmoked air ; the heat was unusual, and heavy storm-clouds were surging up from the west,—the especially rainy quarter.† The hills must attract every mist within their range, and wet weather comes from every direction. All were praying for the " Chuvas de Mangába " (Hancornia)‡ or de Puça (Mouriria Pusá, Gard.),§ the showers which accompany the fruiting of these trees.

Leaving the Saco at 4 P.M., we fell at once into deep sand, with a labyrinth of paths running through the stunted blades of grass. A few yards led to the northern edge of a great Carnahúbal, some four leagues long from north-east to south-west, and large enough to supply the whole river with candles. Every shape and age and size of the palm is here, from the chubby infant a foot high, to the tall thin ancient, whom a breath may fell. The panache gives tremendous leverage, and in parts the trunks lay prostrated by the north-eastern hurricanes, like

---

* This is a local name of the thorny Joazeiro or Zizyphus.

† According to others, the north-east is the rainy quarter by excellence.

‡ St. Hil. (II. ii. 215) mentions two species of Mangába, the M. speciosa (Gomes), and the M. rubescens (Nées and Mart.)

§ This shrub produces a small dark plum.

the long thin alleys which canister cuts through a column of men. In other places water lingered upon the black muddy ground, and the spikey bases of the trees, catching the floating weeds, showed the amount of rise; this has a curious effect when the palms are numerous. Much of the Carnahúbal during the great floods of the year must be crossed in canoes.

After two hours' ride the Carnahúba began to be mixed up with strangers; the Bahú tree, the Mureçi (Byrsonima verbascifolia),* the Puça, and the Mangába. Presently it ceased altogether, and we saw on the right the "Lagôa do Pintor," a green-margined tank, about 200 yards across, with a central islet of lush aquatic plants. During the floods, it is connected with the south-eastern branch of the Ypoeira, and at times it is almost dry. Amongst the trees beyond the water line are a few huts, whose inhabitants seem little aware of the wealth before them. This pond receives from the mountain slope a number of small diamantine streams, and the gems must settle in it. Artificial draining, however, is required, and such operations are far beyond the reach of the present occupants of the land.

Presently we arrived at the hill-foot, cumbered with large and small blocks of stone, which have rolled from the upper heights. This is the western counter-slope of the Serra do Assuruá, a meridional range that prolongs the diamantine formation of the Bahian Chapada. The "ladeira" or ascent was a succession of steps, loose stones and slabs, between which the sandy soil appeared. Reaching the summit of the hog's back, we turned to prospect the "taboleiro" over which we had passed; the large "Salinas," that supply salt to the river, lay upon it in glistening patches, and the Lagôa de Assuruá, about one league in length, was surrounded by snowy heaps of sand, like the "Shrouds" of the São Francisco. This water drains the Serra do Pintor, and its village "Itaparica" takes from it every year £300 worth of fish, here not an inconsiderable sum. The people speak of immense shoals which await exploitation.

Descending the counterslope of the ridge, we saw below us a small Serviço, with a single house and a few thatched huts on both sides of a narrow stony gulley. This Riacho do Pintorsinho flows, like the neighbouring waters, from north-east

* Also written "Murusi;" the bark yields a black dye.

to south-west, and feeds the Lagôa do Pintor. We had no letters of introduction, but we rode up to the doors and introduced ourselves to the owner, Capitão José Florentino de Carvalho, who, after the labours of the day, was reposing under the shadow of his own fig-tree. The fig, by the bye, was a wild Brazilian, which lately took only eight days to cover itself with dense verdure; such is the exceptional fertility of these Itacolumite soils in the rare places where they are fertile. The Capitão and his amiable wife have been diamond washing in this ravine since 1864. He gave us some excellently cooked Surubim, with the usual trimmings of "pirão" and pepper sauce; the Dona sent a cup of aromatic coffee, the hammocks were slung in a room under the fig-tree, and we should have slept like tops but for the heavy rain about midnight, and the tremendous snorting of Sr. Cyriaco Ferreira. I cannot call it snoring, the sound was that of ripping up the strongest new calico. When he did not snort he coughed, and—the place was somewhat close—as the leopard cannot change his spots, so the negro skin, even in a freed man, remains negro. Contubernation with the Hamite does not prepossess one in his favour.

The next morning was warm and pleasant, but it spat, and it seemed to promise rain for the afternoon. Our ungracious guide was salt or sugar, so we resolved to visit Santo Ignacio alone. The cross-path lay over a wonderfully rugged succession of hills, forming prism-shaped ridges, whose crystal waters, delightfully clean and pure, discharged into the Assuruá Lake, and where Itacolumite showed in all its grotesqueness. There were shapes of strange beasts, colossal heads and masques; arches, tunnels, and funnels, worked and turned by wind and rain; huge portals, towers, and cyclopean walls, to leeward smooth and solid, on the weather side seamed into courses of masonry, that showed an imposing regularity. The granular quartz was not so finely laminated as the Cerro formation; some of it was hard, white, and polished like blocks of marble, and at first sight it might have been easily mistaken for limestone, which, as the river bed shows, here and at Diamantina of Minas, underlies the sandstone. It was also more generally stained with oxide of iron, and it had large veinings of quartz, which sometimes formed external layers. Crystallized quartz and ferruginous matter, externally vulcanized, lay about in scatters.

The characteristic feature, also remarkable in the Bahian Chapada to the east, is a bouldery, not pebbly conglomerate, which resembles that of the Scottish Old Red. The huge blocks, many of them weighing several tons, contained proportionate pebbles, some rolled, others angular, here entire and there split, like the halves of almond kernels. The hard paste of sandstone, with nestings of many-coloured porphyry, will be cut into slabs of remarkable beauty.

We crossed the Riacho Largo, a narrow gully heading in a high bluff; its delicious water, the prerogative of the Itacolumite lands, feeds a tiny patch of green grass. Beyond it were three places where the rock wastes to a dazzlingly white sand, and this, in the lower levels where thorns grow, passes into soil, brown with a slight admixture of humus. Then we reached the highest divide; a broad sheet of sandstone shows hollows and holes like the hoof-prints of horses. The vegetation was that of the Cerro, the dwarf Mimosa, and the Ostrich Shank (Vellozia) a few inches high, whereas in Minas Geraes we counted it by feet. On the right the eye plunged into the sandy plains which bore signs of floods, and where other salinas glittered; to the left was an old diamond-washing, from which the people had taken the sand arrested by the big boulders. In front and below us lay the little village of Santo Ignacio, upon the left bank of a Córrego, whose narrow valley was bounded on the further side by a wall of jagged stone, disposed in courses, piles, and peaks. The yellow-green vegetation told the poverty of the soil.

We entered on foot the little mining village, much to the wonderment of its denizens. It had a Rua Formosa, a widening called a square, a miserable chapel, by courtesy termed a church, and men in "Panama" hats, black coats, and white overalls. Every Monday there is a fair, frequented by people from far and wide, and some 150l. or 200l. may change hands. The prices are high: what costs on the coast 0$100 here commands 1$000. We found the shop of a Mineiro from Formiga, who appeared exceptionally civilised amongst the "atrasado," arriéré, race of the Province which still boasts of being the Ecclesiastical Capital of the Empire. The little booth dealt in notions and provisions, red japanned tins of English gunpowder, pots, pans, and bowls; onions, garlic, sardines in cases, and rum in demijohns. The wife being unwell, we could not breakfast, but we drank coffee and ate biscuits under the eyes of brown-faced men, whose principal office in life seemed to be expectoration. This habit is general, as in the United States : perhaps the climate of the New World has tended to preserve it from abolition. Brazilians have told me that it preserves them from obesity.

As far back as 1803, gold was known to exist in the Arassuá Range, and it was worked in 1836. Diamond washing began in 1840, at Santo Ignacio, which was then transferred from the municipality of Urubú to that of Chique-Chique, and the first digging, near the Pedra do Bode, a little down the Córrego, has not been exhausted. In 1841, the Chapada do Coral, some twenty leagues to the south, was found to contain " Cascalho," from which pieces of gold weighing four pounds were taken. In 1842—3, Mucujé, in the Comarca of the Rio das Contas,* became Santa Isabel do Paraguassú, the chef-lieu of its own arrondissement. Presently diamantine deposits were found at Lençoes, so called from the sheet-rocks in the little stream of the same name, the western head-waters of the great Paraguassú River. The place was then a country hamlet, in the Municipality of the Rio das Contas. The discovery was claimed by M. Fertin, a Frenchman, afterwards established at Bahia. It is reported, however, that before 1844 a party of slaves had collected in twenty days some 700 carats, which they offered for sale. These " garimpeiros " were put in prison, but they refused to show the diggings ; they were then let loose, watched, and caught working at midnight. In 1845, Lençoes which had been in the municipality of the Rio das Contas, was made independent. Presently a rush of 20,000 souls took place there, and the city rose to importance.† M. Reybaud, Consul de France, Bahia, calculated from the date of discovery (August 1, 1845), a produce of 1450 carats per diem, and a total of 400,000 carats = 18,300,000 francs.

On return, we walked up the Córrego to visit our kind host's " lavra." The lower part of the bed belongs to another proprietor, who, having water handy, can wash all the year round.

---

* Generally written Rio de Contas, which is, I believe, a corruption.

† It also became the chef-lieu of the Repartição, or Diamantine Department. The papers (bilhetes) issued to the "fais-cadores" were charged annually, at first 0$020 per square braça, now 2$000. They give permission to establish the "garimpo."

Here we found the rock-crack forming the rill converted into a "canôa," or "batador;" the "Cascalho" is thrown in, and the diamonds are arrested by cross-pieces. Following the left bank, we came to a pit some twenty feet deep, where the owner, seated in an arm-chair, with book and snuff-box, was superintending the hands, who, should he happen to go away, lie down to sleep, if at least they find nothing to thieve. Two men, armed with alavanca (crow-bar) and hoe, were loosening a bit of boulder, and were scraping up the desmonte, or inundation sand, which was carried up the pit side by a black girl, a youth, and a boy. The Cascalho must wait to be washed in the rains, and here great inundations or scanty showers are prayed for. The host complained that the increased rate of wages prevented all profit, nor did I wonder: deep works on so small a scale cannot pay. The formation (formação) is here called Pé de Batêa, small dark stones, like iron filings, which settle at the bottom of the pan; there are also the fava, the ferragem, and fragments of light or dark-green clay, unprettily termed "Bosta de Baráta." The Capitão showed us in a Pequá,* one little yellow stone. The gems are mostly small, the largest yielded by this pit was the half-vintem, one grain, or a quarter-carat. The Riacho do Pintorsinho has produced a stone of two vintens, and a neighbouring Córrego four vintens. A diamond of half an oitava (eight carats) had been washed in former years, and the result was a "difficulty," ending in a murder, and in the disappearance of the prize.

We bade adieu to our hospitable host, the Capitão and the Dona, and returned to Chique-Chique with all possible speed. This short excursion had proved that "Cactus-town" has around it lands of immense fertility, salubrious mountains, which as yet have only been scratched and played with for diamonds and gold, and, briefly, all the conditions requisite for a capital. It is connected to the east with the coast viâ Jacobina, Lençoës, and Caiteté,† and to the west with the Piauhy and the Goyaz Provinces.

---

* The Tupy word Pequá, meaning wood generally, is applied to a bamboo-tube a few inches long, from which the stones can be turned out without letting them fall. Castelnau (ii. 343) describes Picoi, "Sorte d'étui fait d'une écorce très flexible." The miners have sundry superstitions about these articles.

† Alias Villa do Principe. The word, written in a variety of ways, e.g. Caiteté and Caïteté, is a corruption of Coa-été, virgin forest, and is thus synonymous with Caethé. In the days of Spix and Martius its neighbourhood was famed for cotton.

We may easily predict that, despite the satirist, some one will presently be proud to—

Ser barão de Xiquexique.

Oct. 11.—We easily ran down the Barra da Picada, which is, however, more tortuous than it appears in M. Halfeld's plan, and after three hours we made the main artery. The left continued to show the containing mounds, here dark with vegetation, there patches of white or yellow sand, and this feature will extend some eleven leagues down-stream. The land is everywhere arid, and the principal features are the " Carrascal " and the Salina. In the afternoon we passed the Arraial da Boa Vista das Esteiras (of the mats), a little chapel-village with some fifty huts on the right bank ; and we presently anchored at a Corôa, known as the Ilha da Manga, or da Porta. Here a rich diamantine "formação" abounded, and the gull, everywhere impatient of man's presence, screamed through the night, justifying Agostinho's epithet " bicho aburrido," * disgusting vermin.

Oct. 12.—We are about to enter a Porteira or funnel, where the stream, after spreading out to five times that breadth, is compressed to 1500 feet. On both sides high lumpy ridges, some bare, others rising umbre-coloured from the green lines of water-shrubbery, either fall into the stream, or form bluffs that face it for some distance. Running down the sand-bar shore we passed with infinite trouble through the first gate. On the right bank is the little village " Tapéra da Cima," with its broad ypoeira. Opposite rises the Pedra da Manga, projecting southwards into the stream a ridge like that of Santo Antonio, prism-shaped, about 100 feet high, by half that breadth, red above and dark below. Here commences the great gisement of magnetic iron, the Itaberite or Jacutinga which we have already visited at Sabará and Gongo Soco ; no examination for gold has yet, I believe, been made. The strike of the metal is north with westing and south with easting,† and it is prolonged on both sides of the São Francisco.

---

* The word is originally Aborrecido, abhorred, hateful, disgusting, the strongest expression of dislike ; it is contracted to aborrido, which is pronounced by the Caipíra aburrido, which, if it signifies aught, means donkeyfied.

† In M. Halfeld's plan, the strike is laid down nearly due north and south. I am probably in error : these formations so "disorient" the needle, that peculiar precautions are necessary.

Below this first portal the river, flowing to the north-east, widens out considerably. The Vento Geral, which had been fitful and fractious at dawn, presently brought a cold wind and violent rain, which made us shiver, though the mercury showed 73° (F.), about the temperature of a comfortable East Indian Club. We made fast to a Corôa till the storm had spent its rage, and then we attacked the second gate. Here again the bluffs on both sides correspond, and both have similar ports, sandy beaches a little down-stream. To the north were the few huts of the Tapéra de baixo, backed by a hill-knob ; and on the south, "As Pedras (do Ernesto)." We landed at the latter, a short row of hovels, and a single block with whitewashed walls. Here the rock chine, prolonging high ground behind, trends to the north-west ; it is broken into blocks, and shows cleavage as well as stratification. Pieces picked up by chance drew the magnetic needle round the compass card, and the substance appeared harder and closer than what we had seen in Minas Geraes.

Again the channel bulged out, as we emerged from the second portal, which ends in a cliff of yellow sandy water on the left bank. And again the grey nimbus in the purple northern sky sent forth howling blasts, and a slanting rain which compelled us to anchor thrice. The pilot determined at last that this *is* the wet season, and somewhat regretted that he had left home. We presently made fast to a sand-bar in the stream, and prepared to night. Far to the west was a blue crest fading in the distance. We are now nearly on a parallel with Paranaguá of Piauhy, on the southern head-water of the great northern Paranahyba River,* and this may be an offset from the dividing ridge between the two valleys, called in maps Serra dos dois Irmãos, and here the Serra do Piauhy.

Oct. 13.—As work was not to be done by day, we determined to try the night in places of minor interest ; the moon also was nearly full, and robbed the snag of a few terrors. Again, the yellow muddy colour of the margin told us that the São Francisco had fallen to the extent of six inches, and we

---

* St. Hil. (III. ii. 250) explains Parana-hyba as a corruption of Pararayba, "rivière allant se jeter dans une petite mer." Sr. J. de Alencar supplies the true derivation: "Para," the sea, "nhanha," to run, and "hyba," an arm, "running arm of the sea," that is, tidal river. Three words in the Lingua Geral are easily confounded,— hyba (hîba), an arm ; ayba (aîba), bud ; and hybá, ibá, ybá or iná, a tree, especially a fruit tree, and often used as a desinence.

jealously watched every symptom, wanting as much flood as possible, with an eye to the Rapids. At 3·10 A.M. there was a mist, or rather a thin rain, the first " Garôa "-fog since quitting the charming Rio das Velhas, and under its influence the river showed a sea horizon. At 7 A.M. we saw over the dark-green right bank the Serrotinho (M. Halfeld's Serrote do Rio Verde), with its two heads of the lightest leek-colour. A little to the south of it enters the Lower Rio Verde, whose mouth is about 230 feet broad, and whose line admits of scanty navigation. Like its namesake, the water is distinctly salt. On the north-east was the Serra do Boqueirão, a long vanishing line of buttresses, forming three distinct bluffs. Upon the left bank rose a little hill upon whose crupper sits the Villa of Pilão Arcado, the end of this highly interesting Travessía.*

* In Mr. Keith Johnston's map, the dotted line of the Rio Verde is placed at some distance below "Pilau," whence it enters the São Francisco, about two miles above Pilão Arcado.

CHAPTER XXIII.

FROM THE EX-VILLA DO PILÃO ARCADO TO THE VILLA DE SENTO SÉ.

EIGHTH TRAVESSIA, 31½ LEAGUES.

PILAO ARCADO DESCRIBED.—RUINED BY PRIVATE WARS.—GREAT IRON FORMA-
TIONS.—STORMS AGAIN.—BAD APPROACH TO THE VILLA DO REMANSO.—
THE TOWN DESCRIBED.—RESUME WORK.—THE GREAT EASTERLY BEND OF
THE RIO DE SÃO FRANCISCO.—THE TUCUM PALM.—LIMESTONE.—AN IRON
HILL, THE SERROTE DO TOMBADOR.—SHELLS.—THE MINHOCAO MONSTER
WORM.—THE WILLOWS.—REACH THE TOWN OF SENTO SÉ.

" The Missouri and the Mississippi Rivers, with their hundred tributaries, give
to the great Central Basin of our Continent its character and destiny."
Mr. Everett, July 4, 1861.

PILÃO ARCADO is still a mere hamlet ; the original settlers here
found a crooked wooden mortar, hence the corrupted name.* A
natural pier of iron-revetted clay projects to the north-east, and
throws the stream to the right bank, where it forms a sack ; the
channel then sweeps to the north-west. The beach shows conglomerate, based upon soft green shale, which is traversed by
quartz veins. Three nameless or unnamed streets, running
parallel with the water, contains about 200 houses, including a
" casa nobre " with wooden shutters. The Church of Santo
Antonio is a mere " tapéra " of bare wattle. The rising ground
behind the settlement shows brown soil, growing tolerable cotton,
and cactus in quantities ; higher up it is scattered with quartz,
white and rusty, and with fragments of various-coloured Itacolumites. Here M. Halfeld places the beginning of the gneiss,

---

* Properly " Pilão arqueado." The
terms do Pilão, or dos Piloës, are often
added to the names of streams, mountains,
and new settlements in the back-woods.
Either a coarse wooden mortar used by the
aborigines was found upon the ground, or
the neighbourhood had peaks, needles, or
cheese-wrings, which the new comers compared with pestles and mortars.

or " gneiss-granite," which will presently pass into true granite.

In former days Pilão Arcado washed gold from its hills, made sugar, which was dark but tolerably heavy and well flavoured, and, being the centre of the Salinas, supplied salt to the settlements up and down-stream.* It became a villa, the chief place of a termo, and the residence of a Juge de droit; presently it lost the privilege—desvillou-se—which was transferred to " Remanso," distant sixteen leagues. The principal cause of its decadence was a private war which lasted for some generations, and which remind us of the days *of the* Percies of Northumberland. Such things were in former times common all over the Brazil as they had been throughout Europe, and traces of the Montague and Capulet system are still to be found in many towns of the interior. Here the rival houses were those of the Guerreiro and the Militão families, names that suited well with their fierceness. The head of the former, in late years, was Bernardo José Guerreiro; whilst the latter were " Captained" by the Commendador, Militão Placido de França Antunes. This distinguished " valentão"† for nine or ten years defied the power of the Imperial Government, here perhaps a unique feat, and he appears to have been like the dreaded " Defterdar" of Egypt, a man of peculiar personal " grit." At the Villa da Barra I saw one of his victims who had lost both hands, and I heard of another whom for a greater offence he had caponized. He died in 1865,‡ aged sixty-two, and, as was said of a certain St. Paul of Scotland, that Militão merited the epitaph, " Here lies he who never feared the face of man." Since the death of this energetic person, who will long be remembered as the " Brigador Militão"—Militão the Fighter—Pilão Arcado and the neighbourhood have known quiet days. It showed as a novelty sails applied to a large ferry boat.

Resuming our work, we found the river trending generally to the north-east, but often breaking to the west, whilst a multitude of islands and sand-bars rendered the course very devious. The channel, in places two miles broad, contained much more dry

---

* St. Hil. (III. i. 293) mentions the Sal do Pilão Arcado, corrupted to Piloēs Arcados, from the Province of "Fernambom," now Bahia.

† I need hardly warn the reader that we must not say, as in the French translation of Koster's Travels, " le valentoens s'age-nouilla."

‡ M. Halfeld (Relat. pp. 105—111) speaks of this brave as one who had departed life.

land than water; the branches were often bigger than the Rio das
Velhas, and in parts, especially on the left bank, a narrow natural
canal, the "Paranamirim" of the Amazons River, had been laid
off by long, thin tracts of insulated ground. A little above the
Upper Remanso (Remanso do Imbuseiro), the stream winds suddenly from north-east to east, with southing. The line now
becomes populous, and on the left bank the fields are fenced in.
The waterside abounds in a lush growth of Capim Cabelludo or
hairy grass, and above it is a wooded wave of ground topped by a
blue-green cone. On the other bank is the Serra do Boqueirão,
the northern extremity of the Serra de Assaruá. The blocks,
separated by low ground, where the drainage passes, were well
defined by the cloud-shadows, and faced the river like cliffs
fronting the ocean. Near the summit are long white lines of perpendicular wall, regular as if fortifications had been thrown up by
the Titans; below them the reddish-brown ramp, apparently
clothed with dwarf bush, slopes at the usual angle. The material
is Itacolumite, based, according to M. Halfeld, on granite or
gneiss (schistose granite).

At the Boqueirão Grande, or Great Gap, between the bluffs, the
river again bends to the north-east, and a little below, off the
Fazenda da Praia, there is a bad rock in mid-stream. Presently
we passed on the left Carauá,* the large white house and
tiled out-houses of the old "Brigador Militão." A "bull's eye"
glared fiercely at us from the east, and an African rain-sun had
warned us to be prudent. We made fast to the north of a Corôa,
called Ilha do Bento Pires, from some huts on the left bank;
and here we found a large barca moored in expectation of the
"temporal." This squall did not come on till dark; en revanche
it lasted through the night.

October 14.—We proceeded cautiously down the channel, which
is here shallow and bristling with crags. The valley is watered
on the east by the Serra do Boqueirãosinho, a prolongation of the
Boqueirão, and on the summit there is a "taboleiro alto," with
fine fertile lands. At 11 A.M. we landed near the Serrote do Velho,

---

* M. Halfeld writes this word Carná.
In Tupy, however, it is Carauá and Carauá-
ta; hence corrupted to Caroa, Caroata,
Caragoata, Gravatá (in the Bay of Rio de
Janeiro), and (Bromelia) to Karatas by the
botanist. In a future volume I shall have
something to say about this most important
genus, whose edible fruit gives spirits and
vinegar, and whose fibre, valued for hammocks and nets, is current as coin in parts
of the Brazil.

the most southerly of three trunnion-shaped buttresses, which we had seen from early dawn looking blue and small. The narrow ledge supported a few pauper huts and bore poor bush upon a red clay, too ferruginous to be fertile without lime. Crossing a foul backwater behind the settlement we ascended the hill-slope; it is scattered over with red Itacolumite, cut and cloven by quartz veins, and with magnetic iron, the hardest possible Jacutinga, black and amorphous. As fuel here abounds, and transport as well as water-power are at hand, it may some day prove valuable.

From the hill top we had a good view of the river, which here narrows, and the gut is rendered dangerous by snags, shoals, and a large central rock. Here again M. Halfeld would control the stream by fascines—a hopeless task. We crossed to the left bank a stony floor remarkably rich in shells (No. 3), which are now common on the river, and which will extend to the Great Rapids; those lying upon the sand-banks were empty, and the animal seems to prefer shallow water near the edges. The storm had now worked round to the south, and the scene looked " ugly " as the mouth of the Gaboon River before a tornado. The sky was hung with purple black, white-grey cottony mists lay upon the earth, and the water gleamed with a sickly yellow. Two men were placed at the helm, and presently the fierce " rebojos "* were down upon us, driving on the " Eliza " with furious speed, and tearing to pieces the surface of the stream. We were compelled to paddle across—always a risky process, as " broaching-to " swamps the raft; tufts of shrub emerging from the water showed where a Corôa had lately been. A bow-shaped ripple to the right hand denoted the bank upon which we grounded; all sprang into the water till the " Eliza," vigorously pushed and shoved, sloped over to the safe side. At the bottom of the reach which runs from south to north, we had seen " Remanso ; " the site is a wave of ground gradually sinking to the deep still water † which gave the place a name; from afar the appearance is striking, but a nearer prospect shows little to admire.

A single barca was being built upon the clay bank, where

* The Rebojo is a gale like the Pampero further south; in the plural it is synonymous with refegas, raffales, gusts.

† At the time when I passed it the "remanso " in front of the town had become a strong stream.

several craft lay decaying. The Villa do Remanso, which till eight years ago was an Arraial or village, extends along stream from north to south. The houses straggle down towards the water, and the suburbs wander over the higher land. It is fronted by a large flat island, and below it the channel is narrowed by sand-bars and shoals. To the west, a lumpy blue rise projects from the dividing ridge between the valleys of the São Francisco and the Paranahyba,* while down-stream are the Morro do Marco and the picturesque Serra do Sobrado, whose crooked cones, quoins, and plateaux form an outline like a crested sea rushing to the north-west.

The houses of the new Villa may number 300, and many of them show a water-mark two feet high. The rains had deposited big puddles in every street, and the damp heat reminded me of Zanzibar. A ragged square to the north still bore the platform of poles erected to hail the return of July 2nd—the Provincial Independence Day. There is another open space to the south, and the Chapel of Nª Sª do Rosario, which appeared so grand in the offing, was a bald little chapel, with its ruined sacristy to the north.

The people number about 1500, more or less. Here men are so incurious that after living thirty years in a hamlet of fifty houses they have never taken the trouble to count roofs or noses. We met with, however, some signs of animation; the tailor was at work, and beer—everywhere the test of civilization—was for sale in the shops. Salinas and good breeding grounds† lie on both sides of the stream. The popular complexion, however, shows sign of dyscratia, and a French "Commis-voyageur," collecting the debts of his Bahian employers, complained of fever, and declared that life at Remanso is "heute roth morgen todt." The "curandeiros" have given some dietetic ideas, and have taught the sick to use bitters rather than sweets. Lieut.-Col. José Cirino de Souza, who acknowledged by a visit my introductory letter, was astonished to see M. Davidson devouring sugar, more Americano, after suffering severely from ague.

At 4 P.M. we set out, and having run a league down-stream, we

---

* The ridge cannot be of importance, as it does not produce any but the smallest influents.

† We here caught the first carrapato-tick since leaving Urubú.

anchored at a Corôa opposite the Serra do Sobrado. Here we seemed likely to rue the night of

Mali culices ranæque palustres ;

and, in addition to the gnats, the mosquitos, which during the day had comfortably housed themselves under the awning and in the nooks of the ajôjo, began to sing and sting. The latter, however, after a few minutes rose and departed ; only a few unusually pertinacious passed with us the night. Presently, as the sun disappeared, hosts of large ruddy bats (noctiliones) wheeled with their jerking flight, aud skimmed the surface of the stream. The thermometer speedily fell to 68°—70° (F.), and the high wind combined with the saturated atmosphere made us tremble with cold. At the same time it effectually silenced the frog concert.

Oct. 15.—This furious weather is, they say, the effect of the full moon, and the wind shows no sign of weariness. On the right bank a block of mountains rise suddenly from the "Baixada," or plain, and prolongs itself down the stream. To the left is the abrupt Sobrado, with cones and outliers. The upper parts were brown, and the lower skirts were already turning green; the hasty drainage probably causes this exceptional phenomenon. M. Halfeld makes the material "Itacolumite with hydrate of iron and pyrites," the sign of auriferous formation. The name is derived from a feature which will be common further on, a tall pile of white stone, emerging from the bush, and not unlike a two-storied house. As we approached (7·25 A.M.) the low and sandy Ilha da Tapéra (do Muniz) an "olho de boi" drove us across the waves, which swept over the raft platform, and in a few minutes we found shelter amongst the shallows to the left. Here we passed the day, imprisoned by the north-east wind. Happily I had with me a few pocket classics, the woe of my youth, the neglect of my manhood, and the delight of my old age, and with Hafiz and Camoens, Horace and Martial, occupation was never wanting.

Beyond Remanso the channel bends round directly to the east, and runs in long reaches, with more or less of northing, but seldom trending towards the west. The wet weather will now cease; the rainy season will break in mid-November, and last only four months; and the showers, which in other parts begin and end the true rains, are often absent. The skies will be clear

ultramarine, and the evaporation excessive; book-covers will
again curl up, and ink will dry in the pen. The sensation was at
first that of a St. Martin's summer,* and, though we had been
threatened with all manner of sufferings from the sun, I judged
the climate to be very healthy. On the other hand we are entering a funnel, a fine conductor of wind, and barcas sometimes take
fifteen days to cover the 108 miles † between this and Joazeiro.
The gale will sometimes last even through the night, and I find
in my journal that every day's Trade is worse than the day before.
The draught increases because the land becomes more sandy, and
there are frequent tracts of rich Jacutinga. Below Remanso also
we miss upon the Corôa the diamantine "formação," and this
suggests that sometimes the supply of the upper bed is not washed
from a great distance. Of the granite and carbonate of lime I
will speak when we reach their limits.

Oct. 16.—Despite the head wind we set out at dawn. Passing
the Ilha Grande do Zabelé, a monster of an island, we saw in the
stream lumps of whitish rock, which proved to be pure limestone.‡
After two hours we were driven to take refuge on the right bank.
Here the land is inundated, and the short manioc must be taken
up before the floods. The plots were defended against cattle with
a wealth of timber. The marshy soil produces the largest and
spiniest "Tucums;" the stems were at least thirty feet high,
double the normal size, and the thorns were strong enough to
pierce a cow's hide. This Palm (Astrocaryum tucum)§ is so unlike
a palm that Sellow would not admit it into the family, and at first
sight the stranger feels disposed to agree with him. It is found
growing upon the seaboard, and extending to altitudes of 1000

---

* The pilots, indeed, called it the "ve-
ranhico," which breaks the rainy season
about December or January. In Peru it
happens about Christmas; hence it is
called "El Verano do Niño"—the Summer
of the Babe. The Spaniards, be it remarked, are far more poetical in thought
and feeling than the Portuguese; it is the
Arab *versus* the Roman. On the other hand,
the Portuguese have produced far better
poets than the Spaniards.

† The pilots who, I have said, always
exaggerate distance, make 40 instead of
36 leagues from Remanso to Joazeiro, and
18 instead of 16 to Sento Sé.

‡ M. Halfeld (Relatorio, p. 117) calls
them "Rochas Vivas," whatever that may

mean.

§ This is the Toucoun of P. Yves d'Evreux (1613). It is mentioned by Piso
and Manoel Ferreira da Camara (Descrip-
çam fisica da Comarca dos Ilheos). Arruda (Cent. Plant. Peru.) has a poor opinion
of the fibre, and his description has been
analysed by Koster (Appendix, vol. ii.).
John Mawe attempted it, and was duly
criticised by Prince Max. (i. 118). In the
Compendio da Lingua Brazilica, by F. R. C.
de Faria (Pará, Santos e Filhos, 1858), we
find that the Tupys called the fruit of the
Tucum, "Tucuma; Mr. Bates (i. 124)
writes Tucumá; and the Peruvians call
it "Chambira."

feet, where it prefers shady ground. Usually the "frêle palmier" is from twelve to sixteen feet in height and five to six inches in diameter. The hard black nut produces an edible almond; the fibre is drawn by folding the foliole and pulling out the nervature of the parenchyma with a peculiar knock. The novice who ignores the twist is sure to break the leaf before the threads are drawn out naked, and a practised hand makes only one-eighth of a pound per diem. The practice is, doubtless, derived from the "Indians," who make their bow-strings of "tucum" fibre, cotton, or Bromelia-bast. Maceration was tried and failed, as the leaf decayed at the end of a week. On the Brazilian seaboard Tucum thread is used for fishing-nets, and bales of the greenish yarn pass as money, with the average value of 2$000 per pound. On the São Francisco River the Tucum is also valued by seine-makers. The leaves when young make good mats and baskets, and when old, thatch. We cut down many of these prickly palms for walking-sticks. They are strong, heavy, and elastic, polishing to a fine dark colour, like those of the Brejahúba palms (Astrocaryum Ayri).

. Here we struck upon and followed a cattle path leading west. The surface was sandy, with platforms of slabs or lumps, compact or scattered, of carbonate of lime, almost marble, ready to make a shell road. Nothing could be finer than the soil, which in places was flooded by the late rains. We were charmed by the vegetation. The "Ingá" Mimosa was hanging itself about with soft white balls, whilst the Juá (Zizyphus) and the Favelleiro in bud gave out the most grateful odour. The Páo Pereira* (a Cassuvia) bore apple-like flowers; it gives wax; the bark is used for fevers; and an extract of it kills, like mercury, the "bernes" that appear in the wounds of cattle. The leguminous "Páo de Colher" (Spoon-tree), a congener of the far-famed "Brazil-wood," turns up its holly-like leaves, as the frizzly fowl does its feathers. The Convolvulus displays especial beauties, and the species of Bignonia (?) known by the general term "Açoitá Cavallo," or "Switch horse," overrun the trees, forming splendid canopies with delicious perfume. One bears trumpet blossoms of the finest mauve colour, and the other, silver-gold with leek-green leaves, is a delight to the eye. We shall often see them down-stream. Many of the

---

* Or Pereiro : it is mentioned by the System.

growths had a spicy odour. The Cactus was everywhere, from the Turk's Fez to the tall Chandelier, nor were the Bromelias less in force. The aloe-formed species (Vellozia Aloifolia) was putting forth long spikes of deep-pink flower, tipped with purple and light blue. Another, called by the general term Carauá (Bromelia variegata), had whitish-green transversal rings upon the dark-green surface, and a terminal spur, sharp as a scorpion's sting, which reminded me of the "Hig" of Somali-land. This species produces the best white fibre for hammocks, and it is stronger when not macerated in water.

Presently we reached the base of the "Serrote do Tombador." It is a detached buttress, now a common feature, and, from different points of view, it appears circular, pyramidal, or cunei-form; it looks higher than it is for want of comparison. The material is magnetic iron,[*] of which traces are found in the clay of the river bank: and it is based upon limestone, its natural flux. The ore was almost pure, and large fragments might have served as anvils; it broke into rhomboids, glittering with finely diffused mica, and it was banded with the whitest quartz, and here and there faced with a paste of pudding stone. The needle was so much affected by it, that we were compelled to take the sun for our guide. Rock crystal, the "flower of silver," was scattered about, and quartz seamed with black mica glittered like galena.

A sharp ridge, striking east and west, crested the hill, which may be 250 feet high; the northern flank is precipitous, but it is easily ascended from the south and from the south-east. The Mimosas and thorny trees become rare as we ascend, and presently disappear, the Bromelia dwindles to three or four inches in length, without, however, any abatement of its inju-rious thorns; the cylindrical cactus is mostly in decay, and from the irregular cleavage of the hill-top, the Macambira raises its tall flower spikes waving in the air. Iguanas and lizards, real salamanders for sun-heat, had here made their homes. We passed the earths of the little Môcô coney, and bleached shells (No. 4), rare below, above common. At this season, unfortu-nately, all are dead, and the young race will not appear till the rains set in. A pair of fine pearl-grey hawks, with white waist-

---

[*] Ferro Oligisto, M. Halfeld (Rel. p. 118).

coats (Falco plumbeus ?), screamed at us, hovered over our heads, and seemed prepared to do battle : probably the nest was near. These birds have a rapid flight, and are said to be good hunters.

From the summit we had a view which disclosed at first glance the gigantic scale of the denudation.* The yellow stream flowed in a broad band at our feet, through a plain subject to floods, and with a minimum breadth of six leagues. It was buttressed by a number of deceptive cones, like that upon which we stood ; some grey-coloured with limestone, others dark with oligiste, and their superior hardness had preserved them from the common destruction. Both sides of the valley were highlands ; to the north the forms were less regular, and the softer portions had been worn away. On the south appeared three long terraces curving into several bays ; below the horizontal surfaces of the upper heights long white lines of perpendicular wall, like sea cliffs, capped their slopes, regular as if laid out by the hand.

Descending the hill, we found the wind breaking the current into backward-rolling yellow yeast. Occasionally taking shelter under a Giráo of four posts with fascined top, we collected the zebra'd snail-shells scattered over the fields. They were met with chiefly in the Maníba,† the dwarf manioc, which ripens in six or seven months. At 2·30 we embarked, but shortly afterwards an opalescent " Olho de boi," crowning a thin column of rain which was falling in little sheets all around, drove us to an anchorage under " As Queimadas." Here the bank, twenty-two feet high, is cut into broad steps by the floods which spread two miles into the country. The people attribute the extensive caving in ‡ of the side, where, by-the-bye, the river forms a gut, to the gambols of the monster " Minhocão" in the days that were. No one, however, would affirm that he had seen the " Worm."

The little settlement contains about fifty thatched huts, the people fish, breed cattle, sheep, and long-legged pigs, cultivate

---

* "They reminded me of Mr. Bates' description of flat-topped hills between Santarim and Pará, in the narrow part of the valley near Almeyrim, rising 800 feet above the present level of the Amazons."

† Usually "Maniba," or "Maniva," is the stalk of manioc, the root is "mandioca," the juice is "manipuera," and the leaves are "manisoba." The latter is pro-

bably the "manacóba" which Gardner applies to a large species of the Jatropha. The root was the staff of life to the Brazilian "Indians," and the civilised race has inherited from them an immense terminology descriptive of the plant : a volume might easily be filled with it.

‡ " Desmoronamento." M. Halfeld (Rel. p. 119) also heard this legend.

maize and manioc, and send to Remanso fruits, oranges, and
limes, grown upon the other bank. Despite the sunset of purest
yellow gold, the high east wind blew all night, and lowered the
mercury to shivering point, 68° (F.). The repose was not comfortable, the tender-canoe bumped unceasingly against the "Eliza,"
and the latter rocked like that great ship which admitted the cow
into the ladies' cabin ; the village drunkard periodically visited us,
asking for fire till the small hours, and the dog Negra received
him with furious barkings.

Oct. 17.—A fine sky and heat-promising sun were perfect conditions for a gale. We passed on the right bank the Fazenda do
Monteiro, a clearing with tiled huts. Behind it is the Morro do
Monteiro ; it is a cone seen from the west, from the east a saddle-
back with smaller adjunct ; the colour is grey, and we picked up
only sandstone and ferruginous quartz. After three hours of vain
struggling, we anchored at Trahiras on the southern bank; here
also is a Morro, which yielded Itacolumite and quartz.* On the
opposite side, the Serra do Pico with the conical Morro do Chifre
form a segment of an arc, whose hollow is to the stream. It is a
low mass with "flancs tourmentés" and cups which, due to
weathering, suggest parasitic craters ; a large ypoeira flows past
its southern base.

Resuming our task in the afternoon, we were soon driven to
the Fazenda do Oliveira, six leagues from Sento Sé. The place
swarmed with negrolings and poultry, amongst which were a
tame Jacú (Penelope) and a peacock, which surprised us with
its melancholy cry. A fine fat pig (capado) was offered for
10$000. The proprietor, Lieut.-Col. Antonio Martins, stalked
about the premises, but did not address us as we brought no
introductory letter ; had he been a Paulista or a Mineiro, we
should have seen more of the inside, less of the outside, of his
house.

October 18.—An awful stillness at dawn was a bad sign.
The river had greatly fallen during the night; we grounded
heavily at the outset, and we had hardly turned the point when
the cuttingly cold east-wind set in, and drove us ashore, whilst
the deep blue cloud-bank threatened to keep us "in quod."
All our attempts to break prison were unavailing till the after-

---

* Here M. Halfeld found veins of chlorite and pyrites.

noon, when the increased heat produced flows shifting to the south. We passed the thatched huts, with here and there a tiled house, called "As Arêas" and "dos Carapinas,"* backed by high waves of white sand. After working five hours to cover nine miles, we were driven to the right bank, near the Povoação da Lagôa. A swamp behind it swarms with water-fowl, and on the northern or opposite bank is a little stream, the Barra das Itans.†

October 19.—This day's weather reflected that of yesterday. We set out at 5 A.M., and were soon forced to lay up under the shelter of a Corôa. On the northern bank, rising from chocolate-coloured bush, was a white-capped dome with a bald ridgey head, and further to the east, the Pico de Santarem, a sharp little cone. Here the crew sold part of their stock to a stout young fellow, the main of whose dress consisted of a bit of leather. He can always catch fish and sell it when caught, and he professed the profoundest indifference for anything but straw hats and sweetmeats. The sands supplied us with an abundant collection of live and dead shells (No. 3).

At 1·40 P.M., when the fiercest gusts had blown themselves out, we again began to wind between the island, sandbanks, and shoals, which rendered steering a difficult task. The right bank, populous with villages and farms, was very rich land; canoes were fastened to the beach, and piles of wood, cut and squared, stood ready for sale. Here the stream was overhung with a shrub, whose homely form we had but lately remarked. The people call it Mangui (here Hibiscus); it is, however, a dwarf willow, which grows in beds, and supplies strong and supple withes. The leaves are spiny at the edges, somewhat like the holly, but by no means so well armed; the rest of the shrub reminded me of the Amazonian Salix Humboldtiana (Willd.), according to Mr. Spence ‡ the only species of true willow known in the hot equatorial plains.

As we advanced the river showed a clear channel, and we

---

* "Carapina" in the Lingua Geral is translated Carpuiteiro; it is possibly an "Indian" corruption of the latter word; but it is popular in Minas Geraes and on the São Francisco.

† Also written Itans and Intanhas. Itán in Tupy means a shell generally.

‡ (Journal, p. 90, R. Geog. Soc., vol. xxxvi. of 1866.) Mr. Davidson remarked that everything is thorny in these lands, even the willow. I did not neglect to collect specimens of this curious shrub: unfortunately they were lost.

passed on the right bank the barras or mouths of two streams,
"da Ypoeira," and "de Sento Sé."* The former drains a
lagoon to the west-south-west, and the latter is fed by the
southern highlands. At 4 P.M., after again wasting five hours
over nine miles, we came to an anchorage—the Porto de Sento
Sé.

* In Mr. Keith Johnston's map we
find below Sento Sé the mouth of a long
dotted line, the "R. do Salitre," which,
with a course of some 35 leagues, drains
the western counterslopes of the "Serra
Chapada Diamantina." The people assured
me that the stream falling in above Sento
Sé is of very limited extent; and, as will
be seen, the Riacho do Salitre enters the
main artery close above Joazeiro. Here
the influents greatly diminish in number
and importance : the flanking ranges
approach the river valley, and render it very
different from the higher stream.

CHAPTER XXIV.

FROM THE VILLA DE SENTO SÉ TO THE CACHOEIRA DO SOBRADINHO AND THE VILLA DO JOAZEIRO.

NINTH TRAVESSIA, 18½ LEAGUES.

SENTO SÉ DESCRIBED.—INDOLENCE OF PEOPLE.—THE PORTO.—THE WOMEN.—
LONG DELAYS BY WINDS.—PRETTY COUNTRY.—VILLAGE NEAR THE ILHA
DE SANTA ANNA.—WE ATTACK THE CACHOEIRA DO SOBRADINHO, THE
FIRST BREAK AFTER 720 MILES.—OUR LIFE ON THE RIVER.—PRECAUTIONS
FOR HEALTH.—REACH THE VILLA DO JOAZEIRO.

> O prospecto, que os olhos arrebata
> Na verdura das arvores frondosa,
> Faz que o erro se escuse a meu aviso
> De crer que fora hum dia O Paraiso.
>
> *Cara.* 7, 75.

THE "Porto de Sento Sé" * consists of fishermen's huts in a row, separated by a tall wooden cross; a few of the tenements are tiled, most of them are thatched, and the walls show a watermark three feet high. All have small compounds grown with shrubs, especially the Castor-plant. The soil is white and sandy, and the floods penetrate deep into the land. It is difficult to understand why the first dwellers did not prefer the opposite bank, where, a few yards higher up, the channel is clean, and there are two undulations which the waters can never reach. We walked to the Villa de Sento Sé, about a mile (1550 yards) to the south-west. The poor dry plain, now coarse yellow sand, becomes during the rains a stream bed : we saw the weeds of the last floods adhering to the shrub-stems. It was sprinkled with

---

* M. Millivet (Geog. Dict.) has grammaticized and nonsensed the word to "Santa Sé," which has been adopted by Mr. Keith Johnston. M. Halfeld, following the pronunciation, writes it indifferently Santocé, Sentocé, Centocé, in the Map Sento Sé. There are many similar names on this part of the stream, as Uruçé and Prepeçé (before noticed). Sento Sé, like Sabará, was the name of an Indian Cacique to whom the lands belonged, and I have followed the spelling adopted by the Sento Sé family.

the Carnahúba palm, which seems to delight in these situations of extreme wet and excessive aridity. On the left of the path was a bit of water which, with its neat border of trees and its central islet, looked artificial; the silent spoon-bill paced away in his delicate rosy coat, and the noisy harlequin plover fled screaming as we approached.

The Villa is at the margin of this "dry swamp;" to the south and west the horizon is fringed with Carnahúbas, showing the course of the stream. About half a league behind it are two lumpy hills gashed into red and grey quarries, and lined and patched with white quartz and sandstone. Here they form cliffs and walls, there they are detached buttresses; the general colour is that of the sunburnt flat, and they seem to reek with heat. This Serra do Mulungú * is, apparently, an offset from the Serra do Brejo, which up-stream showed its white cliff-walls, and which now bends from south-west to north-west. The material is granite piercing through the sandstones and secondary formations; we are fast descending to the rock-floor, the core of the land, and we begin to know without being told that we must be approaching a succession of Rapids.

The entrance to the Villa was viâ the prison, a tiled roof, lath and plaster walls (páo a pique), and iron bands nailed to a window frame. Opposite it stood the Church of S. José, remarkable only for its excellent bricks, and for the "Cantariá" quartzose granite, with spots of black mica in the blue-grey matrix; † with the exception of the wandering block shown to us at the Brejo do Solgado, it is the first upon which we have lighted since we left the coast ranges. Hence it will extend at intervals all down the São Francisco.

By the side of the church, facing north-west, and raised above the floods, are half a dozen tiled and white-washed houses; behind is a scatter of palm-thatched huts, and the only neat tenement is that of the Vigario. The travelled "Menino" bitterly scoffed at this attempt at a Villa, where we found fresh meat and rum, but could not buy even the pepper of the country.

---

* Mulungú (probably an African word) is the name of a thorny leguminous tree with beans of a lively red and black like (but much larger than) those of the Abrus precatorius. They are mashed and applied to the wounds of animals when the "bicho" has entered.

† M. Halfeld (Rel. 124) calls the rock "gneiss-granite," and declares that he found in it pyrites which may prove auriferous.

Signs of a smithy appeared upon the ground,* but no symptoms of an oven ; here they prefer the Pão de Milho, an unleavened " Seven days' bread," of maize-flour kneaded with boiling water. Other favourite dishes are " faróffa," or " passóca," pounded meat mixed with farinha, fubá, or even bananas.

The life of these country places has a barbarous uniformity. The people say of the country " e muito atrasado," and they show in their proper persons all the reason of the atraso. It is every man's object to do as little as he can, and he limits his utmost industry to the labours of the smallest Fazenda. These idlers rise late and breakfast early, perhaps with a sweet potato and a cup of the inevitable coffee ; sometimes there is a table, often a mat is spread upon the floor, but there is always a cloth. It is then time " ku amkía," as the Sawahilis say, to " drop in " upon neighbours, and to slay time with the smallest of small talk. The hot hours are spent in the hammock, swinging, dozing, smoking, and eating melons. Dinner is at 2 P.M., a more substantial matter of fish, or meat, and manioc with vegetables at times, and everywhere, save at Sento Sé, with pepper-sauce. Coffee and tobacco serve to shorten the long tedious hours, and the evening is devoted to a gentle stroll, or to " tomar a fresca," that is, sitting in a shady spot to windward of the house and receiving visits. Supper ushers in the night-fall, and on every possible occasion the song and drum, the dance and dram are prolonged till near daybreak. Thus they lose energy, they lose memory, they cannot persuade themselves to undertake anything, and all exertion seems absolutely impossible to them. At Sento Sé the citizens languidly talk of a canal which is to be brought from the Rio de São Francisco at an expense of £1680. But no one dreams of doing anything beyond talking. " Government " must do everything for them, they will do nothing for themselves. After a day or two's halt in these hot-beds of indolence, I begin to feel like one of those who are raised there.

Returning to the Porto we amused ourselves with prospecting the people. We heard of two elders who could give information, both however were absent, and the nearest approach to manhood in the place was a youth in a suit of brown holland and a wide-awake of tiger-cat skin. We hunted up, however, an intelligent

* The iron, we were told, is brought from the neighbouring Fazenda de Sento Sé of João Nunez, upon the stream of that name.

old Moradora (habitantess) who did her best to enlighten us.
The washerwomen, officially called white, worked nude to the
waist: the subsequent toilette was a shift that exposed at least
one shoulder, and displayed the outlines more than enough, a
skirt and a bright cotton shawl often thrown over the head. The
feet were bare, but the hair, which was admirably thick and
glossy, was parted in the centre and combed out straight to below
the ears, where it fell in a dense mass of short stiff ringlets, reminding one of Nubia. Some women and many of the children
had erect hair, a "Pope's head," a fluffy gloria standing out
eight inches, like the "mop" of a Somal, or a Papuan negro.
One girl had taken for her pet a leaden-coloured, hairless dog,*
whose naked skin had a curious effect when compared with the
head of its mistress. The only trace of occupation was the
twanging of a Jango, or African music-bow, which, in the hands
of a boy, produced a murmur which was not unpleasant.

Before night a small fleet of barcas, which had been weather-
bound, and which the little raft had beaten, came in racing, and
regulating by horn and song the measured dip of their long
sweeps. During the floods they can drop down from Remanso
to Joazeiro in twenty-four hours, now they will have spent nine
days. This is the last trip of the year, and all are anxious to
end it. Most of the barcas had women on board in toilettes as
simple as those ashore. The patrão on the other hand often
wore old clothes manifestly of French build, a sign that we are
nearing civilisation.

October 20.—We set out at 3 A.M., when the barcas were all
asleep; the thermometer showed 78° (F.), which encouraged us
to expect Mormaço, clouded and windless weather. We were not
disappointed in a good working day. On the right, and lying
from south-west to north-east, was the Serra da Cumieira,†
shaped like a vast pent-roof; two days ago we saw distinctly its
snowy-white cliff walls resembling "Palisades" of dolomite,
and terminal ramps slightly concave. It is prolonged by the
Morro do Frade, a similar formation, which takes its name from
a single pike or organ-pipe standing out from the abrupt preci-

---

* Prince Max (i. 219) informs us that
he never saw a specimen of these hideous
canines, which are now not uncommon at
Bahia. He refers to Humboldt (Ansichten
der Natur, p. 90), who mentions them in
Spanish South America.

† From Cume, a top or ridge-beam, thus
we say the Comb of a hill. The Cumieira
(M. Halfeld, p. 126, Comieira) is opposed
to the "Caibros" or rafters, which sup-

pice. The shapes of the mountains now change to the plateaux and quoins, the ledges, bluffs, and uptilted cliffs of a granitic country; these are probably ramifications from the primitive ranges nearer the coast. The river is of noble breadth, 4870 feet, and its right bank about the Sitio da Gequitaia * was pleasant to look upon. Near the water-side, plentiful as Hibiscus on the higher stream, rose in bushes of tender, velvety verdure, dotted with decayed leaves of dull gold, the large trumpet-shaped and mauve-coloured flowers of the Sensitive Canudo, which, however, with all its beauty serves only to poison cattle. On more elevated ground, and sprinkled with the Carnahúba, were fields of the dwarf Maníba-manioc and hay, where ate their fill unusually fine horses and asses. The fences of the wax-palm frond effectually keep out the destructive water-hog (Capivara) and extend to the stream-brink, with passages here and there left open to the water. The countryman is evidently more industrious than the townsman, and I was surprised to see so many evidences of civilisation, where all is supposed by Rio de Janeiro to be a barren barbarism.

Since morning dawned we observed outcrops of rock in midstream, and on both sides; they are probably limestone, which M. Halfeld calls " Pedras Vivas." Near Encaibro is a deposit of calcareous matter to a certain extent quarried. Further down, where we landed for breakfast, the bank was red with iron and mottled with pyrites; along the brink lay bits of calcareous tuff, water-washed into curious shapes, thigh-bones, knuckles, circles, bulges, and spinal processes. Nearly opposite us was the Riacho da Canôa, said to flow near a rich Salina; hence probably the neighbouring chapel, neatly tiled and white-washed like a bridecake, towards which parties of people in Sunday garb were paddling their canoes.

The sun, nearly overhead, waxed hot, and it stung. Yet under the flimsy awning the heat tempered by the breeze never exceeded 87° (F.), and on shore 90° (F.). At 2 P.M. we saw on the left bank the Casa Nova, a large white and tiled house near the left jaw of its "riacho." † It was fronted by a long sandbar,

---

port the "ripas" or thin longitudinal strips under the tiles.

 * M. Halfeld writes Giquítaia, and translates it (Rel. 126) "pimenta soccada com Sal."

 † Above Casa Nova Mr. Keith Johnston places the " R. Casa Nova," which he makes the frontier line between Bahia and Pernambuco, running about twenty leagues from the great river, nearly due west to the long dividing range between the Valleys of the São Francisco and the Paranahyba. As will be seen, the frontier is in the 241st, not the 234th league. M. Halfeld has laid it down correctly.

which the waters had partially covered, and a dwarf vegetation grew apparently from the depths. Below it the bank was green with the sweet Capim Cabelludo : * the Capim d'Agua ; the Taquaril, a thin bamboo used for small pipings and fireworks, and the Zozó, or Sosó, a kind of Pistia, like the P. stratiotes of the Central African lakes. The pebble-banks and the sand-bars are grown with the Angari, also called Járamatáia or Jarumátaia, which springs up even when nearly covered with water; this stiff and woody shrub, resembling a strong osier, will extend as far as the Great Rapids. The wild Guava (Araça) is familiar to us since the mid-course of the Rio das Velhas. About sunset the São Francisco was a grand spectacle, of immense breadth, smooth as oil, and reflecting, like a steel mirror, heaven and earth. The typical formation now appeared clearly developed on both sides ; we no longer see the rule of rolling, rounded hills and waves that characterise the Highlands of the Brazil ; yet there are ridges that continue in many parts to be stone-faced and .white-banded above. In front a distant block, the Serra do Capim, showed behind it a dwarf rounded block which glittered like snow in a Swiss summer. Again, off the Fazenda do Mathias on the right bank, we sighted a low Serrote, lumpy as a camel's back.

This day we had accomplished thirty-three miles in nine hours, an unusual feat, and at sunset we anchored near the left bank above the Ilha de Santa Anna. We prepared for pleasant repose, when the north-east came down upon us, and swept the wavelets over what we called our deck; the only change was from bad to worst *viâ* worse, and *vice versâ* till nearly dawn.

October 21.—The wind was " damnado," as the pilots expressed it, the stream again fell, and, despite the increased velocity of the current, we made no headway. We therefore anchored once more on the left bank, and went forth to " hunt " provisions, which are now becoming scarce with us. The margin showed scatters of granite and lime, with a strew of broken shells, and some good specimens of massive laminated quartz of the purest white. The surface of the river plain is sandy ; and the heavy rains last but four months with two of light showers ; yet the soil, enriched by the calcareous matter below, supports flocks of sheep and goats. Here the convolvulus with fleshy

---

* This useful growth is unknown to the higher stream : it derives its name from the roughness of the stem and of the under surface of the leaf.

leaves, and pink trumpet-flower (Ipomœa arenosa), was a reminiscence of the African coasts. We soon struck upon a bush-path leading to a couple of huts, where good cotton was growing in a fenced field. Yet the people were in rags ; and rags, though we think little of them in England, here startle the eye: the women had not taken the trouble to weave the tree wool almost within hand-reach of their doors. There was a Giráo-garden of lavender and geranium for decorating the hair, but no one had planted oranges or melons, bananas, or vegetables : not even rice was to be had. The country can produce all the wants of life—it bears nothing ; the people should be well off—they are in tatters. I compared their state with those a few leagues higher up, and can explain their inferiority only by some difficulty of communication.

After walking 400 yards we crossed the inundated low land, and reached what may be called the true coast. Here the rise was strewed with water-washed "cascalho" and angular "gurgulho " in regular lines. The soil was drier than usual, and amongst the Cactaceæ towered high the Mandracurú, or Mandacurú (C. brasiliensis, Piso). It is a singular growth, often thirty feet high with two of diameter, and the huge limbs, garnished with stiff thorns, stand bolt upright. The wood is a bright yellow colour with longitudinal white streaks ; it is excellent for roof-rafters (Caibros), and further down it makes the best paddles. The weight, however, renders it unwieldy, and the newly-cut wood falling into the water sinks like lead.

In the evening—anything for a change!—we dropped two miles down-stream to the Santa Anna village. Here it is proposed, during the dry season, to station the steamer which, during the floods, will lie at Joazeiro, nine leagues distant. At present it is a lump of pauper huts raised but little above the bank, whose iron-stained and water-rolled pebbly beds accompany us some way down. For four patacas (1$140)* we engaged a pilot for the Rapids, called do Sobradinho or de Vidal Affonso.† During the last 720 miles we have seen nothing but a wind-ripple ; this is the portal of a new region, and the river will offer ever-increasing difficulties, culminating in an impossibility. We examined

* Barcas pay 4$000, and when lost nothing.

† "Sobradinho" is rock-boulder, generally crowning a hill, and of smaller size than the Sobrado. Concerning the old name, "Vidal Affonso," at present found only in books, I cannot offer the least information.

carefully the lay of the land and stream. Opposite Santa Anna is the Ilhote do Junco, a mere sandstrip, backed by the Ilha do Junco, or de Santa Anna, an inhabited and cultivated island nearly four miles long, by one and a half broad. The channel running from west to east, trends at the end of the Santa Anna island to the south-east, and breaks over scattered rocks for about one league, rendering the whole of the right-hand channel unnavigable. On the left bank the Serras da Cachoeira and do Sobrado approach the stream with a lay from north-east to south-west. Upon the opposite side the Serrote do Tatauhy springing from the south-east completes the head of a broad arrow with the Serra da Castanhera * from the south-west. The latter has a roof-ridge outline slightly bent in the middle, and near the stream projects a white knob, the Serra do Capim. The reefs are nothing but the subaqueous prolongations of these aërial granitic lines.†

Oct. 22.—Despite wind and sun, and " solemn warnings,"—the people caution us against accidents, and I " take blame,"—we shipped the pilot Jacinto José de Souza, and set out at 2 P.M. to attack the Cachoeira do Sobradinho. Running in an hour down the smooth water to the north of the Ilha de Santa Anna, we came to the head of the Ilha da Cachoeira—a thin strip of well-wooded island—about four miles long, with a narrow channel between it and the left bank. The main stream, still flowing on the right, is broken by a number of tufty islets ; the pilot declares that it would be suicide to attempt this gridiron of reefs trending from north-east to south-west, forming the Cachoeira do Junco, and ending in the fierce Cachoeira de Tatauhy.

The navigable Chenal on the left is called the Braço da Cachoeira or do Sobradinho ; the upper mouth, 200 yards broad, presently narrows to half that width, and the general trend is south-east, with shiftings to the south and east. Here the smooth water ends, and the current greatly increases, never, however, exceeding six miles an hour.‡ The first obstacle was

---

* M. Halfeld calls it the "Serra do Sacco do Meio."

† Behind this broad arrow, and forming, as it were, its shaft, is the Serra do Salitre or do Mulato, which resembles in gentle brown ramp and upper white bluff the Serra da Cumieira below Sento Sé. When approaching Joazeiro, the highest part of this range seems to be capped by a bonnet, like the "Pintor" of Chique-Chique.

‡ Of course I mean at the time when we passed it. Even then the six miles may be diminished to an average of four miles an hour.

a pyramid in mid stream, with a platform of rock " en cabochon," projecting from the left bank. The material is a large-grained brown-grey granite, often iron-stained and veined with quartz ; it has large holes, in which the salt-maker evaporates the saline water which he has obtained by straining the mould.

Immediately below the pyramid, the canal is again split by two islets, the Ilhotas da Cachoeira. The upper is of low vegetation ; the lower supports trees ;—and in these places the Joazeiro and the Jatobá, the only growths of any importance, are nobly developed by the exceedingly damp air. In 1857 the head of the second Ilhota was cut off by the current, which also washed away a slice of the left shore proper, upon which were four houses. Unless arrested by the granite, it will go still further, and thus Nature will be her own engineer. The clear way leaves to the left the upper Ilhota, whose head is garnished with lumpy rocks, and strikes, as usual, the apex of a triangle : here two small breaks, passed within four minutes, make the water eddy and boil on both sides. The largest stones are on the right, where an islet is forming, and they might easily be removed by blasting.

Below the second Ilhota is the true Cachoeira do Sobradinho, denoted by a fine clump of " Cupped " trees on the long island to the right ; the left bank shows houses and fences extending all the way down. This chief obstruction is a wall built across stream, with a central breach* where the water breaks in two places. Here barcas prefer cordelling ; they are assisted by the willing country people, who stand upon a low rock on the left side ; but accidents are by no means unfrequent. † We turned stern on, and changing paddles for poles, took, the wind being in our teeth, the left side of the breach. The gap between the two rock slabs, worn into pot-holes, and channelled by water, was so narrow that we almost scraped sides. The sunken stones below this point were easily avoided.

After two hours' work we came to the Cachoeira do Bebedor,

---

* M. Halfeld calls this part the " Caixão : " he makes it 5·70 to 7·1 feet broad, and in the dries almost too narrow for barcas to pass through. The greatest height of the rock above water is 8·60 feet ; the current is 4·17 miles per hour, and the height of fall 3·6 feet.

† Here M. Halfeld's barque, the Princeza do Rio, snapped her tow-rope, and was nearly lost. We read in the Relatorio (p. 132)—

"They informed me that the pilot who guided my vessel during its descent of the rapid perished in the same place." *They* assured me that Manuel Antonio, the pilot in question, had fallen out of his canoe, and had been drowned in smooth water, of course after " liquoring up." Nothing but the greatest carelessness can cause an accident at the Sobradinho.

opposite the hamlet of that ilk; here again the snags and rocks offered no difficulty. The next was the Cachoeira Críminosa, of which the worst part is the name; here, however, the blind rocks are hard to thread, and necessitate frequent passing from side to side. We are now at the south-western foot of the Serra do Sobrado, a remarkable formation which has long been in prospect. Seen from Santa Anna, to the south it is a quoin-shaped mass, with snowy lines sloping to the stream; and it appears to be on the right, whereas it hems the left bank compressing the channel. A nearer view shows the lower three-quarters, invested with tall, thick bush, which dwindles in stature as it ascends; below the crest are two nearly parallel bluffs of bare rock, inclining towards the water, and separated by a thicket-grown level. In the under-cliff appeared the dark mouth of a cave; and further down there is, they say, a larger tunnel.

The mass wears the look of limestone, based upon the granite which outcrops in the river.* The peculiarity of aspect has supplied it with sundry legends. According to the people, a "corrente," or large chain, has been found extending from top to bottom. Our pilot, not an imaginative man, derided the chain; but declared that, at times, especially near the rainy season, the mountain made "estrondos," or loud rumbling sounds, adding that the last had been sufficient to frighten him. As I have remarked, tales of roaring hills are common in the Brazil; perhaps in places the mysterious noises may be caused by the sudden elevation or depression of the mountain.

At the foot of the Sobrado we avoided, by traversing to the right, a succession of small breaks. A little jump was the last obstacle, and at 4·25 P.M. we came to the Boca do Braço, where the south-eastern end of the Ilha da Cachoeira projects a few outlying blocks into the main stream of the S. Francisco, now clean and narrow. Thus we had expended upon the Sobradinho 2 hours 45 minutes, but the wind had always been against us. We landed Jacinto José da Souza on the left bank, and thanked him heartily over a parting "tot:" he is a good man, careful and dexterous, and, wonderful to relate, he works without noise.

This obstruction is in its present state, and at this season, fatal to steamer traffic; during the floods, the only obstacle must be

* The pilot declared the material to be marble. M. Halfeld (p. 133) describes it as "Itacolumite alternating with strata of talcose schist and quartz, running south-south-west to north-north-east, with westerly inclination."

the rush of water.   Canalising through granitic rock is not likely
to pay, and the state of civilisation is here hardly sufficiently
advanced to keep up sluice gates.   Removing the scattered rocks
and bars will draw the water into the central thalweg, and make
a safe passage which, when once made, is not likely to be choked.
M. Halfeld estimates the expenditure at £39,000, which is, perhaps, the minimum, if at least the three miles are to be rendered
navigable for tug-steamers throughout the year.   Altogether, the
Cachoeira do Sobradinho, this furthest southern outlier of the
Great Rapids, is equally interesting to the engineer and to the
geographer.

We ran down the line, which narrows from two miles to a
quarter of that width, and presently we came to another symptom
of rapids, the first rock-islet sighted in the São Francisco.   This
"hog's-back" amid-stream is the prolongation of a Serrote on
the north bank; amongst the broken slabs of the lower part, half
masked by tufty growth, is a cavern with a bad name.   The
novelty of the appearance has, as usual, bred fables; the boatmen, however ugly, will not sleep here for fear of the Siren with
golden hair, who lies in wait for them.   They know it as the
Ilha da Mãe d'Agua; but " serious persons," who " disapprove
of " Melusine de Lusignan, call it de Santa Rita, a saintess to
whom the impossible is possible ; and who, little known in England, is festivated (July 12) in the Brazil with novenas and
rockets which render the day detestable.   At sunset we anchored
off the sandbar do Lameirão: we are now within some 9° 20′ of
the Equator ; the great light is almost overhead, and yet the
weather is cold and gusty.   Five small huts within sight on the
left bank marked the Páo da Historia, the frontier (divisa)
between Bahia on the south, and the Pernambuco Province to the
north.

Oct. 23.—After an hour and a half of paddling, the wind, from
misty clouds, drove us to anchor on the right bank.   Here a
clump of wild figs, tufted with the mistletoe-like Herva de passa-
rinho (Polygonum), and springing from a bed of soft, short, and
green Graminha, the local Bahiana grass, shaded our mats more
pleasantly than any tent.   These delays were inevitable, and the
only remedy was to extract from them as much enjoyment as
possible.   The prospect lent powerful aid.   The lustrous blue
sky deepening through the dark fleshy leaves, was the " glazing : "

the picture was a grand flood, flavous as Tiber, coursing behind the gnarled trunks and the buttressed roots of the Gamelleiras. Life and action were not wanting to the poem. Humming birds, little larger than dragon-flies, red-beaked, and with plumage of chatoyant green, now stared at the stranger as they perched fearlessly upon the thinnest twig, then poised themselves with expanded tail feathers and twinkling wings, whilst plunging their needle-bills into the flower cup, or tapping its side ;* then darted, as if thrown by the hand, to some bunch of richer and virgin bloom. Compared with the other tenuirostres of the Brazilian grove, which are, however, more dainty and delicate than the tiniest European wren, they were Canova's Venus by the side of the Sphinx. And the little bodies contain mighty powers of love and hate—they fight as furiously as they woo ; and no unplumed biped ever died of " heimweh " so readily and so certainly as the humming bird imprisoned in a cage.†

Our day is as follows :—We rise before dawn, and after a "merenda" of coffee and biscuits, or rusks, apply ourselves to writing up journals, and to arranging collections. The crew eat bacon and beans at 7 to 8 A.M. : I reserve the process till 11 A.M., when the neck of the day's work has been broken. The bow of one of the canoes is a good place for a cold bath, and there is no better preparation for the hotter hours. After noon the labour becomes lighter; and the little industries learned by African travel now come into play. For instance, the manufacture of rough cigars with the " fumo de tres cordas," the " three-twist," brought from Januaria. " Reading up " is decidedly more pleasant than writing in a rickety raft upon the mattress stuffed with corn-glumes, which acts table, and the scene-shifting of the river and of the mountains, combined with the subtle delights of mere motion, is an antidote to ennui. When the breeze becomes a gale, we explore the valley for shells and metals, or climb the hills to enjoy the scenery ; or should the demon of Idleness get the upper hand in his own home, we stretch ourselves beneath the trees, enjoying the perfumed shade, and a life soft as moss, an approach to the " silent land." About sunset, we feed in the humblest way, upon rice when there is any, and upon meat or fish

---

* I have often found the hill fuchsia pierced in the lower part of the cup.

† Here the people universally believe that the humming-bird is transmutable into the humming-bird hawk-moth (Macroglossa Titan). Upon this subject Mr. Bates has treated (i. 182).

under similar restricted conditions. When the night-birds begin to awake from their day-sleep, we choose some well-exposed place where immundicities will not trouble us, and "turn in." It is a life of perfect ease, the only fear or trouble is lest the dark hours should be too cold, or the sun too hot, or the wind troublesome; the spes finis is, and should be, the last thing dwelt upon.

During nearly four months' travel down the Rio de São Francisco, with alternations of storm and rain, cold wind and hot wind, mists and burning suns, I had not an hour of sickness. Mr. Davidson, it is true, suffered from "chills;" but he had brought bad health to the river, and he improved in condition as we went. On the other hand, it must be remembered that we did not travel in the bad season, which is here, as elsewhere, near Brazilian rivers, the drying up of the waters. The precautions which I adopted were few, and mostly comprised in my old rule to alter diet as little as possible; it is my intimate conviction that, although the sojourner in foreign lands to a certain extent may, the traveller must not attempt to conform to the "manners and customs of the people." As regards drinking-water, the only necessary care is to wash all the jars every night, and to allow the deposit to settle, which it readily does without alum or almonds. Coffee keeps up the vital heat, and lime-juice corrects that scorbutic tendency which often accompanies the loosened state of the waist-band. On raw mornings, and every night, I "made it twelve o'clock," with a wine-glass of spirits, good cognac (so called), when procurable—Cachaça when there was nothing else. We religiously avoided stimulants, even wine and beer, during the day; and two grains of quinine readily corrected nervous depression. My chief thought was to be warmly clothed when sleeping, a precaution learned from the Arabs of East Africa. The walk and talk were essentially parts of hygiene; but, above all, activity of mind, "plenty to do," contentment, and again, no "spes finis."

Oct. 24.—The night was of a stillness so deep, that an unprotected candle would have burned out. Not so the next morning. We passed on the right the Barra do Riacho do Salitre. The small brackish stream can, during the floods, be ascended for some leagues by canoes.* Here the bank is tall, and white with

---

* Mr. Keith Johnston places a stream far above Joazeiro and another far below it, but none near it.

blocks, layers, and scatters of the finest limestone; the land is well fenced in, and even the Carnahúbal is hedged with dry thorns. Below it we found a labyrinth of rocks sunken and above the surface; no improvements, however, are here necessary. After being nearly swamped more than once, we passed to port the Ilha do Fogo, and found quarters in a baylet * at the eastern end of the " Villa do Joazeiro," defended by a low bush projecting into the stream. Traders usually anchor further west.

* Here called the Ressaca or Resaca.

CHAPTER XXV.

AT THE VILLA DO JOAZEIRO.

> Encrespava-se a onda docemente
> Qual aura leve, quando move o feno ;
> E como o prado ameno rir costuma,
> Imitava as boninas com a escuma.
>
> *Caraïnurú*, 6, 44.

I had long heard of this place as the future terminus where
the great lines of rail were to meet; on the higher São Francisco
it had been spoken of as a centre of civilization, a little Paris,
and the Provincial Government of Bahia actually ordered a
detailed plan of the place to be made and deposited in its archives.
So much for the imagination. Now for the reality.

Joazeiro has a family likeness to the Villa da Barra do Rio
Grande. It is a long line of houses fronting the river, which,
here some 2500 feet broad, flows in a straight line from west to
east. The banks are raised 21 to 25 feet above low-water level,
but many of the tenements show flood-marks. The citizens all
declare that M. Halfeld was in error when he wrote (Relatorio,
p. 140), "the greatest rise in 1792 was of 45 palms" (upwards of
32 feet) "over the usual height, so that on this occasion the
church was flooded 11 palms deep, and so, more or less, were all
the habitations." In 1865, they assert, the inundation equalled
that of '92, and, although it reached the cemetery, it was two or
three palms below the church and the main square.

Some of the houses front the stream, especially in the more civilized western quarter ; the centre shows a ruinous flight of broad stone steps, and here the abodes turn their backs and their yard-walls to the water, which has washed off the plaster and exposed the skeleton of adobe, or palm-wattle, and dab. The sandy soil requires a foundation of the limestone or the freestone, of which the country is a quarry; the streets, however, are totally destitute of pavement, and only the best tenements are subtended by an embryo bottom of brick. A few trees, under whose shade salt is sold, and small transactions take place, are scattered over the beach, which is strewed with pebbles, pudding-stone, and iron-cemented quartz, in the lowest levels. The Villa has but one sobrado, belonging to some fourteen proprietors, and even this has not a sign of glass windows.

At the western end there is a cemetery, with whitewashed and tile-coped walls, including a dwarf chapel. Thence runs the Rua do Mourão, which fronts the river. Behind this thoroughfare lies the Rua do Açougue, " Shambles Street," and, inland of all, the Rua da Recoada, both ragged lines of poor huts, mostly thatched. These streets have the pretension to hoist the white hand and extended forefinger of Rio de Janeiro, directing carriages which way to go, when there is not a carriage within 300 miles.

About midway in the long shallow line is the Praça do Commercio, whose loose sand, spread ankle deep, forms an excellent reflector for the sun : the chief use appeared to be that of an arena for fighting turkey cocks. Attempts have been made to line it with tamarinds, which are now stunted, and with the fleshy-leaved " Almendreira," or Persian almond,* about eight years old, but poor compared with those further inland. Here are the principal stores ; before 1857 there were fifty-two, but many failures reduced the number to fourteen—not noticing the twenty-five " Vendas." That civilization-gauge, the Post-office, is also a dry-goods store; the shopboy permitted me somewhat super-ciliously to inspect the " dead letters," which reposed in a lid-less Eau de Cologne box. The correio is supposed to go out on the 3rd, the 13th, and the 23rd of each month, and to come in on the 2nd, the 12th, and the 22nd ; to-day is the 24th, and it gives no

---

* I have never seen the flower or fruit of this tree, which resembles the Sterculiæ. It thrives in the hot humid atmosphere of Pernambuco, and was planted possibly by the old Portuguese on the East African Coast about Kilwa (Quiloa).

sign, and who cares ?   Here is a single pharmacy, and the Capitão who keeps it prescribes his own drugs; there is no doctor, and, consequently, there is little mortality.   The people are by no means a healthy race; the height above sea level does not exceed 1000 feet, yet catarrhs and pleurisies, fevers and pneumonias, not to mention other diseases, abound.   One of the citoyennes had a nose prolonged like the trunk of a young elephant, and an eye to match; the hideous affection was called Cabungo, or erysipelas.

The head of the square is occupied by the new Matriz of N^a S^a das Grutas, of stone, burnt brick, and lime, of course unfinished.   I had supposed that want of funds was the cause, the citizens declared that such was not the case; probably it is "politics."   Very mean is the original temple, said to have been built by the Jesuits and their "Indian" acolytes.   Above are two open windows, or rather holes; below is a similar pair, railed with thin wooden posts; the belfries, as in Sienna of the Earthquakes, are mere walls, with openings in which the bells are slung, and the quaint finials suggest donkey's ears erect in curiosity.   Beyond the church is the Rua Direita, a slip of a street running off into space.   Here the river is faced by the Rua dos Espinheiros, whose small huts and vendas drive a trifling trade; a large half-tiled shed, sheltering huge wooden screws and new waggons of the oldest style, represents the docks, where the steamer will be launched—when she arrives.

Joazeiro was disannexed from Sento Sé, under whose tutelage it became a freguezia, and was created a villa on May 18, 1833. It is now the head-quarters of the arrondissement (Comarca), and the residence of a Juge de Droit, and of a "Superior Commandant"; it has also a town hall and a jail.   The municipality is tolerably populous, exceeding 1500 voters.   The townspeople were 1328 souls in 1852, and are now about 2000, of whom a quarter is servile, whilst the houses, which have not increased, number 334, subject to the tax known as the Decima Urbana.

The situation of Joazeiro is, commercially speaking, good— a point where four main lines meet—the up stream, the down stream, the great highway to Bahia, and the road to the Northern Provinces.   This central site will secure for it importance in the proposed Province of São Francisco; of course it expects to become the capital, but what is the use of a capital close to the frontier ?   The position will be that of a great outpost, transmit-

ting to the seaboard the produce of Southern Piauhy and of
Eastern Goyaz. It had of old considerable traffic with Oeiras
(eighty leagues), formerly capital of Piauhy, and this continues
even since Theresina, ninety leagues further, became the metropolis. I found in port only two barcas, and the expense of transport greatly injures trade. The "Viagem Redonda," on going
and coming to and from Cachoeira city, at the head of steamer
navigation in the Bahian Reconcavo, has lately risen from 15$000
to 25$000, and even to 30$000 per mule, carrying at most
seven arrobas—about 10s. per 32 pounds. The down journey,
viá Villa Nova da Rainha, occupies ten to thirteen days, the fast
travelling being eight leagues per diem; and it is said that a line
properly laid out would reduce the distance from ninety-two to
seventy leagues.

The lands immediately about Joazeiro, especially on the Bahian
side, are poor, hard, and dry; the rains last from the end of
October till March, and the fertilizing showers of the dry season
are wanting. The price is somewhat high, two square leagues
can hardly be purchased for less than £2000. A little is done in
the way of breeding horses and mules, black cattle, sheep and
goats, pigs and poultry, especially turkeys. Salt and saltpetre,
limestone and sugar with a saline taste, are supplied by the
Riacho do Salitre: this stream rises about Pacuhy, receives the
tributes of the Jacobina Nova and the Jacobina Velha, and feeds
the São Francisco after a course of forty leagues. A place called
the Brejo, distant about four leagues to the south-west, is the
local "celleiro," or granary, and, as this is small, provisions must
be imported from up stream. It produces in abundance pumpkins
and water-melons, especially at the beginning of the rains; the
orange is here small and green, like the wild variety, it does not
find a proper climate, and below Boa Vista it ceases to grow; the
limes are juiceless, and half pips. Cochineal is unimproved, and
there is no tobacco, for which the nitrous soil is well adapted. I
was strongly advised, even by a youth who had lately come up in
three days from Boa Vista, to lay in a stock of beans and manioc,
rice and maize, as nothing was to be found in the starving settlements between Joazeiro and the Great Rapids. The precaution
was taken "on the chance," but, as will be seen, it was quite unnecessary; moreover, it gave considerable trouble. Not a pound
of husked rice was to be bought, the price was high, and the

article was coarse and red, fit only for a Kruboy.* Fish was
abundant, and the Surubim, the salmon of the river, was hawked
about by boys. Some complain that the increased flow of the
stream, the rocky bottom, and the broken waters, are bad breeding conditions, and that the São Francisco is no longer a Mississippi, a father of fish. Others declare, and with truth, that fishermen, not fish, are wanting, that a net is never thrown in vain, and
that the pools, bayous, and ypoeiras produce large shoals.

On the opposite or northern side is the Porto da Passagem do
Joazeiro, of late called Petrolina de Pernambuco. It was a
little chapel, Nª Sª de Tal, and half a dozen tiled houses fronting
the stream, backed by a few huts, and a wave of ground higher
and healthier than the right bank. The two are connected by a
ferryboat which makes use of the " vent traversier," and carries
twenty-five to thirty head of cattle. Each passenger pays per trip
0 $ 080, horse or mule 0 $ 400 (the load and trooper going free), and

---

* In the following list of prices it must be remembered that here the alqueire is four
times larger than that of Bahia :—
1 alqueire beans (in 1852, 11$500) = 20$000.
1 alqueire farinha of manioc (6$400) = 12$000.
1 alqueire salt (12$000) = 24$000.
1 arroba (32 lbs.) of lard (7$680) = 10$000. This was the price which I paid, but
it was nearly 1s. 3d. too much.
1 arroba wheaten flour (0$240) = 14$000 to 16$000.
1 arroba biscuit (10$000) = 16$000.
1 arroba country wax (5$000) = 6$400. Honey is also cheap and plentiful.
1 arroba Carnahúba wax (5$000). It is not made now.
1 arroba Carne Seca (3$400) = 6$000 to 7$000.
1 arroba cotton uncleaned (2$560) = 2$000.
1 arroba cotton, cleaned = 8$000.
1 arroba sugar (7$000) = 4$000 to 5$000.
1 lb. steel = 0$400.
1 lb. bar lead or shot = 0$400.
1 lb. saltpetre = 0$080.
1 lb. sulphur = 0$320.
1 vara (yard of 43 inches) cotton cloth (0$320) = 0$400.
1 vara twist tobacco = 0$160.
1 sugar brick (rapadura) of Januaria (0$240) = 0$160.
1 sugar brick (small and saltish) of the R. do Salitre = 0$080.
Tin sheet (folha de Flandres) = 0$240.
Plank, wooden (1$600) = 2$000.
1 bottle common Barcelona wine (0$640) = 1$000.
1 bottle port = 2$500.
1 bottle vinegar (0$320) = 0$800.
1 bottle corn brandy (of Jacobina, poor) = 0$200.
1 bottle corn brandy (Sto Amaro, the best) = 0$500.
1 bottle Ricinus oil = 0$240.
1 bottle sweet oil (1$000) = 1$600.
Per côvado (cubit of 26½ inches) of Chita cotton cloth averages = 0$230.
Raw hide of an ox or cow (1$280) = 2$800 to 3$000.
Calf, according to size, from 0$800.
Sheep or goat's skin = 0$320.

black cattle 0$300. Matters had changed little since 1852, when M. Halfeld rated the annual movement at 7500 to 8000 souls, 10,500 head of black cattle, and 1300 horses and mules, wild and tame, old and young, intended for the Bahian market.*

My introductory letter was duly sent to the Superior Commandant, Lieut.-Col. Antonio Luis Ferreira, who did not deign to take the slightest notice of it. I then called upon Sr. José Vieira, a young merchant whom we had met up stream : his store was in the Rua do Mourão, or western water street, fronted by a black wooden cross on a pedestal of brick and lime. Of those assembled there none could give me any local information, even the names of the streets. Fortunately I made acquaintance with the Capitão Antonio Ribeiro da Silva, junior, the son of a Portuguese, and born in the place : he had travelled in Europe, and he at once invited us to dinner and chat.

The Capitão spoke of a Gruta, which he described as having a descent to the mouth like the Mammoth Cave ; it extends three to four miles, and is distant nineteen up the bed of the Riacho do Salitre. Here are old legends of silver mines near Sta Anna, and copper at the Fazenda da Carahyba, eighteen leagues to the east-south-east. Our host had found a diamantine formation, covering at least twenty square leagues, in the rich agricultural and coffee country, of which Jacobina Nova is the centre. He gave us some excellent " doce " of the sweet potato, which is here a red variety like beet-root, banded white. His garden contained fine vines four to five years old, trained to a tunnel-work, but almost able to support themselves. This is a grape-country, and nearly every house has its paneiral or arboury : the vines produce all the year round an " Uva durecina," which sells here for 0$240, and at Bahia for 1$000 per pound. Much has been written about the Brazil being capable to produce her own wine. This, I apprehend, will be hardly possible in those climates where the hot season is also that of the rains. The same bunch will contain ripe, half-ripe, and unripe berries, which make a good vinegar. Nor is there any cure for the evils endured by this

Non habilis Cyathis et inutilis uva Lycæo.

On the other hand, where the wet weather begins with the northing

---

* The ferry dues are received by the Villa da Boa Vista, which we shall visit down stream.

of the sun, and where the summer of the southern hemisphere is
dry and sunny, the grape, I believe, is fated to do good service.

My next visit was to Sr. Justino Nunes de Sento Sé, a native
of the town whose name he bears : this gentleman introduced me
to his wife and his fair daughters, who after three months'
experience of Joazeiro, much preferred Bahia, their birth-place.
The father had been chosen by H. E. the Councillor Manoel
Pinto de Souza Dantas to superintend the steamer which was
proposed, even in 1865, to plough the waters of the Upper São
Francisco. Unhappily for the project, Sr. Dantas took the port-
folio of agriculture and public works, whilst his successor, the
Provincial President, was by no means earnest in carrying out
the plans of a predecessor. Sr. Sento Sé complained much of
private opposition. A Joazeiran proprietor, Lieut.-Col. Domingos
Luis Ferreira, had offered for £1600 to receive the vessel from
the hands of government at the Porto das Piranhas, the present
terminus of steam navigation on the Lower São Francisco, to carry
up the sections on horse and mule-back past the Great Rapids, and
then to embark them on barcassas, or lighters. His friends
resented the rejection of his proposal, and spread a report that
the candidate preferred had wasted £6200, that the fragments of
the "Presidente Dantas" were scattered about the Bahian road,
and that an engineer sent from Rio de Janeiro to set up the
machinery had, after four months' waiting in vain, returned in
July, 1867.

Then the steamer, which the papers had made to reach Joazeiro,
and which His Excellency expected to begin work not later than
September, 1867, was, in fact, nowhere. Sr. Sento Sé appeared
to be thoroughly tired of the business, and spoke of raising a
private company for steam navigation of the São Francisco. It is
lamentable to see a great thought thus hopelessly frittered away in
detail by private jealousies and by petty individual interests.
Much as I deprecate the employment of foreign engineers in this
Empire, where natives can be found, there are cases when the
appointment of a foreigner will not raise up against him a hundred
enemies, as will assuredly happen to the native.

We were delayed at Joazeiro until the two men hired at
Januaria agreed, for a consideration, to place me at Boa Vista :
here the people have by no means the best name, and various
tales are told about barquemen robbing their employers, and

leaving them "in the lurch." They drink, and they are dangerous: therefore the men below Boa Vista are always preferred to them. José Joaquim and Barboza, "Barba de Veneno," had always won my gratitude by keeping themselves below: not so the Menino, who at night had returned to the Eliza on all-fours, like one of the lower animals, whilst Agostinho, the slave, was successively drooping with "sea-sickness," and unpleasantly surly. It was a weary time, as are all those enforced halts near towns. The nigger boys splashed in the water around us, and the mulatto youth came to cheapen and wrangle about straw hats, gugglets, and orange conserve. We were anchored amongst the washer-women, who were grotesque objects. One defended her head with a calabash, forcibly reminding me of that Triton song by Camoens (vi. 17): he was very ugly,

> And for a casque upon his head he wore
> The crusty spoils whilome a lobster bore.

In no part of the Brazil had I seen such an excessive display of shoulder: it exceeded the high mode of the Bahian Quitandeira, or black market woman, and it was truly remarkable after leaving the Province of Minas Geraes.

When tired of shoulders I visited the Ilha do Fogo, that small St. Michael's Mount, which we had passed hard above the Villa. It is an interesting feature, and the first of its kind yet seen, a composite river-island of rock terminating down stream in a long sandspit: the level parts were bush-clad, and a splendid Jatobá tree added not a little to their beauty. The northern arm into which it divides the São Francisco is, though navigable, dangerous with sunken rocks: hence probably Joazeiro preferred the right bank.

We landed amongst the blocks and boulders of the western end. The material was a gray granite, coated in places with purplish glaze, like iron that had been exposed to great heat: there were various masses of amygdaloid and veins of quartz, but pyrites did not appear.* It was easy to scale the tower of broken slabs about eighty feet high; certain enterprising sightseers had cut a path through the Niacambíra Bromelias, and had cleared off the Quipá Cactus. The summit commanded an extensive view of

---

* M. Halfeld describes the rock as quartz-veined granite: he found talc, manganese, and pyrites.

the São Francisco, a panorama of plain studded with low hills and dwarf ranges, offsets from the great walls of the riverine valley. East of the Fire-island main heap are two minor outcrops of the same rock, emerging from the thorny brush.

Joazeiro, I have said, is the proposed terminus of two Anglo-Brazilian Railways, that of Pernambuco and its junior the Bahia. Both were offsprings of the law of June 26, 1852, decreeing the concession of the line D. Pedro II. A guarantee of seven per cent. (five from the Imperial and two from the Provincial Governments) easily opened the purses of the shareholders. The reports of a rich and fertile interior, waiting only to be tapped by the Rail, determined the direction from the coast towards the Rio de São Francisco. Works were undertaken with a recklessness characteristic of great expectations. No general commission was organised to arrange the system upon which the great trunk road should proceed. A staff should have been appointed to make serious preliminary studies of the ground: this was neglected, and in the Brazil I have seen calculations for cuttings and embankments based upon a flying survey, whose levels were taken with the Sympiesometer. The result was what might be expected. The lines were laid out and built with almost every conceivable defect; they began at the wrong places, and they ran in the wrong directions; they were highly finished where they could have been made rough; they were dear where they should have been cheap; they had tunnels where the land was to be bought for a song. Thus the estimates were shamefully exceeded, and the seven per cent. became a snare and a delusion. The branch roads and feeding lines were not made: hence complaints and recriminations; the shareholders were losers, and the Government found itself saddled indefinitely with a huge debt, which it had calculated to pay off by the increased yield of the railways. Here, and here only, has the steam-horse assisted in uncivilizing the country by unsettling the communications which before were bad enough, and are now worse. Here, and here only, the mule can successfully contend against machinery: anti-Brazilian writers compare the progress of the country with that of the sloth, and truly at this rate it will be behind even Canada. Finally both these main trunks stopped short within a few miles of the Provincial capitals, where they had commenced and built their last stations, either in the virgin forest or in Campo ground, little more productive than the favoured regions about Suez. At this moment railway enterprise in the Brazil may be said to stand still, and the Empire has suffered in the money market of Europe for a maladministration whose blame attaches chiefly to foreigners.

On the other hand, steamer navigation has prospered, and from Joazeiro downwards, we shall find that the weekly arrival of a little craft at the Porto das Piranhas * has galvanized the whole country as far as Crato in Ceará, a radius of 270 miles. Leather-clad men, who would never have left their homes, are now loading their animals with cotton, and are making purchases of which a few months before they would not have dreamed. In 1852 M. Halfeld remarked, " by reason of the great rapids on the São Francisco, both above and below the town of Cabrobó, fluvial traffic has been little developed." The description is obsolete in 1867, showing how vitalizing, even in these thinly populated regions, is the effect of improved communication. I hope to see the Bahia Steam Navigation Company (Limited) † increase her

* The first commercial steamer left Penedo, August 3, 1867, and reached Porto das Piranhas, August 5, 1867.

† This Company was organized in 1861. Its Articles of Association were approved by the Imperial Government in 1862, and it dates its proceedings as an English Company from June of that year. The capital is £160,000, of which about £150,000 has been paid up. The subsidies granted by the Imperial and Provincial Governments amount to £20,000 per annum, equivalent to one-eighth, or 12½ per cent. on the whole capital. The contract actually in force extends till 1872, and an Imperial Decree (No. 1232 of 1864) authorises the Government at the end of the above period to revise and extend the convention and the subsidies for a term of ten years. The obligations of the Company comprise communication with the chief ports on the Brazilian sea-board, extending north-wards from Bahia to Maceió, and south to Caravellas or São Jorge dos Ilhéos ; likewise the internal navigation of the Reconcavo, from the provincial capital to the cities of Cachoeira, Santo Amaro, Nazareth, Valença, and Taperoa, touching at the intermediate villages ; thirdly, the navigation of the Rio de São Francisco, from Penedo to the Porto das Piranhas ; and, fourthly, the navigation of the Lakes Norte and Manguába, in the Province of Alagôas. The floating property is represented by the following sixteen steamers, six of which are employed in the coasting navigation, and ten in the internal, or bay and river navigation :—

		registered tons		horse power
1.	S. Salvador	280 ;	,,	150
	Dantas	295	,,	165
	Gonçalves Martins	298	,,	126
	Sinimbú	312	,,	126
5.	Santa Cruz	178	,,	103
	Cotinguiba	195	,,	103
	São Francisco	153	,,	60
	Dois de Julho	261	,,	50
	Jequitaia	250	,,	60
10.	Santo Antonio	153	,,	40
	Boa Viagem	153	,,	40
	Itaparica	62	,,	30
	Lucy	30	,,	12
	Victorina	3	,,	3
15.	(building)	200	,,	75
16.	,,	200	,,	75

small fleet of sixteen to fifty vessels. She has taken the right line, and with energy and economy she must prosper.

As regards fixed property, the Company possess at the city of Bahia workshops, &c., for the repair of the fleet, and suitable stores for materials and coals. At the city they have recently completed the new landing-piers and receiving-houses for cargo, and they have constructed suitable landing-places at all the Bay Ports.

This information was given to me by Mr. Hugh Wilson, of Bahia, the energetic and progress-loving Superintendent. I have only to hope that his views will be adopted with its usual liberality by the Imperial Government, and that a tramway will presently connect the Porto das Piranhas with Joazeiro. Evidently this should have been the step first taken ; but should it be the last, we shall not complain.

CHAPTER XXVI.

FROM THE VILLA DO JOAZEIRO TO THE VILLA DA BOA VISTA.

TENTH TRAVESSIA,* 22 LEAGUES.

GENERAL REMARKS ON THIS TRAVESSIA, THE GARDEN OF THE SAO FRANCISCO.
—THE "TWO BROTHERS."—THE CACHOEIRA DO JENIPAPO.—THE VILLA DA
BOA MORTE, ANCIENTLY CAPIM GROSSO.—ITS ORIGIN.—ITS SCANTY CIVI-
LITY.—RESUME WORK.—PRETTY APPROACH TO THE VILLA DA BOA VISTA.
—THE CANAL PROPOSED.—ALSO ANOTHER CANAL.—ARRIVE AT THE VILLA.
—THE COMMANDANT SUPERIOR—RECRUITING OF THE CONSERVATIVES.—
ORIGIN OF THE VILLA.—ITS PRESENT STATE DESCRIBED.—ENGAGED A NEW
CREW, THE PILOT MANOEL CYPRIANO AND THE PADDLE "CAPTAIN SOFT."
—MADE NEW PADDLES FOR THE RAPIDS.

> Terra feliz, tu es da Natureza
> A filha mais mimosa ; ella sorrindo
> N' um enlevo de amor te encheu d'encantos.
> *(Poesias B. J. da Silva Guimaraes).*

WE now enter a country which has left upon me the most
pleasant impressions. Between Joazeiro and Boa Vista is the
lower garden of the São Francisco, perhaps a finer tract
than that about the Pirapora. The stream becomes swift,
averaging four knots an hour, and though the sunken rocks
present some risk, the travelling is much more pleasant, and the
swirling and boiling of the water show that it has a considerable
depth. On both sides there are farms and fields, each with its
scarecrow frightening the capivaras and the robber birds, and
there is no drought, though the air is intensely dry, the effect of
evaporation. The dew is heavy, and the dry winds carry off the
watery particles to form rains on the higher bed. The sloping
banks are all green with manioc, maize, beans and wild grasses.
The valley is studded with pyramidal hills, of which as many as
five are sometimes in sight ; they are backed by waves of ground
covered with thick or thin bush ; these Catingas Altas † will con-

---

* Formerly this Travessia extended
twenty-nine leagues to the extinct town of
Santa Maria (276th league), the terminus
of barca navigation down stream ; it is
now reduced to Boa Vista (269th league).

During the floods boats run down from
Joazeiro to Boa Vista in twenty-four hours.
† The term is applied to the ground, as
well as to its vegetation.

tinue till Varzéa Redonda. The Cajueiro and the Cajú Rasteiro are now common;* the principal growths are the cactuses, the gigantic Mandracuru, the Facheiro (faxeiro), whose dry wood serves for torches; the echino-cactus, Cabeça or Corôa de Frade; the Xique-Xique, or cylindrical plant, the common flat band nopal, and the dwarf Quipá. The bush or undergrowth is chiefly the Araça (psidium), and the Tinguí (Magonia glabrata, St. Hilaire). The larger growths are the Páo Pereira (Aspidospermum); the leguminous Carahyba, whose large green bitter pods are loved by goats and deer; the leguminous Catinga de Porco† whose leaf resembles the Barbatimão acacia; the Salgueiro,‡ and the Páo Preto, whose black trunk appears scorched with fire.§ In many parts fuel is wanting near the stream. We have left behind us the diamantine "formação" and the iron fields; here we find pyrites, traces of gold and large outcrops of limestone. The winds are furious at the present season, but they will have no power below Boa Vista; here the trees and grasses are bent up stream by their persistency and power. We were told to expect windy nights, and hot still days; we shall have wind night and day, cold and furious by night, hot and furious by day. The mornings are cold and cloudy, but the sun begins to sting at 10—11 A.M. and lasts till late in the afternoon.

Friday, October 25, 1867.—We managed to set out at 11 A.M., and dropped past Joazeiro Velho on the right bank; the place has become superannuated since its desertion by the channel. The trade wind was moderate, but tourbillons of sand and dust-devils (Redemoinhos), coursing over the broad river-plain, made us furl the awning. Of the five hills in sight only one block, white and bushy, approached the river, which the many islands, sand-bays, and islets divided into sundry independent streams, never less than two. Red and purple glazed rocks scattered in the bed, again gave the familiar sound of the Cachoeira.‖ We grounded once by hugging too fondly the left shore, and for a few

---

* The tree when I went down the river was not yet in fruit.

† It has a powerful smell, which, however, hardly justifies the harsh name "pig-stink."

‡ It produces a useless fruit: the strong hard wood is applied to the "Cavernas," or ribs of barcas.

§ The wax exuded by the bark makes candles which are exceedingly hard, and if a melted drop fall upon the hand, it removes the skin.

‖ M. Halfeld would remove them at an expense of £340 and £500, and in three others £680, £170 and £720, or a total of £2,410 in twenty miles. This may be done in due course of time; at present it is useless to expend one milreis. A good pilot can steer clear of the difficulties, and we went down safely with men who, if they ever knew this part of the river, had quite forgotten it.

minutes we hung poised upon the crest of a sunken rock which gave no sign. The banks were green with the spiky Capim Cabelludo, which is planted for dry season fodder; if not drowned by the floods it lasts, they told me, twenty years.

After sunset we anchored off the huts of Mato Grosso on the right bank. Here the course of the São Francisco is north with a very little easting, and the bed is no longer so broad as above Joazeiro. Opposite us, or nearly due west, is a fine landmark, the Pico da Serra do Aricorý, or Ouricorý,* attached to a lumpy line, whose trend is north-west. Though distant five miles the features are clearly distinguishable.

October 26.—The crew, eager to advance, began work at 5 A.M., and we shot rapidly past the Ilha da Manisóva † and other unimportant features; we were, however, driven to anchor from 9 A.M. to 1.30 P.M. whilst the world was airing and warming. The third league showed us the Fazenda do Pontal; here on the right or southern bank a line of scattered cones drives the stream from north-east to south-east. Opposite it enters the Riacho do Pontal, ‡ and below it stretches the remarkably long island of the same name. Further down the Bahian side bears the little Arraial da Boa Vista and its chapel of Nª Sª dos Remedios. For many an hour we saw in front the peculiar Serrote dos dous Irmãos, twin pyramids with gentle, regular and equal slopes on both sides; their cliff facings of white stone were thrown out by the now greening "bush;" and, after sunset, a shadowy grey colour stole over them. When the gloaming began we sped by the Cachoeira da Missão, an unimportant break to starboard, and presently we landed on the Pernambuco bank, at a place called the Pontalinha, opposite an islet of the same name. I had given a passage from Joazeiro to a young fellow whose home was here; three women came down to the landing-place and carried off, on their heads, with much coquettish recusancy, the few bricks of sugar and the dozen greybeards of gin which he had brought as

---

* Also written Aricori and Ouricori (the name of a palm): in the Lingua Geral the terminations "i" and "y" are equivalent and are used indifferently, as Tupi or Tupy, Guarani or Guarany.

† This is doubtless the Manacóba or large Jatropha of Gardner. Generally Manicóba is the Seringa or Caoutchouc-tree. The minor features are the Fazenda de Paulo Affonso and a few rocks at the Barra do Vieira, which do not require

removing.

† "Pontal," like "Começo," is applied to the head of an islet, especially when the point is bluff. The Riacho do Pontal comes from the Catingas Altas, and though much broken, it is ascended by canoes during the floods. My informants gave it a length of thirty leagues. Mr. Keith Johnston makes the "R. Pontal" drain the dividing ridge.

a stock in trade. They were wild-looking beings, their very small faces were set in a frame of hair, and their beady eyes peeped out from the profusion of unkempt, witch-like locks.

October 27.—Passing the Two Brothers we were driven to anchor at the Começo da Cachoeira do Jenipapo, a small break some eight miles above the main feature of that name. Delayed between 7 A.M. and 2 P.M., we saw at 3 P.M., on the Bahia side, the Barra Grande* do Curaçá; the mouth is about 230 feet broad, and the right jaw projects into the main artery a large dome of stone. The view up-stream discloses a pretty vista of lively verdure. About three miles below it is the Cachoeira Grande do Jenipapo, with houses on the right side and rocks dotting the stream; we found, however, a clean central way. Far to the north-east appeared a lumpy hill range with a brown, green and white-streaked surface; the bank to starboard showed, alternating with grey granitic schists, large snowy blocks of laminated lime-stone, whose scatters we repeatedly mistook for human habitations.†
At the Barrinha, a little stream and village further down, two broken reefs of projecting rock run parallel with each other along the bed from south-west to north-east. The wind tossed us about fiercely, the current ran very fast, and we were nearly dashed against a hard head by the pilot. He complained of ague, attributing it to the rising and falling of the stream; the fact is, he was suffering from over coffee and Jacúba.

As the sun sank low we sighted from afar, on the right bank, a picturesque village, the Villa do Senhor Bom Jesus da Boa Morte, whose vulgar neighbours persist in calling it Capim Grosso—big grass—the old original name. Fronting to north-west and towards the stream, a white-washed and red-tiled church in the Bahian fashion, with pinnacles instead of towers, and a façade sparkling with imbedded fragments of glazed pottery, displayed itself upon the crest of a ground-wave. Along the river were two Sobrados and a line of white houses backed by brown huts. The field-fences extended to the water side, and on the sloping bank

---

* Mr. Keith Johnston ignores it, and I do not think that, despite its fine name, the stream can claim any importance.

† M. Halfeld (Rel. p. 147) says that the rock is white and ash-coloured, with block veins traversing the strata in wavy bands (*bichas onduladas*) "of primitive forma-tion resembling marble, and as their breadth is sufficient for the saw, they may serve for works of taste, tomb-stones, &c." Those which I examined were an excel-lent building material.

These are the features which gave rise to the common local names for hills and mountains, "Sobrado" and "Sobradinho."

were two tall shady trees* which looked gigantic by the side of the
thorny shrubs. Half way up the range, and high and dry since
many a day, was an old barca, there beached by the last floods.

We anchored in a sheltered place below the rock-pile fronting
the church; here however the river is broken by two islands, the
Ilha do Torres to the south and the Ilho do Giqui (Jequi) hard
by the left bank. We had scarcely made fast, when a report
spread that a steamer had arrived. Rushed down the bank a
posse comitatus of notables, mostly "bodes" and "cabras," in
black coats, paletots (a word which here becomes "pariatóca"),
and white etceteras. Only one man approached whiteness; he
was probably the Professor of First Letters, and he squatted,
Hindu-like, upon a stone, washing his face with both hands, and
towelling it with his pocket handkerchief. The disappointment caused by the "Ajôjo" elicited peals of laughter, and the
smallest jokes bawled in the loudest and coarsest of voices. I
seemed to hear once more the organ of African Ugogo. Exceptionally in the Brazil all ignored the presence of strangers, and
they made unpleasant remarks about the certainty of such a
craft never reaching Varzéa Redonda. I have, however, been
threatened with drowning ever since leaving Sabará. Presently,
hearing that a bullock was to be slaughtered, all rushed away,
eagerly as a flight of turkey-buzzards.

Capim Grosso, which deserves to be entitled Villa Grosseira,
was, till 1853, an Arraial; it rose to township by the Sup-
pression of Pambú (283rd league). The houses may now
number 70 and the souls 350. The broad streets are not badly
laid out, and the thoroughfare running parallel with the river is
cumbered with hard talcose slate and quartz-banded granite,
which will readily supply building material. The prison, crowded
with recruits for the war, peering from behind its wooden grating,
was guarded by four soldiers, and the Cámara was denoted by
the papers pasted to the door. The church, of burnt brick upon
a foundation of gneiss, was out of all proportion to what met the
view. In the usual square we found a few shops, and an "Aula
Publica Primeira." We then walked along the deep, sandy path
to the little cemetery and its shed-chapel behind the settlement.
Hereabouts began the thorny Catingas Altas, where, however,

* They are called by the people "Moquem:" this is all that I could learn about
them.

cotton seemed to flourish. The ground was strewed with pebbles and quartz-blocks of all sizes and colours, and the stone appeared to be auriferous. This place commands a fine view of the Serra do Roncador on the other side, where the wind is said to "snore" furiously. About a league and a half to the east is the Serra da Capivara, a long broken lump, which all declare to contain gold, although the metal has never been worked. Hence, probably, the auriferous pebbles.

Capim Grosso is the wildest place that we have yet seen; it did not show a trace of hospitality, or even of civility. Yet the people seemed tolerably "well to do." Many of them were on horseback, the saddles were made "country fashion," with strong cruppers and poitrels for riding up and down hill. The Caipíras wore, for protection against the sun, ugly "Sombreiros," and the swells cocked up one side of the broad brim, and the flap fastened by a large metal button made a three-cornered hat. These "Chapéos de Couro" are of goat, sheep or deerskin; the latter is the best, but any will serve, and they look like the "babool-stained" leathers of Western India. The women greatly out-numbered the men. We had inadvertently made fast near their bathing-ground; after dark they disported themselves in the water all around us, and debated, giggling, about the advisability of doffing the innermost garment. The site of Bom Jesus da Boa Morte is nature-favoured, but this was the only merit that we recognised in the place. I hope that the next travellers may be justified in giving a better account of it.

October 28.—The Januaria men here found relatives, and this delayed us till 6 A.M. After about two and a half leagues we came to a break, the Cachoeira das Carahybas; the river had again risen, the water had become exceedingly clear, and we easily found the safe line near the natural stone jetty on the right. The rains at this point are expected soon to break; the weather, however, has been dry since September when there was a short and copious fall.* On the left was the Serra do Curral Novo, re-markable for its rounded summits, platform and high demi-pique saddle-back. The lands on both sides of the stream were of exceeding fertility, presenting a most amene and riant appearance. At the Fazenda de Goiaz, a neat tiled and whitewashed house, the river began to turn from its northerly course to due west-

* Here called "Manga," or "repiquete de Chuva."

east. On the left bank at some distance appeared the Pedra Branca, a wave of bushy highland with a block of white limestone conspicuous upon its flank. Below it was a brother formation, the Morro da Boa Vista,[*] apparently two hills, but really three lumps disposed in a triangle with the base towards the stream. The nearer rise was capped with white near the stream; the further was thinly clad with Catingas Altas, like a head becoming bald. At the south-south-west of the latter rose the Villa da Boa Vista, our destination.

On the left, and about half a league above the town, we passed the Ilha do Icó.[†] The bank, a "baixada," or low land, is broken by the Barra Grande da Boa Vista. Here M. Halfeld (Rel. 149 —150) would begin the great canal proposed by Dr. Marcos Antonio de Macedo,[‡] and other "educated men." The waters of the São Francisco are to be drained through a channel to the Riacho dos Porcos which falls into the Riacho Salgado, an influent of the Jaguaribe River, traversing oriental Ceará from south-south-west to north-north-east. It is a "gigantic project:" it would effectually lay the horrible plague of famine, and awake from their profound lethargy the people of inner Ceará and their neighbours of the Parahyba and Rio Grande do Norte provinces. Unfortunately, at a distance of some forty leagues, the line is cut by the Serra do Araripe, the dividing line between Ceará and Pernambuco. M. Halfeld highly approves of the idea if a pass (baixada) can be found through the range. The people of Boa Vista had never heard of Dr. Marcos or of his canal, and when I read out to them the Relatorio, they laughed. The projector still lives, it is said, at Crato, in his native province of Ceará, which should be truly grateful to him for his good intentions. Even were the canal to fail, the strong current of currency which would be generated even by the attempt would doubtless bear fruit.

As I am speaking of canals, it is as well to say that others have been proposed. Perhaps the boldest idea is that which

---

[*] Aliàs "Dos Deus Irmãos," although there are three : the people ignore this name.

[†] The Icó or Ycó (Colicodendron Icó), which gives its name to a city on the Jaguaribe River of Ceará, and which will become common on the São Francisco, is a shrubby tree with an edible fruit, the latter resembling the common Ameixa or yellow plum of the Brazil, where it has long been naturalised. The leaves are injurious to cattle, producing inflammation of the intestines and of the kidneys. The System refers to kitchen salt and castor oil as remedies.

[‡] This name is mentioned by Gardner : I do not know if it be the same person.

owes its origin to Lieut. Eduard José de Moraes. This officer was apparently encouraged by "M. Emmery's" report on the Hudson River and the Lake Champlain Canal, and by the brilliant picture of prosperity which M. Michel Chevalier portrays as the result of canalisation in the United States. He would simply take the waters of the Rio Preto, the main affluent of the Rio Grande,* and throw them into the Paranaguá, or Parnaguá lake, near the city of that name on the headwaters of the Gurgeia † River, the great central affluent of the Northern Paranyba. The distance between the streams is only twenty leagues, which, it is reported, might be reduced to fifteen; but unfortunately the dividing line bars the way. This difficulty is most naïvely alluded to,‡ and it is confessed that "le Rio-Gurgeia n'ait pas encore été exploré." "Un inconvénient (!) se présente cependant dans le tracé de ce canal, c'est l'existence d'une chaîne de montagnes entre la vallée du San Francisco et celle du Parnahiba, et qui a été pour cette raison appelée das Vertentes § par le Baron d'Eschwege, qui la trouve la moins élevée de tous les autres systèmes de montagnes du Brésil. Il est donc naturel de penser qu'une partie de ce canal pourrait être souterraine, cependant rien ne vient prouver ce fait puisqu'une reconnaissance n'a pas encore été faite dans ce sens; peut-être existe-t-il une gorge, une dépression où l'on pourra le faire passer même à ciel ouvert." And to attempt such chimeras as these, the author would tax the English gold-mining companies in the Brazil, which have never yet been able to support the smallest impost.

Compelled to cross from the right bank, through a little break above the town, we were nearly upset by the violence of the "raffales." We succeeded, however, in making fast behind a rocky projection, and I sent without delay my introductory letter to the Commandante Superior Sr. Manoel Jacomi Bezerra de Carvalho. He at once called upon us and undertook to find a pilot and paddle-man. We talked of the railroad projected from this point to the Porto das Piranhas, thereby defeating all

---

* See chapter 21.
† In Mr. Keith Johnston, "Grugeia R."
‡ Rapport, &c., p. 29.
§ The "Serra das Vertentes" is some 1260 miles to the south: we passed it at

Alagôa Dourada. The Rio Preto is supposed to arise in the Serra dos Pyrenéos, which M. Gerber and others extend from the headwaters of the Tocantins to the western valley of the São Francisco.

the Rapids; our visitor declared that the line was sandy and
without hills, whilst its tortuous length can be reduced from
seventy to sixty leagues. Neither he nor any of his friends had
seen the neighbouring Niagara; they have often when riding
down to the port passed it within a few miles. The latest newspapers dated from early September, and yet we are here only 200
miles from steamer navigation. The Commandante presently
left us in a prodigious hurry, having to superintend the ironing
of ten whom he called twenty recruits. They were sent to head-
quarters at Tacaratú, and we met the returning escort of fourteen muskets who had escorted them. They were wild-looking
fellows, servile as well as free, and only the chief man had a
horse; the dress was old-fashioned shirts and tight smalls of
strong homespun cotton, leather hats, waistcoats and sandals.
In the evening I saw a wretched "Conservador" pursued through
the bush by mounted men who presently captured him for the
war. No wonder that these places look like the ruins which the
slave wars have made on the Lower Congo.

The Fazenda da Boa Vista, some five leagues down-stream,
and belonging to the Commandante's grandfather, José de Carvalho Brandão, was originally an Aldêa, or settlement of Indians,
and the head-quarter village of these parts. Presently a church
was here built, and the huts gathering round it took the name of
"Arraial da Igreja Nova," which is still preserved by the riverines. In 1838 it became the Villa da Boa Vista, the head of a
Comarca, and the residence of a Vigario, a Juiz de Direito, a
Juiz Municipal, and other requisites for self-rule. Its two fre-
guezias, Santa Maria da Boa Vista, and the Senhor Bom Jesus
da Igreja in the Povoação da Cachoeira do Roberto on the left
bank of the river, number 6000 souls, an estimate founded upon
the fact that a single parish has 1000 voters.* The town may
contain eighty-five houses, and, at festivals, 500 inhabitants.
They support themselves by breeding cattle, and agriculture, and
they want but little here below; we found fresh meat for sale, but
absolutely nothing else, not even a water-melon. Many spoke to
us of the Serra Talhada, distant some fourteen leagues from the

---

* A rough but ready way of estimating
the population in these outstations is by
the number of voters, which every one
knows. In some parts a tax is paid upon
doors and windows, but this again leads to
errors in counting roofs, or as households
are still termed in wigwam phraseology,
"fires."

left bank. It is said to contain alum and saltpetre, but not a single specimen was to be procured. One man brought me a match-box full of iron pyrites, which being bright and brassy was for sale as gold: it is said to come from the western country.

The town has, as may be expected, little to show. We visited the natural pier at the western end which fronts south-west and runs back to the north-east. The substance is talcose slate, containing much quartz distinctly stratified, with cleavage lines trending from east-south-east to west-north-west, or nearly perpendicular to the direction of the beds. The harder parts can supply large blocks ready cut for building; in places it is soft and is worn down by the footpath which descends it in steps. Further to the west large fragments have slipped into the stream. At the eastern end there is another outcrop with strike to the south-east and dip north-west 35°; and in parts it is spread without regularity over the steep bank of stone, sand, and stone dust. It is mostly banded with white quartz, and has embedded lines of amygdaloid. Near the stream its surface is revetted with a coat of the darkest chocolate, the usual ferruginous glazing; here, however, iron is not found, and must be brought from down-stream. The highest floods, even those of 1857 and 1865, the worst on record, did not extend half-way up the pier. The general belief is, that the inundations are diminishing, and with them the fevers.

We visited the church of Nª Sª da Conceição, a typical shape, tall, narrow like the people; its only charm is its site, a rocky platform forming the highest ground in the settlement, and fronting up-stream. A whitewashed cemetery appears to the north or inland, separated by a depression, into which the floods enter without, however, insulating the settlement. South of the church is the town showing a single row, the Rua da beira do Rio. With the usual unwisdom here customary, the people have fronted their houses towards the glaring temple and the hot stony hill, whilst their back-yard compounds and plots of pomegranates and flowers enjoy the charming view, and the breezes floating up and down stream. Looking south the Serras da Capivara and the Curral Novo break the horizon, and the broad river, with rocks above surface and rocks below water, serpentines through its subject valley. To the south-east are the Serras do

Piriquito and do Estevão; of the four pyramids one is remarkably acute-angular, whilst further east three knobs denote the
Serra dos Grós. There are no glass windows even in the richest
tenements, and the jail, at the eastern end, is a house like the
other houses.

Boa Vista is the terminus of barque-navigation ; at this season
only ajôjos and canoes go to Varzéa Redonda. Here I dismissed, with an additional gratuity for extra service, the pilot
José Joaquim de Santa and Manoel Felipe Barboza, aliàs das
Moças, aliàs Barba de Veneno, and of late usually known for
shortness as "Manoel Diabo." The latter, having quarrelled
with an angry father, had fled his family, settled a few leagues
down-stream, and had not seen it for fourteen years. He contented himself with writing a letter from Boa Vista, and he set
out contentedly with his friend in a small canoe which will take
at least a month to make Januaria. We separated well satisfied,
I hope, with one another.

There was no difficulty in finding men.* The Commandante
directed the pilot Manoel Cypriano to hold himself in readiness :
the tariff is 25$000—not bad for five days' work in these regions,
and the new man presently came to see us. He was a dark
senior, dating from 1817, but looking at least sixty-five ; he
declares that his premature old age has been brought on by a fast
life, and that he has long passed the time when men begin to
die. He has a queer dry humour, he delights in chaffing the
people upon the banks, he twangs the guitar, he takes snuff as
most boatmen do, but requires a snuff-pocket like our grandfathers, and he has a private bottle of country rum wrapped up
carefully as if it were a baby. He never works except when half
seas over, and I should fear to trust him when dead sober ; he
is slow to excess, taking five minutes to don his coat and to slip
his feet into his ragged slippers. Yet he is the only real pilot
that I saw upon the river, he knows it thoroughly, he *will* be
master on board, and he slangs a recusant paddle with the
unction of an Oxford coxswain—in my day. Certainly no beauty
was M. C., but he was stout-hearted and true. We soon
learned to confide in his nerve, force and precision. There was

---

* M. Halfeld (Rel. p. 61) says that here
it is hard to find watermen for rafts and
canoes, on account of the Rapids. The
sole obstacle is the extreme laziness of the
people. It is, however, only fair to confess
that I lost but a single day.

something more interesting even than beauty in his danger-look, when, working his paddle like the tail-fin of a monstrous fish and firmly planted in the stern canoe of the rocking and tossing raft, he bent slightly forwards, steadily eyeing with straining glance the grim wall upon which we were dashing at the rate of twenty knots an hour, and, by a few ingenious strokes of the helm at the exact moment, brought round the bows and almost grazed the reef.

I gave Manoel Cypriano carte blanche to choose his oarsmen, and this was a prime mistake. Like almost all his countrymen, he had a certain amiable defect, a constitutional inability to say "No," which is often worse than a moral incapacity to use "Yes." Thus when he was set at in due form by one José Alves Marianno, he objected faintly, he held with him long palavers lying on the bank, and he ended by engaging him. All this time he knew the man to be a noted skulk, whose nickname on the river was Capitão Molle—Captain Soft—and whom no one would engage.

Marianno is, he tells us, a son of Petrolina, by no means a good locality. His immense curly head-mop of jetty colour, proves an African maternity, and the legal saying "partus sequitur ventrem" is true in more ways than one. He sings well, he has an immense repertoire, and, as a repentista, he is known to local fame. Ergo, I presume, he has taken the poetical and Arcadian name Mangericão (Ocymum basilicum) which he pronounces "Majelicão," and which soon becomes Manjar de Cão—dog's meat. He is hopeless, he drinks like a whirlpool, he eats like an ogre, he pretends to faint if pushed to work, and, if undue persistency be applied, he loses some of the tackle. He loves to "put on a spurt" where the stream is swiftest, so as to make a bump fatal : in still water he lolls back, snuffs, chats, chaffs, or chaunts. The worst is that I cannot be seriously angry with the rascal ; he is abominably good-tempered, and he seems to look upon himself as the greatest fun in the world. Yet it was a relief when he received his 16$000 and showed us his back.

The next day was a forced halt. The Escrivão of orphans, Sr. Felipe Benicio Sà e Lira, kindly allowed me the use of his house and his desk, which made the hours move more nimbly than they otherwise would. The wind blew strong and contrary.

The pilot had paddled away to his house down-stream, and was laying in small stores. We wanted large paddles; yesterday the only carpenter in the place had been engaged in ironing the recruits, and till that important operation was concluded he could not go to the bush and cut down a Mandracurú-cactus. These paddles look something like action, rude and heavy, but strong and pliant: they are perfectly straight, five feet long, and with a leverage of 2 : 1—the little paddles used up-stream are nearly equally divided, and the effect is like using a large kitchen-ladle. When the work was done, he asked about four times its worth, and he took the opportunity of offering for the Eliza 100$000. Had he paid as he charged, he would have said 1:000$000.

CHAPTER XXVII.

FROM THE VILLA DA BOA VISTA TO VARZÉA REDONDA.

Eleventh Travessia, 45 Leagues.*

The Rapids and the Smooths.

"Les Brésiliens avant la conquête de leur pays par les Européens étaient au dégré le plus bas de la civilisation."—*Prince Max*, ii. 396.

Section I.

THE GOOD RAPIDS TO CABROBÓ.

A LITTLE below Boa Vista, the river, after a short and tolerably clear northern sweep, returns to the eastern direction, and enters upon that Cordilheira of breaks and rapids which will last for some thirty leagues. Earth here begins to show her giant skeleton bare. The bed broadens in many places to a league, and is worn down to its granitic floor; it is a mass of islands and islets, all bearing names, of reefs and rocks sand-scoured, cut and channelled by the waters, which glaze them to a grisly black. As a rule, the bed is too winding for the winds to form high waves, but this is by no means always the case. The rocky quartzose highlands, disposed apparently without any system, approach the channel and throw across it broken walls of stone. The Cachoeiras offer some risk to those descending, but more during the ascent.† There are many and sundry triangles of water, and the old rule of the Rio das Velhas, namely, to make for the single apex, will hold good here; in some places we must get into broken water to avoid sunken stones, and sometimes we must run straight towards a rock, and rely upon helm and current to escape it.

---

* M. Halfeld, by some curious oversight (Rel. p. 6), makes this eleventh Travessia thirty-eight leagues, and reckons the distance from Santa Maria, instead of from Boa Vista. The pilots stretch the distance to fifty-two leagues; namely, sixteen to Cabrobó, and thirty-six to Varzéa Redonda.

† The people declare that accidents never take place, but we shall find two wrecks *en route*.

The only really bad part will be passed on our sixth day ;* it has nine rapids, two whirlpools, and two shallows, which form, during the space of five leagues, obstructions as serious as the whole course of the Rio das Velhas. Here a committee of pilots could point out the best line, which might be cleaned, marked,† and rendered passable; it would be far better, however, to abandon all this part of the stream, and to run a tramway to the Porto das Piranhas, distant 70—72 leagues along the channel.

The beauty of the banks still continues, and houses, farms, and fields extend down the whole way. Gold cascalho, talcose slate, and quartz, frequently appear on both sides. At this season the vegetation is much burned, and the finest trees are upon the comparatively damp islands. The nearly total absence of palms gives to the scene a look of the temperate regions. Agriculture and stock breeding are the main resources of the people, but where the stream is low sunk, they have no idea of the Persian wheel or the windmill. The banks, especially the right, are much broken by Alagadiços or swamps, and by Ypoeiras, which here take the Tupy name Igarapé, or Ygarapé. The influents, known by the bright green grass at the mouths, are mere nullahs, owing to the increased narrowness of the river valley; at this season their short narrow courses are either dry or slow strings of pits and pools (Caçimbas and Poçōes), during the rains they roll terrible torrents.

We were told that during the windless nights a candle might be used naked; this was the case only once. The Serras to the north-east, Araripe and Borborema, obstruct to a certain extent the thorough draught. The Trade changes with the direction of the stream, and at this season it invariably comes up-stream. In the morning we have catspaws, the wind blows strongly during the sunny hours, and woolpack gathers in the afternoon. There is an immense evaporation, causing a constant thirst, and crumbling tobacco to dust: the whole of this section is a laboratory that distils a copious stream for the higher river. The rains are mostly from the north, sometimes from the south ; showers, violent only in March and April, extend between November and January, further down they are fiercest in February and March.

---

* Between the 295th and the 300th league of M. Halfeld.

† " The channels are so intricate that we find, at the bifurcations, bits of sail-cloth hung on the bushes to guide the navigators on the route to Pará." (Lieut. Herndon, p. 333).

We now enter the head-quarters of the extinct Jesuit missions, a land of ruins strange in a country so young; and we see with astonishment that more than a century ago the neighbourhood was much more advanced than it is at present. The company, it will be remembered, was denaturalised, and departed with confiscation of property from the Portuguese dominions by the celebrated law of September 3, 1759. The Jesuits—abstraction faite de leurs institutions vraiement nuisibles, et du mal résultant de leur domination—certainly taught their converts the civilisation of labour, and now the " Aldeâdos," or villaged Indians, have allowed their chapels to fall, and are fast relapsing into savagery. Finally, the place of the old Fathers has been but poorly filled by the Italian and other missionaries, who, of late years, have been thinly scattered amongst these out-stations.

Wednesday, October 30, 1867.—With infinite trouble we managed to set out at 11 A.M. The old " Menino " was drunk, and well nigh incapable, and the new paddle, " Herb Basil," after a very short spurt, began to droop in all save in the matter of singing. We dropped down between the left bank and the large Ilha Pequena ; it shows fenced fields and thatched roofs on four poles, under which the shepherd shelters himself from the broiling sun. The sheep and goats are poor and lean at this season, and the owners ask 1$000 for a bag of bones. On the bushy hills to the north are many sobrados, the usual house-like lumps of white limestone ; below are scattered tiled huts, with here and there a large tenement, and the negroes are singing over the task of bush clearing. The banks are of fine quality ; from the raft we see no bottom to the soil, and the tap-root strikes straight down.

This Roça channel, which we have taken to avoid the furious Cachoeira do Ferrete on the right hand, recalls to memory the Rio das Velhas below its confluence with the Paraúna stream. Presently we find on the right the Ilha da Missão (Nova), and its broken fane, one of the most southerly establishments of the Jesuits. It runs west-east, with a convexity to the north, and it is at least three and a half miles long. Sundry islets rise between it and the left bank ; there are few breaks, but the many sunken rocks require careful piloting.* The Serrote do Páo Torto on the left

---

* Probably M. Halfeld surveyed this part of the stream when it was low. He talks of sundry Cachoeiras and " Cachópos " (shoals here called baixios, or bancos de Arêa), which are mere " Corridas," or runs. Between the Ilha Pequena and the bank he places the Cachoeira do Fuzil, which, when I passed, can hardly be called a rapid.

gave us a taste of its quality; further down the Morros dos Grós, the three hillocks seen from Boa Vista, form a bluff, and approach the channel, which is compressed still more by the Serra do Estevão* on the other side. Both banks project natural piers of rock, which make the stream dark and swirling. Above the village "Os Grós" is a lump of stone; from this enchanted ground the barquemen have often heard the sound of the drum and the song, and tramp of crowds passing along. As the dangers of the bed increase, so will grow the belief in things unseen, till at last almost every bluff will have its own superstition.

At this point we turned from east to north, and passed between the left bank and the Ilha da Missão Velha. It contains a ruined chapel, Nª Sª da Piedade, fronted by a cross; formerly it was populous and cultivated, now it is inhabited by only one Morador. "Captain Soft" determined that he had worked enough for that day—it was then 3 P.M.—and as I declined to put into the left bank, where he had friends, he neatly let slip the new paddle of Cactus, and managed, perforce, to effect a halt. It was useless to attempt the rapid which we heard roaring down-stream without all our gear perfect. I made him dive—he swam like a fish—but the current was strong, and the heavy timber was, doubtless, soon rolled far down-stream.

We halted on the left side, opposite the Missão Velha, and Manuel Cypriano set out at once to cut down a Mandracurú. Huts and clumps of noble Joazeiro and Quixaba trees gave the bank a pleasant aspect. The Zozó, or Pistia, formed bright beds in the water, especially at the mouths of creeks, and in places the tall Ubá† was apparently planted by the people. Ledges some two feet square upon the water slopes, were laid out with onions, mint, that made excellent juleps, and the Merú, an edible tuber, with an Arum-like leaf; whilst the forks of low trees bore pots of lavender and flowers for the women's hair.

We were presently hailed by a familiar voice from above, and we recognised, despite certain borrowed plumes, the jolly face of "Manoel Diabo." His brothers, hearing that he had left Boa Vista without visiting his home, indignantly pursued him, and brought him back, nolentem volentem, to receive his mamma's blessing. He had "loaned" his friend the pilot's black coat and

---

* M. Halfeld calls it the Serra do In-hanhum, from the large island at the bend of the bed.

† This Saccharum is probably the "fre-cheira," or arrow-cane, of the Amazons River.

slippers, and he cast them both off when he led us about the
"Fazenda." Here the bank is flat, and subject to an excess of
droughts and floods. It is backed by a grey hill of talcose slate,
veined with and passing into quartz below. The cotton shrub
grows admirably, and each "foot" is said to produce thirty
pounds; a little has been exported, but the old "lavrador" complained of a blight which had lately appeared: the plant probably
wants new land. Most of the cottages here have looms, which
are, however, superior in nothing to those of Unyamwezi. Cattle,
sheep, and goats looked tolerably thriving, and the crew found
abundance of birds; the flights of wild pigeons are described to
resemble those of the United States. In the evening the fatted
calf was killed, men and women complimented the truant in
extempore verse, to which he replied with interest; and the drum
was not silent till sunrise. We heard for the last time the Whip-
poor-Will, his wrongs are taken up by another volatile, who ever
complains, like the West African bush-dog, that the fire has
gone out.*

October 31.—The old pilot worked hard at his carpentering
under a shady tree, and even "Majelicão" bore a hand—I had
deferred breakfast till the paddle was ready. At 10·30 A.M. we
shook hands all round, and pushed off towards the spot whence
the roaring came. This upper Cachoeira da Panella do Dourado,†
the first below Boa Vista, has been descended by barcas even
during the dry season, but it is perfectly capable of doing damage.
We ran to the north of the Ilha da Missão Velha, and, poling up
against a strong current, we passed between it and its northern
neighbour, the Ilha do Serrote. Then turning poop on, we
dashed down the usual channel,‡ with the Angicos Island on the
right, and the Cabras on the left, and we escaped without anything more serious than a long graze. It was a wild and haggard
scene, a series of rivers within a river, a tortuous labyrinth of
currents formed by seven large and a multitude of smaller rocks,
through which the "eau sauvage" ran straight as an arrow. The
broom-like shrub, Jarumataia, or Angari, brown below and green

---

* The cry is supposed to say, "Fogo
'pagou" (apagou).

† "The rapid of the pot-hole of the
Dourado," a fish of this species having
been caught in some "boil" by the first
travellers. "Panella" signifies either a

pot-hole worked by the water in the rock,
or a small whirlpool, a conical swirling
depression in the surface.

‡ There is another channel to the right
of the Ilha do Serrote, but it appeared to
me very dangerous.

above, grew in clumps upon the islets and in the shallower waters, and heaps of drift-wood were thrown upon the convenient corners and ledges. The rocks, banded with snowy quartz, glazed like the pigeon's wing, ruled in straight lines by the several levels of the water, and in places bored into basins, appeared singularly characteristic. Further down, where the flood reposes in depths studded with foam, and where the current wheels round in lazy circles, we came to the cause of all this disturbance. On the left bank, without correspondence on the other side, a Serrote 80—90 feet high, and projecting to the north-east, sent a rib of rock clean across stream from north-west to south-east. The bluff showed strata of hard sandstone striking to the south-west, and split into brick-like cubes by the perpendicular cleavage; the face was lined with thick and thin ramifications of snowy white quartz,* which everywhere lay in fragments upon the surface. From the south it assumed a quoin shape, with a bushy hog'sback declining to the west.

Beyond the Ilha das Marrécas we fell into the main stream. We had not seen it united since we coasted the Ilha Pequena, and now we found it flowing like Arar "incredibile lenitate." On the left of this reach, some four miles long, opened the mouth of the now dry Riacho do Jacaré† and its island down-stream. Below it the channel passes between the right bank and the Ilhota do Serrotinho, a spine of hard sandstone and white quartz raggedly covered with trees; the tail end has clearings and cultivation.

Presently we turned almost due east, and sighted ahead another mass of obstructions. They are caused by a number of stony cones on the right bank, and on the left by the Serra das Carahybas. This is a block and outliers of rock, with waves of bushy ground (Catingas Altas), which, contrary to rule in the Brazil, show no tree fringe on the top. At 1·30 P.M. we passed the Ilha Grande, where M. Halfeld gratefully inserts the residence of his pilot, Cyriaco, whose dexterity and courage he greatly praises. Curious tales are told of the old man, who seems to have inherited from his "Indian" ancestry a coolness of head, a clearness of vision, and a strength of arm and will quite exceptional. The boatmen

---

* M. Halfeld makes the bluff to consist of quartz, chlorite, mica, iron, and titanium.

† According to the pilot, it comes from As Queimadas, distant thirty leagues. M. Halfeld shows a very narrow embouchure.

In Mr. Keith Johnston it drains the eastern slopes of the western dividing ridge, heading near the sources of the Caniudé, the river of Oeiras in Piauhy.

declare that he knows every stone in the river, and that he can travel by night over the wildest dangers, especially when "tomado," or slightly "sprung." We shot down a rocky run between the Ilha Grande and the "Ilha de Villa da Santa Maria," formerly the end of the tenth and the beginning of the eleventh travessía. Two ajôjos, laden with salt bags, and raised hardly four inches above the water, were painfully poling up-stream.

We landed on the island to inspect a ruin which we had seen from afar. The soil is of immense fertility:* it bore cotton in small quantities; manioc wherever men had taken the trouble to plant it; the pinhão bravo, or poison croton-nut, which feeds the tenantry of the landshells; and fields of Icó trees, whose ancestors were probably planted here by the Jesuit Fathers. The people, with lank hair and broad yellow faces, showing indigenous blood, were better clad than those up-stream, and inhabited the same miserable huts. After a walk of a few hundred yards to the south-west, we came upon the temple fronting west towards the right bank and up-stream, where is the finest view. Monastery, church, and chapel were all a mere shell, and the latter bore inscribed upon the entrance

RESVRGE
NT IN NO
VISSIMO DIE
1734.

The material was the finest brick, and the maximum size was two feet square. It was almost as durable as the ashlar of talcose slate with which it was mixed, and the chunam, probably shell lime, was of the best quality.† One of the belfries had fallen, and cactus grew upon the walls where roof and ceiling had been. The dimensions of the church were 100×25 feet. There were remnants of an arch under the throne (for the Host), and a line of stout, square piers forming an aisle or sacristy to the north. The lizard and the pigeon were the only inhabitants of the grim old ruin. I left it in sadness. There is something unpleasantly impressive about these transitory labours, upon which the lives of men have been

---

* Below Sta Maria the lands will become sandy, less light and rich.

† Hence the pilasters are called by M. Halfeld (Rel. 156) Columnas de Pedra. He remarks, "In the above-mentioned church they still inter the defunct, but with so little piety, that the corpses, hardly covered with loose earth, exhale an insupportable fetor." The hint has been taken, and we had not to complain of the atmosphere.

wasted. The whole scene reminded me of the once renowned City of Wari (Warree) in the Kingdom of Benin.

Resuming our way, we ran down the Island of Santa Maria, and presently we came to another great stone-river coursing through the roots of the Serra do Orocó. These hills form a hollow crescent between the Serra das Carahybas and the water. The appearance is made peculiar by two knobs, which a curtain connects, and an outlier, the Upper Orocó proper,* approaches the left bank. The stream flowed like a sluice, and in the Cachoeira de São Pedro, where, despite the manful slanging of the pilot, the paddles preferred looking behind them to working, we struck heavily. We then threaded our way down the mid stream, though a land of islands.† On paper the channels look like the blue fissures of a glacier. The Ilha de São Miguel showed a deserted temple on a knob of ground. The Ilha de São Felis disclosed through its dense trees a whitewashed and tiled church, with belfry and two terminal towers; here the saint still resides.

As we emerged from this "belt" the water fell smooth as a metal plate; but the prospect was not less wild. On the right bank the Serra do Aracapá crouched like a sphinx with jubated neck, and from down-stream the head will seem distinctly traced. We then paddled between the Aracapá Island and the left bank, where enters the Riacho da Brisoda (Brigida).‡ The back water flows up the green mouth for about two leagues, and the rest of the bed at this season is a string of pools. We found good anchorage ground at the Porto do Aracapá, near the Fazenda of that name fronting the island. The people here breed horses, mules, and black cattle for the Cabrobó market, and a good ox is sold for 20$000. It was a still, quiet evening, favourable for mosquitos. The hoary eastern clouds at nightfall threatening

---

* There is an Orocó debaixo on the left bank, about two miles down.

† To starboard were successively from west to east, the Ilha das Almas, do Juá, de S. Miguel, and da Piedade, with a mass of others, especially the Ilha Comprida, between them and the right bank. To port lay the Ilha de São Felis, which the pilot called "de São Pedro:" it had about midway a billock (outeirinho), upon which the Chapel is built, nearly due north of the rising ground that supports the ruins of São Miguel. Between S. Felis and the left bank is the Ilha da Tapéra; east of S. Felis is the Ilha do Aracapá, about four miles long. Its channel is much broken, and Cascalho appears upon its left bank. Here the broadest part of the river is nearly two (geographical) miles.

‡ Mr. Keith Johnston calls it "R. Bregido," and makes it drain, correctly, I believe, the southern slopes of the dividing range between Pernambuco and Ceará.

wind, and the jacaré splashed around us, while the flute and the song came loud from the shore.

Nov. 1.—We set out at 6 A.M., but the gale soon drove us to take refuge on the western side of the Aracapá Island, where we found a few huts belonging to moradores and fishers of the Trahíra. The low lands are often flooded, but there is a dorsum of higher ground to which the people can retire. I utilised the delay by engaging another paddle. The liveliness and the free, swift motion of the rapid after the smooths above were rather enjoyable than otherwise; but the process seemed somewhat a "tempting of Providence" in the now crazy Eliza, manned by a crew that would not work. The pilot, who knew the dangers better, was far more anxious than we were; and he presently returned with a barqueiro who, moyennant 10 $ 000, agreed to accompany us. Antonïo was a stout, dark youth, with a heavy shoulder and muscular arms. He justified all the good things said about the people of the river below Joazeiro. Having received a small advance, he crossed the stream to fetch his sheepskin, and he took his paddle at 11.30 A.M., the old "Menino" being placed in the stern to amuse himself with his kitchen ladle.

We coursed down the end of the wild Aracapá Channel, passing an islet on the right and the Ilha do Taboleiro* to the left. Presently we shot through a gate formed by several rocks to port, and to starboard by an enormous block which had assumed the domed form into which granite masses are so often weathered. The colour was cinereous brown above, and below it was japanned to the semblance of a meteoric stone. Here, as elsewhere, the colouring matter does not penetrate the surface except through fissures. The coat varies but little in thickness, and when broken with a hammer the fragments show that the glaze is easily to be removed from the stone. I have before alluded to this phenomenon of coloration, which is common to both hemispheres.† A long series of observations is required before we can answer the question—"Does the river hold the oxides suspended like sand and other earthy substances, or are they found in a state of chemical solution?"

Beyond the gate with its grisly tower was a remanso, half boil,

---

* In M. Halfeld's map it is called the Ilha dos Bois.   † Chap. 15.

half dead water, which taxed the paddlers' arms. We then flew through a "violent rapid" formed by rocks between the eastern end of the Aracapá Island and the left bank. It is known by the descriptive name of Desataca Calçoës "—loosen your breeches—which is all that requires to be said of it. We then passed on the left bank the Serrote da Ponta da Ilha da Assumpção, where an ypoeira setting off to the north-east insulates a tract of ground three and a half leagues in length by five-eighths of a league in extreme breadth, more than double the width of the river water.* The Serrote is a lumpy, half-bald hill, with grey bush scattered over a whitish surface. The upper part is banded with scattered rocks of lighter tint running parallel with the bed, and tailing off down stream.

Perforce we took the main channel to the right or south of Assumption Island. The bank is mostly of sand based upon hard clay, and its "Cascalho" extends to the water edge. The low-lying land has Catingas Altas, where the people fly from the inundations. They remember in 1838 an exceptional flood, which rose 32 feet. The soil is said to be good; horses are bred in the island, and black cattle are said to have run wild. The mountains of the "terra firma" in front form a picture. The broken line of the Serra do Milagre contrasts with the lumpy mass of the Serra da Bananeira, upon which is said to be an "Olho de Agua;" and whilst the valley is bone-dry, its highlands are fed by rain. Far to the left two pyramids, regular as if cut, breed reminiscences of Cheops and Cephren in a certain valley of the Old World. Seen from the south-east, these hills lose their venerable appearance, and become a saddle-back, banal even as is its name, "Serrote do Jacaré." †

At 2 P.M. we entered the "spuming rapid" of Cachauhy, called da Assumpção to distinguish it from two others down stream. It is formed at the end of the Ilha das Vaccas and other complications by the Serrote do Salgado, a knob on the right bank. We went down the breaks where the water danced about the dots of rock,

---

* We shall ascend the easternmost part of this ypoeira to make Cabrobó. The direct route would be by the west, but at this season it is impassable. The principal rapids, going down the line, are : 1. Born Successo ; 2. Cachauhy ; 3. Tucutú ; 4. Camaleão ; 5. Urubú ; 6. Cauam, or Cauan ; 7. Fouce ; 8. Catarina ; and 9.

The Cachoeira do Gavião, or do Portão. The latter, about half a league above the town, is described as a drop (despenhador), which can be passed only in the height of the floods.

† None of these names is given by M. Halfeld. The Jacaré appears to be his Serra do Bendó.

and safely accomplished the always delicate operation of crossing. We landed on the Island of Assumption to see the church, part of whose brickwork cumbers the shore in large masses. Nothing can save it. In 1852-54 it was 51 feet from the stream, which has now laid bare the southern side. It was built in 1830 by a citizen of Cabrobó, whose name is already forgotten; and the style as well as the material are a long way behind those of the Jesuits. It is to be hoped that the next traveller will find a little more attention given to the dead who are buried in the roofless enclosure.

The people collected to see us. Apparently inclined to be "saucy," they came with knives and small bird-bows and arrows. The old savages have all died off, and these are mostly a mixed breed, whose curly hair comes from Africa. The pure blood showed the well-known signs—big, round Kalmuk heads, flat Mongol faces, with broad and distinctly marked cheek-bones; oblique Chinese eyes, not unfrequently bridés, rather brown than black, and dwelling upon objects with a fixed gaze; dark and thick eyebrows; thin mustachios fringing the large mouths full of pointed teeth; and small beards, not covering the long, massive chins. The hair, brought low down over the forehead, was that of the Hindú, jetty and coarser than in the pure Caucasian. The nose had an abominable cachet of vulgarity, small and squat, with broad fleshy nostrils; in fact the feature was all that an Arab is not. They were well-made men, except that the trunk appeared somewhat too long and large for the legs, and the shoulders seemed to project horizontally just below the ears. The extremities showed that delicacy of size and form which has passed so remarkably into the Brazilian blood, and the skin was brown-yellow and ruddy only where exposed to the light and air.

A glance down the river from the tall bank discloses a grisly sight. There was the rich golden glow of the unclouded sun now slanting west, and many a silvery line of stream to suggest

Den Silberbach in goldne Ströme fliessen.

But a purple nimbus with a long grey lappet ahead threatened a gale of wind, and the richly tinted surface was fanged with murderous black stones. Here the Serrote da Lagôa Vermelha* runs

* Some call it the Serra do Milagre; others the Serra da Lagôa Dourada.

parallel with the right bank, and extends to it many a hill spur that reefs the stream. We easily traversed the bed, and stumbled through the Cachoeira da Pedra do Moleque, which breaks and boils right across; but at the next place we nearly came to grief.

Here the channel bends to the south-east, and dashes at a hill of stone and red clay. This Alto da Lagôa Dourada deflects it almost at right angles to the north-east. The water flows down hill, and we distinctly feel and see the angle formed by the raft platform. It is a violent torrent, pouring at a rate of 10 to 12 knots an hour over the rocks, swirling around them, and producing a complication of currents. In the runs there is a visible convexity of surface, the waters being heaped up as it were by the compression of the sides; and between the torrents are smooth boils which seem as though produced by underground springs. As we were entering the worst part, the strong east wind struck us, and in a minute we were thrown helpless upon a rock. I had taken the precaution to secure everything on board with ropes; had not this been done the surges which swept us as we heeled over would have cleared the deck. The pilot exerted himself desperately; the men kept their presence of mind, and the current, in whose power we were, beneficently sent the Eliza's head down stream, with no further injury but a scrape and graze. Enough for one day. The storm set in with fury. We managed to pass the Serrote da Lagôa Vermelha, and we anchored on the right bank, a little below the extinct town of Pambú.

This place of unintelligible name lies in a sack of the southern bank, and to the east of its unimportant Riacho. The site is a level at the feet of a thicket-clad ridge. The offset from the high wall to the south-west and the huts may number a maximum of thirty-five. The church, built and dedicated to Santo Antonio by a rich proprietor of Cabrobó, has indulged itself in an architectural eccentricity. The façades are double, whitewashed in the rear, brown clay in the front, and the effect is that of a man with two heads.

We passed the night at the Pedra do Bode, fronting Pambusinho Island, which was backed by the great Assumpção. A clean patch of sand was dotted with the Oití da Praia (Pleragina odorata, Mart.), a bush here considered useless; with the wild Icó, bending under the weight of its fruit; and with the Piranha, a scaly tree, which is green and lively above, whilst the lower parts

supply good dry firewood. Behind the beach rose the rocky and
clayey hill, scattered over with quartz pebbles and red silex that
resembled Rosso Antico. Passing through the bush we were
attacked by a carrapato-tick, now a novelty, but none the more
agreeable.

Nov. 2.—We ran down half a league to the tail of Pambusinho
Islet, and then turned north-westwards into the Braço do Tucutú,
the channel parting the mainland of Pernambuco from its subject,
the Ilha da Assumpção. The latter is here kept in position by
lines of rising ground, which face to the east and to the south.
It was a delicious morning; the air was sweet and rain-washed,
and the temperature that of Cairo in the cold season. How
much would be paid for such a day at such a season upon the
banks of the Thames! All creation looked its best, and the birds,
unusually numerous, sang gaily in the bush, especially the tame
and familiar red-headed songster of many names.* The ashen-
grey maracaná † with the long cuneiform tail, was trooping from
the forest to plunder whatever maize was to be found; and the
fine large blue alcedo,‡ a king amongst the kingfishers, crossed
the stream with his " vol saccadé," or sat upon the spray of pale-
green glaucous verdure, looking out for what he could devour.
The brown-black nimble plotus shot swiftly past us; the ichthyo-
phagous crauna,§ with dark plume and yellow beak shaped like
the curlew, heavily flapped its long wings; and the Socó boi
(Ardea virescens), so called from its bull-like bellow, looked
twice at us before it would take the trouble to fly. Plundering
seems here to be the fashion; even the pigs were necklaced with
wooden triangles‖ to temper their love of manioc.

At this season the lower channel is clear; but during the
height of the dries the Cachoeira da Boa Vista, alias da Boca do
Braço, must be troublesome. The scenery was the usual pistia
and hairy grass near the water, thin Mimosa growth higher up,
and plantations upon the more elevated lands. Fish was plenti-

---

* It is called Cabeça Vermelho, Gallo
Campina, and Tico-Tico Rei. The "Menino" declared that he had sold for
10 $ 000, at Rio de Janeiro, a pair (casal)
of these birds, which are prized for their
song.

† There are two species : Psittacus
Macavuanna and P. Guianensis (Linn.).

‡ The people call it "Socó," and de-

clare that it is fat and good eating.

§ The word is doubtless a corruption of
"Guaraúna," the "blackbird ;" but it is
pronounced as above, and many places
upon the river are named from it. The
people praise the flesh of this bird, after it
has been fried in fat.

‖ Locally called Canga (a yoke), or
Cambão, a rustic Portuguese word.

ful, but the fishermen asked for it exorbitant prices. After four hours of very lazy poling we turned a corner from south-west to north-west, and came upon a clump of huts and a large compound wall facing towards the stream. A little above was the Porto, where a ferry plies between the island and the main. It is a broad green boat, with a short mast made fast to a bench stepped in front. Here we found the usual scene—women washing, men filling their "odres" (water-skins), and "borrachas" (leather bottle-bags, with wooden corks), and children splashing and catching the Piaba and the Piau. There were many horses, and the clean-limbed cattle fed upon the heaps of cotton-seed which had been thrown upon the banks. The other live objects were very lean pigs, prowling dogs, and poultry, which here includes turkeys and guinea-fowls.

Ascending the bank I found unexpectedly a large place without any of the sleepiness which had characterised Joazeiro and Boa Vista; the site is the mainland, in the Comarca of Boa Vista, Province of Pernambuco. At present it is a very dry land, the evaporation curls up the leaves of the orange tree, whilst the tall stout papaws seem to enjoy the temperature. And at times it is very wet; the floods enter the settlement, deluging its floor of sandy clay, and driving the people to the Catingas Altas, which we see scattered about. The main of the town, which may contain 125 houses and 700 souls, is formed of a large street, or rather square, running north to south, and containing the dismantled church of Nª Sª da Conçeição. The houses are unusually low and massive, and they use shutters instead of glass windows, declaring that the road to Bahia is 140 leagues long, and that many of the stages want water. On the north of the settlement is the cemetery. The centre shows the new Matriz and inevitable cross, the work of a rich devotee, D. Brigide Maria das Virgens, whose husband built the now ruined fane on Assumption Island; both are in the same style, and this bears the date 1844. The interior is unfinished, showing a ceiling of naked rafters; there are, however, two pulpits, an organ loft, and carpets upon the floor, which show that it is in use. The Vigario recites mass every morning, and all the "respectables" of both sexes are "expected" to attend with a regularity which reminded me of the Mosque. Here and there are some decent shops, and I bought without difficulty ʼeat and poultry, rice and water-melons, salt and liquor. A tiled shed represented the market-place, which was crowded with leather-clad men from the interior, chaffering over their cotton bales* and broacas of rapadura and farinha, which will here be exchanged for wet and dry goods.

Here, after some months, I again saw "the Eagle zin" at work; the material comes from the inner highlands to the north, where yesterday's rain fell. This is a country of great fertility, and extends north to the Serra de Araripe,† distant from Cabrobó thirty leagues. The range is described to be a succession of mounds of rich red clay, across which there is an easy road, whilst behind it is the stony Serra de Borborema, which inosculates with the Ibyapaba Range, separating Ceará from Piauhy. At the southern foot of the Araripe is "Ixú," whose Villaship has been transferred to "Granito." On the northern counterslope are Crato and the Villa da Barra do Jardim. In this chalk range Dr. Gardner first found the Ichthyolites which now go by the name of "Penedo Stones." The nodular and rounded lumps of impure fawn-coloured limestone, when split down the middle, display the skeletons of the Mesosaurus, and fishes belonging to the recent Cretaceous epoch.‡ The people know of their existence, and some are still sent to the coast as curiosities.

We at once see the cause of prosperity at Cabrobó. The land road between the Villa da Boa Vista and the Varzéa Redonda runs by it, and is met by the highways from Ouricory, Crato, and the Cairirys§ to the north and north-east. The cotton bales are embarked on rafts or carried down by horses, to the Porto das Piranhas, distant along the river 55 leagues (165 miles). Then after long wandering they find a steamer which ships the exports

---

* The bales averaged five to six arrobas : they were unpressed, but made up neatly enough with "tie-tie."

† Gardner has described this chalk formation. The name Araripe has been wholly omitted in Mr. Keith Johnston. Sr. Candido Mendes de Almeida has not forgotten it : he does not, however, show it backed by the Borborema, which, properly speaking, includes the two Cairirys. Of the latter more presently.

‡ Recent cretaceous fishes have been lately found by that excellent traveller, Mr. William Chandless, on the Rio Aquiry, an affluent of the Great Purús. Most of them, according to Prof. Agassiz, occur between S. lat. 10° to 11°, and W. long. (Gr.) 67° to 69°, in localities from 430 to 650 feet above sea-level. Here the latitude of Araripe is about 7° south.

§ This name is locally applied to the country about Crato and Jardim. Cairiry, also written Cariry (Carirys), Cariri, or Kiriri, was the name of a Tapuya tribe, the ancient possessors of Itaparíca Island, in the Bay of S. Salvador. In 1699 a Jesuit missionary, P. Luis Vincencio Mariani, published at Lisbon his "Arte da Grammatica da Lingua Brasilica da Naçam Kiriri." Many places hereabouts bear the name of Cairirey ; they were doubtless localities to which the old savages emigrated. There are two principal ranges, the Cairirys Novos, in the Province of Parahyba do Norte, and the Cairirys Velhos, in Pernambuco.

to Bahia. In 1852—54, I have said, all was languishing, where now we find life and energy. A good rolling road, but more especially a tramway, would give a mighty impulse to trade by facilitating it; and the many men relieved from the carrying trade would at once become producers.

I called upon the civil young Delegado Sr. Bertino Lopes de Araujo of Parahyba do Norte, who had married and settled in this place six years ago. During that time he did not remember a single assassination, although, of course, quarrels had taken place. Neither he nor any of his neighbours could explain the word " Cabrobó," also written " Quebrobó;" all they knew was that the old Indian name had been given to a Fazenda which presently became a Villa. The Delegate warned me, as others had done, to make everything snug on board the Eliza, as we were soon to be in difficulties.

SECTION II.

THE BAD RAPIDS TO SURUBABÉ.

Nov. 3.—After manifold delays—the Delegate was writing letters for us, the pilot attended mass, and " Majelicão " hid himself in the nearest brothel—we ran down the narrow arm, safely passed its central " Camboinha," and, after an hour's work, sighted the Banco d'Arêa,* on the Bahian side below Pambú. At this point the Rio de São Francisco begins the great south-easterly trend, which it will keep, with a few insignificant varia-tions, to the end of its journey. The north-eastern Vento Geral now becomes a side wind, and sometimes blows almost from behind. The sun is decidedly hot, clouds gather to the east and to the west, we see from afar symptoms of a " repiquete," or violent shower, and we therefore expect a gale, if not a rain-storm.

On the left side a sandy islet hid from us the mouth of the Riacho da Terra Nova, or do Jequi (Giqui), a nullah of some importance.† Beyond it we entered the Passagem do Ybó, the

---

* It is a clump of huts above a large sand-bar or beach, known as the Corôa do Bom Jesus.

† It is said to head about 30 leagues from its mouth in the Ararine Range, near the place called " Cairirys Novos." Mr. Keith Johnston calls the north-eastern fork " R. Terra Nova," and the north-western, " R. S. Domingos."

narrowest part of the São Francisco, where people can talk from side to side. The formation is a deep gorge in the valley line, which, however, shows no especial features; the banks are sandy, the right is not flooded, whilst the left is swept, and a low rib of loose rock stands up in mid-stream. The water, at this season 95—100 feet deep, swirls in palpable domes, and foams in shallow "pots." A little below the Fazenda do Ybó, and a point projecting from the right bank, the 770 feet of stream spread out to more than a mile. The total breadth of the river below the narrows is three-quarters of a league, but the greater part of it is occupied by the Ilha da Vargem, fronted by the main artery, and backed by its own little branch of the São Francisco. Well inhabited, and with fertile soil, this island, shaped like the letter L, with the angle pointing south-west, is one of the largest, each limb being about a league and a half long.

Easily passing the narrows, we ran along the left bank between it and the Ilha do Estreito.* This is the only line passable. Beyond this island the left bank projects a rounded point towards the concavity of the L, and fills the river with rocks and rapids; the heights are apparently limestone, and again we see along the brink iron conglomerate in dark ledges. At the apex begins the Cachauhy de Antonio Martins, the second of the name. The roar of this rapid is worse than its bite; the foul channel, however, is compressed on the right by the Ilha do Cachauhy, and further down by the high and sandy Ilha do Caruá.

We then crossed the river from west-north-west at the tail of the Ilha da Vargem, to the "Largo do Brandão" on the east-south-east, a long reach of deep smooth water which appeared a "Remanso" after the swift stream higher up. A gaunt island, the Ilha dos Brandoës,† here defends the bed from the rocks of the left bank, whilst the right bank protects us from the wind. Opposite the head of the island, and on the Bahian shore, is the mouth of the Riacho da Vargem, which is said to run twenty leagues from a height called the Tombador. Rice fields were on its borders, and boys were pelting the greedy birds with loud cries of "diabo." At 4 P.M. the pilot said that we must anchor, as

* The Ilha da Boa Vista in M. Halfeld's maps.

† On the left bank are three Fazendas, called Brandão, probably from some family that first settled here. At the Brandão do Meio there was a neat white house and a clump of cocoa-nut trees.

there is no safe ground amongst the Rapids, which will extend ten leagues down stream. This is by no means the case, but Manuel Cypriano's eyes are not now of first-rate quality, and he does not like to pass broken water either in early morning or in the evening shades.

We made fast below the Fazenda do Abarê,* opposite the head of the Ilha Grande, a thin strip about two and a half leagues long, immediately succeeding to that of the Brandoẽs. The bank is here lined with nodules of lime. The little settlement of tiled and tattered houses had its chapel, and we met no difficulty in buying a pig and poultry. The crew reproached me for not having killed a harmless water-snake, and amused themselves with bullying an unfortunate frog of large size, which is popularly supposed to swallow sparks of fire. The boatmen have tales of the " Sapo " getting to Heaven by the aid of the birds, and the animal seems to hold in these regions the position of the spider on the coast of Guinea.†

Nov. 4.—This is the critical day—the acme of our rapid-troubles; we shall pass nine bad places in 6 to 7 leagues. The breadth of the stream is a constantly varying quantity, but generally it is unusually narrow, the effect of increased slope. The left bank is a long line of little hills, whilst the right side is mostly flat and bushy. The profile of the bed is an inclined plane of rock and gravel, divided into sections by level spaces. Long islands and short islands, rocks and reefs, sandbars and shoals, cumber the bed, and the former bear bits of noble forest. There is something majestic in the aspect of the São Francisco, whose turbid waters, here building up, there lieing low, now flowing in silent grandeur, fanned by the gentle breeze, and reflecting the gold and azure of the sky, assume an angry, sullen, and relentless aspect when some obstacle of exceptional importance would bar its mighty path.

Rising with the dawn, but not pushing off till 7 A.M., we took the channel formed by the Ilha Grande to the north, and pre-

---

* A little below this point is the Barrinha do Abarê. Both are reminiscences of the Jesuits, one of whom was called Abarê bébé, the "flying father," because he was always on the move. They, as well as the Prelates, took the title of Pay Abarê Guaçu, the Pope being known as Pay Abarê oçú etê (biggest of all). The friars of Saint Anthony were called Abarê tucúra, father locust or grasshopper, because the hood reminded the savages of the "gafanhôto."

† M. Halfeld (Rel. 215) mentions the Calborge, a singing and amphibious toad, which covers itself with froth. It has also its legends.

sently scraped over the shrub-grown stones as we passed down through the middle of the great "Tubarana" Shoal. Its site is at the head of the Ilha da Missão, where this landstrip, also long and thin, lies parallel with the Ilha Grande. On the right was the Barra do Tubarana, alias da Fazenda Velha, another nullah with a pooly bed. At 10 A.M. we coursed down the middle of the Cachoeira do Imbuseiro, formed between highland in Bahia, fronted in three tiers by the islet "do Meio," and the islands da Missão and Grande.

Twenty minutes then took us to the "impetuous Cachoeira of the Rosario." This is a break right across between Bahia and the head of the Ilha do Serrotinho. We hugged the right bank, and shot an incline of water, which made us sit back in the raft as upon a horse landing after a leap. The channel is smooth, lucent, and visibly lower—now a general feature—than the stream, which breaks with a railway rush on both sides. A heavy bump was the only damage done; here the rule is a bump and a scrape at least once a day.

After the Rosario we took the narrow channel made by the right bank and the Ilha da Barra, a lumpy island, one of a group of three disposed in unicorn—the others being the Ilha do Meio and the Ilha da Patarata. Near an affluent, known as the Barra do Mucururé,* the awning was taken down, and the thermometer showed in the sun 114° (F.), which made my companion suffer; even the black boys on ashore crouched and cowered under their little awnings of yellow straw. At the tail of the Ilha da Barra was a narrow presently "flaring out" into a bay. Looking back through a gap to the north-west, we saw the whitewashed Church of Belem—another missioner name—upon a dried-up plain, backed by a range of wavy hill.

11 A.M. brought us to our third trial, the "furious Cachoeira of the Cantagallo." It is a "long Sault" of half a mile, with two distinct breaks. The lower is by far the worse. We rolled down the mid stream through boiling glassy water, fringed by rows of surge flowing noisily. At the bottom we shaved the left bank of the Ilha do Cantagallo, a pyriform plain of sand, with a small rocky "Serrote." On the right of the channel is the third

---

* M. Halfeld calls it Barra do Tarraxi. It is said to rise at a place called the Imburanas, at the Ponta da Serra, and to measure forty leagues.

Cachauhy (do Pianoro, P. N.),* which is always avoided. The river, obstructed in its course, there breaks into waves which dash with thundering violence against the broken reef, and rush between the jags of rock in sluices of dazzling velocity.

We have now a clear league ahead without rapids, but requiring great care. Stones, shallows, and many little runs which the pilot calls simply Pedras, stud the bed. On the left bank is the Serrote do Papagaio, which has been visible since leaving the Ilha da Barra. The profile, seen from the west, is the "Phrygian bonnet," generally known in the Brazil as the "parrot's beak;" from the stream opposite it is a vertical ridge of bare rock. † Here begins the upper break of the second Cachoeira da Panella do Dourado, which the pilot facetiously calls O testo da Panella, the pot-lid. Below is the "famous whirlpool and rapid" of that name; the only sign of a maelstrom was cross-waves from the left or north-west, but on the downward side of a rock-lump by which we ran, we were struck full on the beam by a current flying rather than flowing, and we were once more nearly, but only nearly, swamped.

After a short halt for baling, we resumed. Number 5, the Cachoeira do Boi Velho, was not of much importance; it gives a fair way to the right, leaving the heaviest break on the left. Again the stream became clear, and the banks were lined with settlements; prairie fires, a symptom of expected rains, and burnings for new "roças" appeared all around; but they were of small extent, as the people want grass for their stock. The air becomes even more arid than before, and the surface of the land is mere dust. The right shore showed the Arraial da Missão de São João Baptista de Rodellas, more curtly called "As Rodellas;" it was a village of "Caboclos," pauper huts gathering about a large and well-washed church, backed by a wave of high ground. In 1852 the temple was in ruins, but a Capuchin Missioner, Frei Paulino de Lusione, collected alms, and reconstructed it. The pilot told me an ugly story about some ghostly man, here stationed, who showed a pronounced propensity for "Caboclas" (the feminine), under the age of twelve. One of his

---

* M. Halfeld gives at this place a Cachauhy de Cima and a Cachauhy de Baixo.

† In the Relatorio (p. 168) it is called Serrote da Pedra; the trend is made from north-east to south-west, and the material is stated to be "gneiss granite" (unstratified gneiss), granite and quartz.

victims ran away, and complained to the Delegate of Police, who at once imprisoned, and finally compelled the Reverend to quit the country. There were other tales of debauchery, cloaked by sanctity, especially one of holy water, which proved to have a pestilent taste of gin; true or not, they prove that the moderns do not secure the respect paid to the ancient Jesuits.

We rested on the left bank opposite Rodellas, and the boat-men bathed to prepare for the finale, an ugly stretch of two leagues. The channel widens out, for the last time, to nearly three-quarters of a league, and bending from the south-west almost to the south, becomes a mass of islands. Of these eight are considerable tracts of wooded ground.*

At 3.15 we put off from shore, and easily passed through the Cachoeira do Urubusinho, which is some hundred yards in length. On the right was the hill-island do Urubú, a kind of Careg-Luzem-Kuz, which from up stream appears like a monstrous elephant, with white ear and head partly averted, lying down amongst the trees; its spine is a bristling crest of bored and hollowed stone.† To the left of this "hoar rock in the wood" lay the long thin Ilha da Viúva; hence we passed directly into the "furious Cachoeira of the Fura-olho," or gouge-eye. I confess to having felt cold hands at the sight of the infamous turnings, the whirlpools which the Relatorio calls the terror of navigators, and the pot-holes some fifteen inches deep in the water. Head on, we dashed at the rocks—here bare, there shrub-clad—and more than once we prepared for the shock; often, too, the pilot giving the raft a broad sheer with the sweep of his heavy and powerful paddle, carried us safely through places where we could almost touch death on either side. It was a wild scene; the Eliza swayed and surged to and fro, as she coursed down the roaring, rushing waters that washed the platform; the surge dazzled the eyes when it caught the sun, and on the smooth depths the beams were reflected as by a mirror. "Shout, boys, shout!" cried the old man, in his Cachoeiran element; "I love

---

* The islands, beginning from above, are—1. Ilha do Cuité; 2. Ilha da Viuva, or dos Cubaços, the latter name confounding it with a smaller feature to the south; 3. I. da Tucurúba; 4. I. do Jatobá; 5. I. de São Miguel; 6. I. da Crueira (sic in map, Cruzeira?); 7. I. do Espinheiro; and 8. I. do Surubabé, in the Relatorio called "Sorobabé" and "Zorobabé." Besides these the Plan shows some thirty-five islets of larger or lesser size, not including rocks.

† M. Halfeld (Rel. 169) says that the formation is granitic, and he places the channel on the right, whereas we passed to the right of the "Elephant Islet."

to hear the shout in these places!" "Hé Fura Olho!" they exclaimed, with their glapissant voices, calling upon Nossa Senhora, and crying, "Ô bicho feio!" to the whirlpools and the ugly-headed, black rocks, whose faces glistened like the hippopotamus fresh from the deep, and whose necks were cravatted with bands of rushing white water, a thin and semi-transparent gauze. We managed "Gouge-eye" in fifteen minutes, and pronounced it very pretty shooting—when it was ended.

The ground-swell below, not a little like the "Gallops Rapids" of Canada, bore us down between the Ilha da Tucurúba,* and its outlying rocks on the north-west, to the Pernambuco bank. We now enter upon the 299th league, which is said to be the worst upon the river, but we found it less formidable than that preceding it. The course begins with a stony break between the left shore and the Ilha dos Espinhos;† a mass of Mimosa tasseled with pink flowers, and well-armed with thorns. Presently it passes a small nameless river-holm on the left, then the line hugs the bank to avoid sunken rocks and shallows; once more it winds amongst the islet-rocks, above the head of the Ilha do Sorobabé; and, finally, it returns to the left side. The tide flowed like a mill race, and in parts the speed would have distanced any steamer; but we had often to hang back, and the total of two miles occupied us twenty minutes.

Then, as the sun began to slope behind the Imbuseiro trees, we heard just ahead the roar of Surubabé, the ninth and last trouble, where ends this upper Cordilheira of Cachoeiras, which preface the Great Rapids. Manuel Cypriano, whose motto certainly should be festina lentè, proposed reserving it for to-morrow, but the day was only 4.40 P.M. old, and for old reasons, I at once negatived the measure. During the floods between December and May, which, however, are very uncertain, Surubabé is shot by canoes, and even by small barcas, the only danger being the rapidity of the run, which dashes them to pieces if they touch. During the dries a portage for merchandise is always made. The river has now risen from five to eight palms, ‡ and thus our difficulties are greatly lessened.

---

* On the left bank the Riacho da Tucurúba, a mere nullah, falls in.

† Others called it "do Espinho." M. Halfeld's plans name it the I. de S. Antonio.

‡ At Varzéa Redonda the people declared that it was four to five fathoms above low-water level.

The Surubabé, also called the " Cachoeira do Váo,"* began with our ugly rapid between its island and terra firma. Here the São Francisco "fervet immensusque ruit." Passing this, we landed on the left bank of the island, above the great obstruction, a wall of granite, extending right across from east-north-east to west-south-west, which might easily be opened. The greater part of it has a clear fall of two feet, and a reconnaissance determined us to attempt the right side of the ledge, where the shoot slopes like the places to which we have been accustomed. The chief danger is the impetus which drives the craft upon the " Váo," or granite-bed just below and in front of the fall; the dashing water curls back in waves two or three feet high, and would bury the intruder.

The pilot and two men out with their poles, whilst four of us manned the rope, making use of the trees where the tow-path was foul with slippery water-grass, dry shrubs, and tough fig-roots. Down went the raft headforemost, dipping deep her platform, and grazing a boulder on the right side. When she had reached the bottom of the fall, Manuel Cypriano and his men stopped progress with their poles, we sprang on board, punted to the left of the " Váo," poled back to the island, and, after another little difficulty, which also required cordelling, we exchanged poles for oars, we ran to the left bank, and landed at 5.15 P.M. Our day's work had covered twenty-seven miles instead of fourteen, the average since leaving Boa Vista.

Then we passed out of the gloom and torment of the Rapids into the calmly flowing stream, whose light blue was stained with the gorgeous red of the western sky. Thus satisfactorily ended all my troubles with cachoeiras upon the Rio de São Francisco, and the sensation was certainly one of great relief. We passed a pleasant quiet night upon the water-grass and the iron-stained Cascalho that banked the smooth channel; under a " dome of steel lit up by the stars;" and within hearing of the dead monotonous crash of the Rapids—perhaps my prejudiced ear did to it injustice—underlying the music of the breeze. There was not a trace of dew, which partly explained the burnt-up look of the land.

---

* Of the ford, or shallow. M. Halfeld gives Vão, which means an empty place, or desert.

SECTION III.

THE SMOOTHS.

PARROT HILL.—CHALK FORMATION SIMILAR TO THAT OF THE AMAZONS RIVER.
—DIAMANTINE DEPOSITS.—ROCK INSCRIPTIONS HITHERTO NEGLECTED.—
END OF RIVER TRAVEL.

After the "Thousand Islands" and their accompanying
terrors, the São Francisco is a sightly stream; here, in the pilot's
phrase, you may fasten a branch to the bows and float safely
down stream to Varzéa Redonda. The scenery somewhat
suggests the valleys of the Nile and the Indus when they reach
the dry country; but the artificial glories of the far richer
Brazilian river are all to come. The area of drainage is narrowed
by mountain-ranges on both sides; there are few influents, and
none of importance; the breadth of the bed greatly diminishes,
an immense evaporation ever sucking up the waters, and reducing
their volume where we expect to see it increased. On the other
hand the depth is more considerable, and the flow if not swift is
steady, making up for want of size. Hence the ypoeira becomes
an unimportant feature, and we miss the long chain of island and
islet, built up by the waters in the shallower portions. The
climate becomes exceedingly dry, and the three-months' rains do
not suffice for the sandy thirsty land, rich only in thorns. It
wants, however, only water to become fertile as Sindh, and the
canalization of Egypt will be much facilitated by the compound
slope of the lands about the stream. Agriculture, and even
population, are confined to the banks, where the crops thrive by
capillary attraction through the porous soil; not a gourd of water
is ever bestowed upon the growth, and digging a deep trench,
with a dam to preserve the supply during the dry season, is far
beyond the power of the present generation.

Here we change the wild, stiff, upright scenery of a granitic
country for the soft, amene, and rounded lines of the sandstone
and cretaceous formations. The right side shows " Catingas
Altas " at a short distance, and at times dwarf bluffs facing the
stream; the left is low, and excepting a few scattered lumps, it
stretches uninterrupted to the Serra de Araripe, not visible from
this point. The water margin, as far as Varzéa Redonda, is

frequently lined with " Cascalho " of all sizes; some of fine conglomerate, which fractures easily, others of jasper and the various forms of silex, mossed with black lichen, or stained with iron, and giving, when struck, a metallic sound. Here and there the formation shows points of gold, and the people know it as " Gurgulho brabo." We shall spend three days over the fourteen to fifteen leagues along the river which separates us from Varzéa Redonda. The São Francisco makes a great bend to the north, covering seven leagues, when across the heel the line is hardly five miles. I had no reason to lament the loss of time; this section unexpectedly proved itself the most interesting part of the voyage.

Nov. 5.—We found the united stream to measure only 300 fathoms (braças), and its comparative narrowness was set off by a dorsum swelling on the right side, here a normal feature. I was surprised to see so many signs of labour, cultivation extended to the water side, and long lines of hedge ran down the gently sloping bank. It was a peaceful pleasant scene, where nought jarred upon the senses, save only an old negro who was paddling a broken canoe, and cursing like a Celt, because he had lost his hat. Wherever there is irrigation, maize and sugar can thrive, onions and ground nuts yield abundantly, and the sweet potato attains an unusual size. The peach-tree abounds, but here as elsewhere in the Brazil, as far as my experience extends, it is hard and tasteless, fit only for stewing. As on the uplands of the Rio das Velhas the characteristic colour of the flowers is a laburnum yellow; even the Carahybeira now changes its mauve-coloured trumpets for gold.

On the left bank we passed the little influent known as Riacho do Páo Jahú,* and presently we struck the great northerly bend. This round turn in the bed is subtended by the Serra do Penedo,† a long and regular ridge with outlines of sandstone grit. On the north side a dwarf cliff buttressed the stream: the material is coarse arenaceous matter, almost horizontally stratified with perpendicular fracture, tinted red and yellow, and in places black with iron glaze; it was riddled into holes by the water, and displayed long straight lines of imbedded conglomerate, which

---

* At present it has no stream, and only water wells are in the bed. The people declare that it drains the northern dividing ridge. M. Halfeld (Rel. 171)

writes the word as it is pronounced, Pajaú. Mr. Keith Johnston does the same, and makes it a considerable stream.

† M. Halfeld gives "Serra do Penedinho."

seemed to have been deposited in a calm lake. Our Manuel
Cypriano, who had complained of fever during the night, was not
himself to-day, and we humoured the laziness of the men by
dropping down in mid-stream. A fierce wind presently came up
from the north-east, raising the waves in a few minutes, and
nearly lost us in the safest part of the journey, not an unfrequent
accident with shipping, British and other.

After some difficulty we made the left shore and baled. Here
the tall bank was white with marl, and in the upper levels cotton,
all unheeded, spread the ground with snow. When the gale had
somewhat abated we struck across stream to the Serrote do Pico,
whose regular dorsum and sliced cliff suggested rapids, but none
were found, the bed being too deep. This lump rises abruptly
out of a sheet of sand crumbled from itself; the height is about
110 feet, and the material is the now normal coticular sandstone,
iron-glazed below, and of brick-red and grey-yellow in the upper
parts. The summit is a bluff, and about the middle the slope
assumes the natural angle, growing a few trees; the strata dip
easily to the north or up stream, and the perpendicular fracture
forms at the corners columnar blocks. As we climbed its knee-
cracking sides, the little Mocó-coney came out of his home to
stare and bolt back. From this point offsets the direct road to
Itacutiára, an hour or two's ride, whereas the stream-way will
require from us three days.

We took the right of the Ilha da Tapéra, the only island as yet
seen below the Rapids; flat, green, and wooded, it was remarkable
near the arid red ground, and the thin dry bush of the bank.
Farther down the Bahian side showed us the Povoação da Tapéra
do Valentão—the "Village of the Ruins of the Ruffian." The
pleasant name is descriptive of the old inhabitants, a race of
bullies muito "enthusiasmados"—as the pilot magniloquently
expressed himself. The tall well white-washed and belfry-
boasting church, that does not set off the little hovels, is said to
be a deception, tumble-down inside.*

Dust-devils flew about in front of us, and sheets of distant rain
gave us a hint to take the bank; we made fast in the "nick of
time." Whilst everything around us, even the pinnate leaves,
was profoundly calm and still, rose the roaring of a mighty wind

* M. Halfeld (Rel. 173) speaks of it as a "vistósa igreja."

from the north-east, and columns and mists of brown-yellow silt
came charging down upon us as though we had been in the Valley
of the Indus. Then the gale tore through the woodland, plough-
ing the smooth surface of the water and rushing violently up-
stream. The meteor, which brought with it only a few thin drops
of rain, appeared to be, like the African tornado, merely local.
An independent squall was seen further down. It took an hour
to work round, again striking us from the south-west at 4 P.M.
We were then, however, securely embayed in a shallow bight protected by reeds, near a little settlement called Sabuicá. The night
brought wind and violent rain, which kept the mosquitoes quiet ;
our crew, however, seemed to fear them less than the "Besouro
Grande," a large black and yellow insect like our bumble-bee,
and they declared that its sting causes fever. To-day we saw
for the first time under the Gurgulho bravo, agates and onyxes,
banded with red and yellow.

Nov. 6.—We resumed our way down the right bank, which
was lined with ledges of dark Cascalho. Presently the stream
began to bend from north-east to east, the effect of ground
waves on the left bank, especially the Serrote do Ambrosio, whose
white ridge and light greenery, seen through the morning mist,
were easily to be mistaken for a giant tree. We then made a
"travessa braba,"* rendered fiercer by the sunken rocks, to the
Riacho dos Mandantes on the left bank ; the dry nullah at whose
mouth the grass had been cut for fodder, becomes an ypoeira
during the rains. Here the channel bends gradually from east
to south-south-west; the cause is the Serra do Papagaio, a
block through which the stream appears to have broken, and
which was formerly continued to the Serra do Penedo, passed
yesterday. From up-stream the "Parrot's Range" looked like a
"Castle Hill," with a tall, ruined tower on the right, connected by
a curtain with a smaller donjon to the left, and trending from east-
north-east to west-south-west. We went down cautiously under
pole, and presently landed to examine the chain ; at the same time
a furious south-easter came up and rendered progress impossible.

At the foot of the Castle Hill is a nullah flowing in from the
east, and formerly it supplied the banded stones (pedras lavra-
dinhas), for which the place is celebrated ; now, however, it is

---

* A "dangerous crossing."

choked with sand. We walked to a stony slope further north, and found on the riverward face, specimens of flint and coloured quartz that soon filled our bags. The most common form was the red and yellow-banded pebble, like those so common about Cambay in Western India; a few were striped white and black. There were also well-stained blood-stones; onyxes fit for cameos; cats'-eyes, as in the streams of Ceylon; "water-drops" (quartzum nobile); crystallized quartz, fragments of rock crystal, and a coarse opalline formation.* Formerly the valued stones were in great abundance, but for years they have been carried off, and we met a rival collector in the shape of a Brazilian youth.

I then worked my way to the Castle Hill, crossing sundry ridges that were crested with upright slabs, like the vertebræ of monster snakes. The broken surface bore nothing but stones and thorns, the usual species of Cactus and Bromelia. The ascent of the lower tower gave us some trouble; even Shakspere's Mark Antony, in matters of physique the beau-idéal of a traveller as of a soldier, would have complained of shaky knees and short breath after a two months' diet of manioc, rice and fish. The material is a friable grit, breaking almost with the hand, pierced by small holes, as if worm-eaten, and too coarse for whet-stones; in the higher parts the particles are smaller and more closely disposed. Lines of harder material ramify over the surface, and rise in alto-relief, forming irregular compartments; but even these crests can be knocked off with a stick. The soft places have been weathered into pot-holes and caverns that from afar resemble a dove-cot. The lower part shows a slightly green dis-coloration, which at once suggests our upper greensand overlying the stiff blue "gault;" and the higher walls are grey, red and yellow, doubtless a ferruginous tint; in fact, signs of iron every-where appeared. Lightly imbedded in the arenaceous matter, and evidently deposited by still waters, were horizontal bands of pebbles, smaller sized than those strewn below; hence, doubtless, comes the diamantine "formação" which we shall find in such abundance further down.

Approaching the summit, where the goats had trodden a smooth.

---

* I have already alluded (chap. 21) to existence of the opal, the only gem which as yet art has not learned to imitate. José Bonifaccio (Viagem Min. p. 29) found near Ipanéma of S. Paulo, the "common opal very like those of Tel-cobania in Hungary." I have not yet seen in the Brazil the quartz with the harlequin play of prismatic colour, which is still so valuable.

path upon the friable grit, I worked across the crumbling curtain. The latter is a "facão de Morro," a narrow spine with a fall on both sides, and in process of rapid degradation, soon to be level with the plain. It ends in the taller donjon, where a large, upright, and striated block of sandstone, whiter than usual, looks in the offing like a quartz "dyke." The cold, damp sea-breeze and windy clouds interfered with the prospect from the summit; it showed, however, that the lowlands were sandstone flats, from which rose many little buttresses similar to that upon which I stood. This formation we first noticed at the narrows of Ybó. Downstream I could see on the left bank the Serrote dos Campinhos, a sister block; and the substance extended with outcrops of granite and alternation of limestones to the city of Penedo, on the Lower São Francisco. It was noticed by Gardner at Crato in Ceará, and vestiges of the cretaceous period have been traced from Maranham to the Upper Amazons.*

Nov. 7.—From the Serra do Papagaio a road strikes southeast towards Varzéa Redonda, distant four leagues by land and seven by water. The pilots calculate respectively five and eight, the normal exaggeration of distance, which is measured by the laziness of the crews and the pace of the wretched nags. We set off at 4 A.M. in a dark, cold drizzle, at times lighted by the gusts, and, after a league, we passed on the left bank the Serrote dos Campinhos (de Baixo); here the sandstone rises bare and it forms outlying single pillars, weathered to cheese-wrings, and sometimes resembling "logan stones." The place is known by an ugly two-headed rock projecting from the river. The next feature is Icó on the left bank, backed by its Ypoeira, which is said to breed shoals of fish.† High in front rose on the right bank the Serra da Itacutiára, backed by "Catingas Altas," and fronted by a similar formation, the "Guixaba" on the opposite side.

The roaring wind again arose and drove us for shelter to the left side; on the bank women smoked their long reed pipes, with

---

* Dr. G. S. de Capenéma, a Brazilian savant, was of opinion that Gardner's discovery of immense cretaceous deposits about the north-eastern shoulder of the South American continent, might be reduced to "Tauatinga," or degraded felspar. The journey of Professor Agassiz has, however, set that question at rest. The signs of the cretaceous period are ferruginous sandstone deposited upon a lower greensand; marls and limestones soft and compact; thick beds of a finer grained soft and coloured sandstone; and lastly, a great dissemination of chertz, silex, and true flint. It is apparently barren in organic remains.

† In this part of the river fish is caught easily during the dry season, and especially about the rapids. The people shoot, as well as net and hook, the prey.

small clay heads, and fetched water, whilst the men scraped fish, which they refused to sell. None were in rags as about Joazeiro. The popular skin was yellow rather than sallow; the features were regular and sometimes handsome, the hands and feet were well-formed but large, showing Portuguese blood, and the long, lank hair was "Indian," whilst the pointed teeth probably came from Africa. All were armed, and some carried pouches of the Maracajá, a wild cat spotted like the ounce, and very destructive to poultry and kids. Those who passed by on horseback had shoes with long front leathers, over which the spur strap ran; they used halters and not bridles, and the stirrups were provided with swivels above the instep. They were not uncivil, but independent as their ancestry of the wild, and, perniciously frugal, they ignored the wants of civilization. Yet the land is good, producing in abundance maize and manioc, beans and ground-nuts, sweet potatoes, pumpkins and onions, melons and water-melons, sugar-cane and rice, in the places where it is not flooded by the stream, whilst cotton rotted, as usual, on the uplands.

Upon this "praia," at the turn of the stream from south-west to south-east, we again recognized for the first time after an interval of ninety-three leagues, the true diamantine formation. Along the water were strewed lines of the white and black cattivo, the jetty ferragem, the square Santa Anna, the Agulha, here very large, the snowy Ovo de Pomba, the straw-coloured Siricória (chrysolite or white topaz), and the feijão, the fava and the many kinds of polished "caboclos," whose bright lustre is held to be a good symptom. Further down stream we met with it after turning up the large pebbles (gurgulho brabo), and under the superficial humus it is also spread in a thin sheet. These deposits will continue as far as the Cachoeira da Itaparica, eight leagues below, and there it will again be noticed. The people have never beheld a diamond, and their cattle tramp over what may prove to be a mine of wealth. When they saw us picking up shells and pebbles they lamented their "backwardness," but in the present state of things, exploitation is hopeless. The place is only 90 to 93 miles from the highest station of the steamer, and it is my conviction that it should be carefully examined.

Two young fellows, Rufino Alves de Sà, and Francisco Maria de Sà, of the Engenho Novo lands, were loitering about and asking us the usual questions, *e.g.* if the English had a King. I examined them upon the subject of a " letreiro," or inscription, of which we had heard up-stream ; they declared that they knew the place, and the sight of a Milreis note easily persuaded them to become my guides. They shipped on board the raft and assisted us across the Itacutiará break, which can hardly be called a rapid.

Here the river, sweeping round to the east, passes between the long dorsum of the Itacutiára hills and the bluffs of the " Guixaba; " the two connect by a ridge of iron-glazed sand-stone. On the right is a clear channel up and down which boats can pass even by night; in the centre is a peculiar mush-room-shaped rock, and between it and the left bank the bed is very foul. As we approached the reef, and were rushing at full speed with the water, " Captain Soft " let slip the lashing of his paddle, fell upon his back, and remained there grinning like an idiot. The strangers prepared for a cold bath by loosening the band which held their short cutlasses ; fortunately, however, the old pilot, furiously working the stern paddle, and using the while language of the most energetic description, drove us safely through the upper break.

We landed on the right bank at the Sitio da Itacutiára to the north of the hills, and walking through a manioc field we reached a sandstone wall, locally called a " Talhada." It bears south-south-west of the upper break, and forms an angle whose arms face to the east and the south-east, thus obliquely fronting the stream. The material is coarse sandstone with lines of conglo-merate, reddish-yellow above and below, glazed as though the river had once washed it. Between six and seven feet from the ground there is a roof-like projection, and above it the rock is piled up in blocks. The highest strata in the mountain mass are cut for querns and whetstones. Under the roof the whole wall is covered with characters, varying in size from a few inches to two feet in length, and they extend about twenty feet on each side from the apex of the angle.

I was delighted with my trouvaille, the first of the kind which I made in the Brazil, and which here has not before been noticed.* Jacinto Barbosa da Silva, the farm-owner, declared

---

* The Relatorio does not refer to its existence. I shall recur to these inscriptions at the end of the present Chapter.

that it was a roteiro or guide pointing out where treasure is concealed, and such is the general opinion touching these inscriptions. An Italian traveller in the days of our interlocutor's
grandfather had found that it directed him to a hole in a neighbouring nullah, and by dropping stones they found the cavity to
be deep. Slaves were sent to work at it, but presently the waters
came down and the spot was lost for ever.

We then resumed work, and easily finished with the Itacutiára
break. On both sides there are little settlements called " Ao
Pé da Serra : " * opposite these there is a heavy swirl and a
string of small whirlpools which have a dangerous look. A clear
channel, however, is on the right, and boats go down a hollow in
the water with raised rims and strong lateral shoots. A little
below on the Bahian bank is a cliff of red-yellow sandstone, a
" written rock," resembling, but somewhat smaller than, that
first visited; it projects across the stream a similar dark ledge,
much grooved and turned by the floods. There must be some
risk in ascending as well as descending this break when the
winds are violent; and we observe upon the banks that the Cana-
fistula trees, bent almost at a right angle up-stream, rest for
support their leeward branches on the ground. At the Pé da
Serra of Pernambuco, a line of red sandstone bluffs faces the
river with outlines of pillared fragments and rocking stones,
whilst a low plain of their own wastage separates them from the
bank.

The next feature of importance is called Morros do Sobrado,
because supposed to resemble a house. On the left bank below
a large Corôa of sand, thicket and stunted trees, extending
across three quarters of the bed, are twin bluffs, tall and yellow,
separated by a sandstrip, 400 yards long. Stratified and with
cleavage they show " lócas " or caverns of unusual size, whose
black mouths look as if iron faced; they are favourite nesting
places for birds, especially the large grey-coloured hawk (F.
plumbeus ?) which does so much damage to the young of the
flocks. Large blocks have fallen into the water, and have received, like the granites, a coat of glaze. On the right bank a
mass of glistening black " Marumbés " runs into the stream like
a bed of fresh lava, contrasting strongly with the red hills, the
loose yellow sands, and the brown Catingas Altas.

* Thus we have Saint Magnus ad pedem pontis, &c.

Presently turning to the north-north-east we sighted one of the most picturesque reaches in this picturesque valley. The river, now of noble dimensions, bulges out and narrows with graceful curves, and the view down-stream is closed by the long low ridge of Tacaratú. The banks, gently shelving, have their slopes divided by hedges of dry thorn, and bear upon the ridge-tiled houses; here they are sandy, there they are green with grass and corn. To the right is the hamlet of Casa Nova, consisting of some twenty houses, and faced by three magnificent Cashew trees, whose domed heads of verdure extend their leafy locks almost to the ground. On the opposite side is the Porto de São Pedro Dias da Varzéa Redonda *—our destination. The thundering roar of a rapid below tells us that we have now finished our voyage.

Here then is the great terminus of navigation on the mighty Rio de São Francisco, down which we have floated some 309 leagues, nearly thrice the length of England. I felt the calm which accompanies the successful end of a dubious undertaking, whilst the beauty of the site and the splendid future which awaits it, supplied the most pleasing material for thought.

I now return to the inscription.

These "written rocks" appear to be common on the Lower São Francisco. In this part they are found at Icó of the Ypoeira, at Itacutiára, and at the Pé da Serra. Below this I heard of them at Salgado, two leagues from the Curral dos Bois ferry (320th league); and upon the Brejo, a breeding Fazenda belonging to the Capitão Luis da Silva Tavares, opposite the Porto das Piranhas and distant six or seven leagues. The people have stories of "Estrondos" and superhumanities which wait upon these indications of buried treasure; at the Brejo there is an "Olho d'Agua" where the clashing of steel rods is heard.

Such inscriptions were known to the old travellers. Yves d'Evreux, speaking to an acolyte, said of "Sainct Barthelemy," "Tien, voila ce grand Marata qui est venu en ton pays . . . c'est luy qui fit inciser la Roche, l'Autel, les Images et Escritures qui y sont encore à present, que vous avez veu vous

---

* All writers, including M. Halfeld, call it Varzem Redonda. I can only say that the people do not. Varzem and Varzéa, however, are synonymous, signifying water meadows or land occasionally flooded.

autres." His editor, M. Denis, refers to the "grand voyage pittoresque" of M. Debret (i. 46), which does not want a certain interest; the rocks are upon the mountain do Anastabia near the Rio Yapurá, in the Province of Pará. Long before him Koster (ii. Chap. 3)* mentions "a stone in the Province of Paraïba upon which were sculptured a great number of unknown characters and figures, especially that of an 'Indian' woman." The rock, which was of great size, lay in the bed of a nullah, and the people who saw the draughtsman at work told him that there were many similar features in the environs, and named the localities. The Count de Castelnau copied inscriptions from rocks on the Araguaya River which were pointed out to him by the Capitão Mór Antonio Rodrigues Villars: † he found them (v. 113—114) at Serpa, i.e. "pierre gravée," ‡ on the Lower Amazons, and he alludes to the carved figures on the rocks of the Rio Negro, and to the rock inscriptions of the rivers Orinoco and Essequibo. On the Upper Paraguay the huts of the "Indians" and the neighbouring tree trunks were covered with "singular hieroglyphs" of very varied form, but the traveller could not determine whether they were mystical writing or merely copies of marks which the people had found upon stolen cattle. H. I. M. D. Pedro II., a most diligent student of Brazilian antiquities, has collected all the current information upon the subject of these "incised rocks," and told me that he held them to be the work of Quilombeiros or Maroon negroes. I cannot accept this view, as the African at home ignores every species of inscription.

The glyphs found upon the São Francisco were much less European in form than those published by the Revista Trimensal of the Brazilian Institute.§ The symbols show considerable monotony, the most remarkable forms being the hand, the hoof

---

* He had his information from a priest who had visited a friend in the Parahyba Province ; and he was prevented from copying the sketch by his leaving Pernambuco more hurriedly than he had expected. Southey alludes to this inscription.

† They were seen in 1774 during an exploration by the Ouvidor Antonio José Cabral de Almeida. Cunha Mattos (Itinerario de Rio de Janeiro ao Pará) would trace the inscriptions to the Jesuits.

‡ According to Mr. Bates (i. 308) the name of Serpao in the Tupí language, "Ita-couatíara," signifies striped or painted rock, from the prettily variegated Tauatinga clay and conglomerate.

§ The reader will find in the Appendix a translation of this curious document. Its allusions to the Great Rapids of Paulo Affonso are evident, but the tale of the deserted city is popularly supposed to be a romance. A Bahian Padre dedicated himself for a score of years to the re-discovery, and died before he effected it.

with a vertical line or lines bisecting it, and the old Gothic double-looped ⚭. My kind friend Dr. (D.C.L.) A. Moreira de Barros, President of Alagóas, and M. Carl Krauss found other characters upon the Rio da Agua-Morta at the village Olho d'Agua do Casado, near the Porto das Piranhas, and about one direct league from the São Francisco River. The site is a grot from three to five metres in breadth, with perpendicular walls of hard, massive granite (syenite?), from which the mica has almost disappeared, and dyed red by oxide of iron. M. Krauss believes that the inscription was made with iron tools. I would remark, however, that the jade hatchets of the natives were with savage perseverance capable of dinting the hardest stone.* Mr. C. H. Williams, of Bahia, who ascended the Panema influent of the Lower São Francisco, found, two leagues up the bed, characters traced in red paint upon the under part of a rough granitic slab. It is much to be desired that all these ancient remains may be photographed before they are obliterated ; at present every Caipíra, instinctively it would seem, digs his knife-point into the " letreiro " as if in revenge, because it will not betray its secrets. The interpretation will light up a dark place in the pre-historic age of the Brazil,† and the mere mention of them shows that the traveller is wrong to assert " Au milieu des rochers et des arbres gigantesques de ces forêts qui défient les siècles, il ne se trouve pas d'hiéroglyphes ou aucune espèce de signes gravés sur la pierre." ‡

---

* It is not easy to understand how the savages worked refractory substances. Almost everywhere, however, man has invented the rudiments of a file by means of sand adhering to a gummed thread. In India nephrite was treated with corundum or diamond dust.

† The inscriptions on the following pages are those found by Sr. Moreira de Barros and M. Krauss, to whom my gratitude is due.

‡ Prince Max. in 1815—1817 (ii. 314).

This inscription, twenty to thirty metres up stream, shows only about half of what was probably the original size. In this as in letter A, the arrow points along the stream, and seems to indicate a certain point under the sands where possibly there may have been old diggings. It is certain that gold was taken from this place in old times. M. Krauss found nothing there, but his visit was hurried. He considers this carving to be a plan of the stream.

The straight line shows a fissure in the rock.

$\frac{1}{20}$th of the actual size.

A.—Horizontal Projection.

Igrejinha.
(Caldeirão).

Rio Agoa Morta.

These characters are found at the bottom of a natural Caldeirão or pot-hole, which the people call Igrejinha (little church). It is about 3 metres in diameter, 4 deep, and $2\frac{1}{2}$ above the actual bed of the stream.—$\frac{1}{40}$th of the actual size.

B.

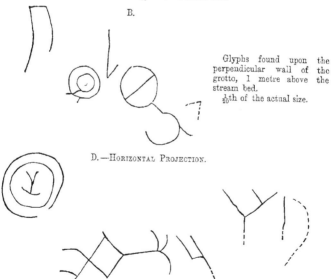

Glyphs found upon the perpendicular wall of the grotto, 1 metre above the stream bed.

$\frac{1}{10}$th of the actual size.

D.—Horizontal Projection.

This is also judged to be half the original size. It is at the bottom of a little cave whose plane is some two metres above the stream, and which can conceal two persons.—$\frac{1}{20}$th of the natural size.

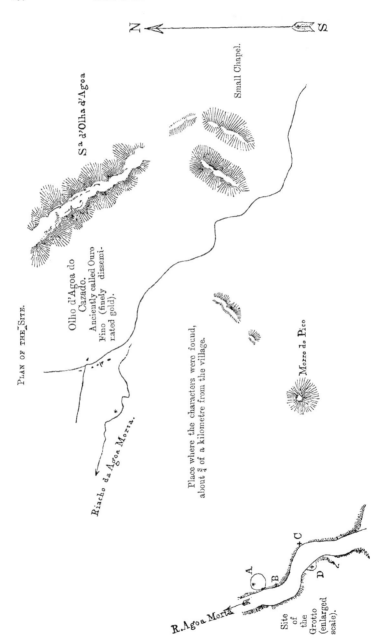

PLAN OF THE SITE.

N ⟵   ⟶ S

Sᵃ d'Olha d'Agoa

Small Chapel.

Olho d'Agoa do Cazado.
Anciently called Ouro Fino (finely dissemi-rated gold).

Riacho da Agoa Morta.

Place where the characters were found, about ¾ of a kilometre from the village.

Morro do Pico

Site of the Grotto (enlarged scale).

R. Agoa Morta

A
B
C
D

Mr. C. H. Williams favoured me with a copy of the characters, which he traced upon the "Panema," and these are the most remarkable forms.

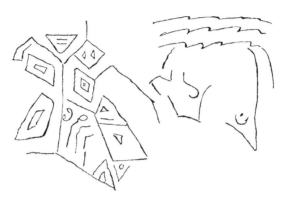

The subjoined are the normal types of what I observed at the Itacutiára.

CHAPTER XXVIII.

TO THE GREAT RAPIDS.—PAULO AFFONSO.

VARZÉA REDONDA DESCRIBED. — DISMISSAL OF CREW AND CONSEQUENT
RELIEF. — THE MULETEERS OF PERNAMBUCO. — GREAT RAPID OF THE
ITAPARICA.

"Then" (when 'Brazil's best customer and most natural ally,' Uncle Sam,
sends a few thousands of his energetic children) "might Brazil, pointing to the
blossoming wilderness, the well-cultivated farm, the busy city, the glancing
steamboat, and listening to the hum of the voices of thousands of active and
prosperous men, say with pride and truth, 'Thus much have we done for the
advancement of civilization and the happiness of the human race.' "—*Lieut.
Herndon*, p. 372.

I LOOKED at Varzéa Redonda for a city, or for a thriving town,
and at best I found only a wretched Quarteirão, which may contain
a score of houses. The population is confined to a slip of ground
along the river, and for want of water, with millions of gallons
flowing within cannon-shot, the uplands are utterly neglected.
The whole of the left bank, from the Serra do Papagaio to the
Varzéa Allegre, belonged three generations ago to a Brazilian
landowner, Manoel da Souza. When he died, the eleven leagues'
length was split up into the various fazendas do Atalho, da
Varzéa Redonda, and further down, da Varzéa Allegre. All
are still occupied by the multitudinous descendants of the original
proprietor. The law of genesis, or development, is here carried
out with a peculiar vigour; the sole métier is apparently père de
famille—a man who has not his dozen is considered a poor devil.
The women bear per head from ten to twenty-five children, and
rare is the hut that does not show a great-grand-parent. The
effect is due to a healthy climate, abundant, though coarse provision, scanty occupation of the body, and yet scantier of what
is popularly called the mind. And I must notice, that even as
Bahia was found inferior, as regards social life, to Minas Geraes

and S. Paulo, so the interior of Pernambuco lay arear of Bahia, whilst Sergipe and Alagôas will be behind all. The latter two, indeed, might easily be thrown into one, but for the political necessity of keeping up as many "government places" as possible.

After arrival I called upon Sr. José Manoel da Souza, ex-justice (inspeitor de quarteirão), who lives at the Porto do Atalho, a few yards above the main landing. He freely offered us the use of his house, but wishing to make some last arrangements before paying off the crew and dismantling the "Brig Eliza," I wished to sleep on board, and regretted my resolution. The night was furious, and the wind raised waves that nearly beat the old raft to pieces. My men having reached the end of their work, had the usual boatman's "spree," hard drinking, extensive boasting, trials of strength, and quarrelling, intermixed with singing, shouting, extemporizing verses, and ending in the snores and snorts of Bacchic sleep. "O Menino" swore that we could not, and should not, advance a step further without him, which ensured for him the "sack." "Majelicão" complicated matters by stealing all the provisions, metal, and loose woodwork that came handy. Next morning my ruffians shed tears of contrition, and cane-rum. The former received 150$000 for his two months of work-shirking, and, complaining that he feared alone to face the fierce strangers, was permitted to take with him the dog Negra, who had also ended the voyage. The brute combined the unpleasant qualities of cowardice and savageness. It could not be trusted near children and small animals, whilst it would fly from the charge of an angry porker. Another sensation of deep rest followed the last glimpse of my crew's backs as they disappeared en route for Boa Vista. The only face which we regretted to see fade in the afar, was that of the good old pilot. I would earnestly recommend him to the future traveller.

The ex-justice showed us the country round about, and our first walk was down stream. The banks near Atalho are a better site for a settlement than those below, where the Ypoeiras produce extensive insulation, and whose levées can hardly remedy the evil, especially when the streams fall in from the hills. There are extensive scatters of Cascalho, that yielded gold some years ago. The experiment, however, was not repeated. We saw a few agates and a hepatic-coloured silex, here known as fígado de gallinha—hen's liver—which is hard enough, they say, even to wear away steel. Amongst the usual Catingas Altas appeared the Shady Mariseiro tree, whose fruits were scattered upon the ground;* the Quina-Quina, with convolvulus-shaped flower and pointed leaf, and the Embira, whose bark is used for fibre, and whose ashes make good brown soap. The country, according to our host, is full of game, ounces, deer, and wild pig (peccari). We had the usual tales of the man who single-handed slew the "tiger" with his clasp knife, and of the "Cabolláda" that eat everything, hawks and lizards.

The higher parts of the bank showed us on the opposite, or Bahian side, "Olho d'Agua," pronounced "Oìdá"—a fine hill-block some 800 to 900 feet above sea level.† To the south, and on the left side, is the Serrote do Brejinho, here called the Serra da Itaparíca. It is distant three to four leagues along the stream. A mile and a half below the Porto do Atalho begins the second and the greater Cordilheira of Rapids, which will extend over some twenty-five leagues.‡ The Cachoeira da Varzéa Redonda, the portal of the unpassable region, is formed by stony hills on both sides of the stream. Black rocks appear on the right of the bed, and boats have passed it, but in great and lively fear of the panellas, or little whirlpools.

Close to this rapid, and on the left bank, is the Varzéa Redonda proper; its small chapel, under the invocation of São Pedro Dias, bears the date 1862. It is a ground-floor affair, with a verandah and two shuttered windows, pointing to the west. The Vigario resides at Tacaratú, five leagues to the east-south-east. This is a market-town, which supplies Varzéa Redonda with provisions. A fair is held there every Saturday, and the country-people visit it from afar, riding in to make their purchases and to attend the next day's mass. It is also connected by a good road with Bahia, the metropolis, said to be distant 110 leagues.§

In 1852, Varzéa Redonda boasted only of eight to ten houses: the number has now trebled. Nothing is easier than building. The river bed gives the best materials for tiles, and hard woods

---

* The fruit, when boiled, is said to taste like almonds.

† It is on the Bahian side, and belonging to the Freguezia of Curral dos Bois. The people speak highly of it, declaring the soil to be excellent, the water abundant, and the air healthy.

‡ The muleteers stretch out these 75 miles to 27, and some to 30, leagues.

§ That is 90 to Alagoinhas, and 20 by railway to S. Salvador da Bahia. Many still prefer this line to the steamers.

are abundant. Sandstone, of which every hill is a quarry, splits up into natural bricks, ready-made ashlar. The finer kinds are good whetstones. I saw many specimens, in which a thin stratum of sparkling grey-green colour was contained in two layers of brown ironstone: the latter, which soon wears down tools of steel, is readily knocked off with the hammer. About eight leagues to the north, at a place called Poço Cercardo, lime is sold at the rate of 2$000 per alqueire. Thus, here again, the calcareous overlies the arenaceous matter.

The climate of Varzéa Redonda is famed for salubrity. We have again slept in the wind and moonlight, in the rain and dew, with rather improvement of, than injury to, health. Here, as we approached that vast ventilating machine, the Paulo Affonso, the wind comes from all quarters. The north brings tornados of thunder and lightning; the south, "inverno, " i.e. wind and rain; the east, light showers, which are considered desirable, and the west is a dry draught. The rainy season opens with storms in October and early November; and the heaviest downfall is about the close in February and March. This is also the rule at the head of the lower Rio de São Francisco, whereas, as I have remarked, the coast rains of the same latitude begin when here all is drought.

I dismantled the " Brig Eliza," which had now been " home " for the last three months. The planks were given away to our host. The anchor from Morro Velho was left in his charge,* and the two canoes bought from the Piaba of Sabará, were here sold for 120$000 to the host's uncle. The next step was to procure animals, which were then rare, being, we were told, engaged in transporting cotton. The charge for making the Porto das Piranhas by the Tacaratú road, was properly 6$000 a head. I vainly offered 8$000, and was asked 10$000 instead of 7$000 for the journey along the stream to Piranhas. The fact is, the ex-justice, remembering that blood is thicker than water, determined that we should come to terms only with his cousins. Neither he nor they, nor any of the neighbours, had seen aught of the Great Rapids, save the mist-cloud which canopies it, and the uncle had told me that it was all a " peta," or " do."

---

* The iron was of excellent quality, and much valued by the people. The ex-justice promised to remit the value to Bahia, but I have not heard that he has done so.

Curiosity here cannot overcome the obstruction of a few hours' ride. The muleteers were not ready till the third day.

The party consisted of the worst men, the worst beasts, and the worst equipments that I had seen in the Brazil; and the disappointment was the greater as Fame has long spoken loudly in favour of the "tropeiros" of Pernambuco. If these be fair specimens, two of them are not worth one Paulista, or Mineiro; and, during the march, there were many unavailing regrets for the troopers Miguel and Antonio, and for the well-fed mules of Morro Velho. The horses were more stubborn and headstrong than mules; they lagged behind; they strayed to enjoy grass and shelter; they rushed forward to prevent being the hindmost, and sometimes they lay down with their loads. Hence some of my collections were lost; hardly a bottle remained unbroken, and the best water-proofs were pierced by the villanous packsaddles. The only attempt at correcting the hoof was to place it upon a plank and to cut down with a formão or chisel. The overcloths of the saddles were loose, and the stirrups just admitting the toe tips, rendered mounting anything but a pleasure. The quadrupeds were weak from want of forage, and the owners would not buy grain for them. Moreover, they were barbarously treated; and for the first time I saw cruelty done to animals by a Brazilian.

The human beings were two and a half, the moiety being represented by a small boy, known as Niger Quim, short for his name, Joaquim Gomes Lima. He looks like twelve, but claims fourteen years of age; and his gruff voice is in his favour; a strange mixture of man and his father; he carries tobacco, flint, and steel; he knows all the local chaff; he is "up to" every adult vice; he offers drink to women thrice his date, and yet he plays with wild fruits, and he climbs up the cruppers of the horses as young Bedouins mount their camels in play. The adults Ignacio Barbosa da Silva, and João, popularly known as "João Caboclo," combine almost every fault of the trooper— intoxication excepted. The difference between them is that Ignacio has a merry eye, which does not belie his nature, whilst very vile is the temper of the Caboclo.* Both are extra lazy. In the morning I have to turn them out of their hammocks, and

---

* Mofino como Caboclo (poor devil as a Caboclo) is an old Brazilian proverb.

they sleep in the bush when they should be collecting their animals. During the first night they allowed (as if they had been Somal) a dog to plunder their meat-bag, rather than take the trouble to hang it up. They must drink water after every two hours; they rest after three; they put their head into every cottage, and they halt to chat with chance-comers on the road. They squabble about carrying half a pound of each other's cargoes, and they use foul language, here by no means a common practice. The principal amusement is to couch the staff like a lance at rest and to dash at the cattle feeding near the path, this "making a fox," always produces a scamper that is ever enjoyed. On the morning after our start, the Caboclo found his temper, and loaded his two beasts to return, demanding, when he knew me to be at his mercy, an additional sum before he would continue the journey. His beard was in my hand after we had reached our destination, but I contented myself with making him yellow with fright for the benefit of those to come, and with *not* paying the money unjustly claimed. A similar proceeding at the Island of Zanzibar, after my return from discovering the Lake Regions of Central Africa, proved strongly to me the absurdity of "public spirit." Yet mere calumny will never deter me from doing what I there and then did. Travellers will never be well treated as long as their predecessors act upon the principle—or rather non-principle—of forgive and forget at the journey's end, because it is the journey's end.

*   *   *   *   *   *   *

The cream of the expedition was now to be tasted, but the enjoyment began with a succession of bitters.* Payment by the host's uncle delayed departure till noon; we could not be allowed to go fasting, and the sun neared the horizon as we mounted our wretched nags, and cast a last loving glance upon the graceful curve, and the "cupped trees" of the fair "round reach."

The path lay southwards along the left bank down the old riverine basin, over which the waters were wont to spread as far as the rolling-ground on either side. It was deeply cut by

---

* The approximate distances of the march were as follows :—

			time		miles		in the league
1.	Nov. 10,	Varzéa Redonda to Itaparica,	3h. 30',		9,		in the 317th league.
2.	,,	11, Itaparíca to Barra do Moxotó	,, 6h. 0',		,, 15,	,,	324th ,,
3.	,,	12, Barra do Moxotó to PAULO AFFONSO	,, 2h. 0',		,, 5,	,,	326th ,,

Total   11h. 30', miles 29

"tip-overs," which during the rains form islands; these must resemble the higher sites in the Egyptian valley, when "pingui flumine Nilus" floods. Between the waters are dwarf table-lands, sterile enough, except where the stream has flaked the sand with hard-baked mud; the richest parts are the Brejos and Brejinhos, little swamps which produce an abundance of cane, cereals, and oil-seeds.

The violent "Vento Geral," here much deflected from the north-east, struck full in our faces. I blessed it for the first time since the beginning of the last month. During this portion of the journey, it generally rose with the moon, and blew itself out after a few hours. The climate suggested that which Bruce called the "hottest in the world," 61° (F.) at sunrise, 82° at sunset, and at 3 P.M., 114°, rising even to 120° in the shade.

When the moon hung high in the air, we reached the "Brejinho de Baixo," and were well received by the owner, Manoel Victor da Silva. His little sugar-mill stands near a swamp, which feeds an old but still luxuriant clump of cocoa-nut trees; the fruit was in the best condition for drinking, and the pleasant sub-acid and highly cooling milk revived memories of Fernando Po.

In front of the house rose the Serra da Juliana,* with a prominent knob called Nariz Furado, or pierced nostril; this is the eastern limit of the ancient bed. We passed an unusual sight within hail of a rapid which we had not shot, and which we were not to shoot.

Before the shade had cleared away, we were aroused by the local alarum, the cry of the Papanshó bird, and as the Rosiclér, or morning light, dawned over the hill-tops, we went with our host to examine the rapids of the Itaparíca.† Here the stream, whose glossy smoothness we had yesterday admired, falls suddenly into a convulsion; a little bay on the Pernambuco bank shows where canoes find the "ne plus ultra," and a few yards beyond it, the São Francisco dashes at a grim ridge of black rock and, splitting into three foamy white lines, disappears from sight. The gate is formed by a rounded hill, the Serra do Padre ‡ on the left or

---

* So called from an ancient Moradora. "Morador," I may remark, does not always mean "habitant isolé;" here it is mostly applied to a class for which we have a name, "peasant-proprietor," but whose existence is sadly wanted.

† This name also may be an ornithological index; in the great bay of S. Salvador (Bahia), the Long Island is called "Itaparíca."

‡ M. Halfeld (p. 178) calls this staple the Serrote do Brejinho.

north-eastern bank: and on the opposite side by the Serra da Itaparíca, a long, straight curtain of bluffs, disposed almost perpendicularly to the bed. In former times they were parts of a dam, which, intercepting the stream, contained a lake; the waters found a " soft place," burst through the wall, and formed the present Cachoeira.

The material of the ridge is sandstone, lined and pudding'd with large and small water-worn quartz pebbles, often passing clean through the blocks. On the banks are found iron-pasted cánga, and large blocks of amygdaloid and boulder conglomerate. The base is a fine, pink syenite, like that of the Nilotic Cataracts. Where the violence of the waters extends the rock is iron-plated with the usual glaze; it is black, as if coated with tar, and the aspect is grizly in the extreme.

On the right, or Bahian side, the channel has a perpendicular fall,* which has dashed to pieces all the rafts and canoes drawn into it by accident; this feature would be best seen from the slopes of the Itaparíca bluff. A tall, longitudinal hogsback, of dark and slippery sandstone conglomerate, rises between the south-western and the central channel. The latter is divided from the third or north-eastern branch by a "mal-paiz" of polished black rock, radiating heat and fire afar, not unlike a lava-field. During the depth of the dries, the jutting-stones approach one another, and a man with a leaping-pole might cross dry-footed the whole Rio de São Francisco.† At such times, also, the site is excellent for a bridge; but in the floods the whole bed becomes a furious, dashing torrent. The rocky ridge extends to some distance on both sides, and here it becomes plain that lateral canalization should not be attempted.‡ It will be plainer still when travelling along the stream, whose banks, alternately rocky and sandy, here dry and there much inundated, will drive the tramway to the skirts of the northern containing ridge.

Nowhere had I seen such gigantic " caldeirões," pans and potholes, turned by the water-lathe; some of them were fifteen feet

---

* The height of the drop varies, according to season, from 6 to 32 feet : the united altitude of the fall, within half a mile, is not less, during the dries, than 55 feet.

† All the Brazilian rivers which I have visited show these extraordinary narrows, when the water, after having spread out for perhaps a mile in breadth, is compressed to the dimensions of a brook. Of course they are always fatal to navigation.

‡ M. Halfeld justly deprecates a line of 72 geographical leagues, mostly cut in the rock, and requiring 108 flood-gates ; at an expense of about 100,000,000 francs

deep by half that diameter, and the sides and surface were jetty as the Itacolumite rock turned into a hole. Mr. Davidson, whom the guides used to call " old lizard " (lagartixo velho), from the ease with which he swarmed up the smooth ridges, and stopped upon the wall-crests, where they would not venture, found in these natural wells the finest crystals and the best diamantine formation. Many similar cavities are doubtless covered over with thin slabs of rock, easily broken into by crow-bars. These places should be carefully searched for gems, and some favourite of fortune will probably make a " pot of money " in a few months. The only use to which at present the open cauldrons are applied is for tanning, their lips are white, and the pits are full of tainted liquid.

I then walked to below the rapids. Here the syenite, the proper material for sphinxes and obelisks, crops out of the white and tree-scattered sands in smooth, bold, and rounded heaps. This point shows the meeting of the waters, which, spuming and whirling from the prison-walls of iron, rush roaring into one another's arms. There is nothing of grace and little of grandeur in the spectacle; all is dark and lurid as a river of the " Inferno."

> Tudo cheio de horror se manifesta,
> Rio, Montanha, troncos e penedos   *   *   *
>
> *Claudio Manoel da Costa.*

The first six miles after the Itaparíca led over a land like that of yesterday. Then we came to the Riacho do Mouro,* where hill-spurs, abutting upon the stream, afforded us only road, narrow gullies, with steep walls, and gutters paved with rolling-stones—in fact a " Caminho perigoso." This bad bit may, I was told, be avoided by following the " Desiro do Bom Querer," about half a league away from the stream. After working for two whole hours, the troopers would halt under a tree, ragged as are all on this bank, at the Porto opposite the Passagem do Jatobá. The river now becomes generally repulsive, it narrows to the size of the Upper Rio das Velhas, and the dull, yellow waters sullenly swirl, boil, and foam around and against central and side jags of

* M. Halfeld (p. 180) calls it the Riacho do " Murro," and in his map, " Muro ;" he makes both to be corruptions of Morro. I give the popular pronunciation. In the Brazil, Mouro or Moor, and Mourería, a " Moorery," mean a gipsy, and the quarters of the cities or towns to which they were confined by law.

rock, whose black or tawny skins contrast unpleasantly with patches of chalk-white sand. The "Passage" is comparatively safe, and there is a ferry to the Curral dos Bois, on the highway to Bahia. Some fine trees appeared upon the far side, sheltering the village, which showed a scatter of huts,* with a chapel dedicated to Santo Antonio da Gloria. The ferrymen sat and stared at us, bare-footed and half-savage beings in foul leathers, and chaplets suspended to their necks. All held, disdaining the support of a belt, sheathless, plate-handled knives, equally ready for a friend or an ounce; formerly the ecclesiastics of these regions never went abroad unweaponed. An "Eu sei?"* long drawn out, as by an Essex boor, was the question that answered every query, and "Au shé,"—Oh, pshaw!—was the token of dissent.

We resumed the march under what Sr. Ignacio was pleased to call "sol macho," a "male sun" of ungentle beams. The road was now deep and sandy, better for man and worse for beast, cut by dry nullahs, scattered with pink quartz and silex superficially streaked and banded; and obstructed by blocks of syenite and porphyritic granite—the Olho de Sápo, or toad's eye, of São Paulo. Trees which up-stream bore fruit three weeks ago, were here in flower awaiting rain, and the river presently resumed the dimensions which it had at Pirapora; the depth, however, was great, and this, combined with its swiftness, and the great loss by evaporation, explains the shrunken dimensions.‡

The moon had risen when we descended the vile bank, and threaded the stagnant pools of the Moxotó or Mochotó, a stream which rising near "Cairirys Velhas" to the northwards, divides the provinces of Pernambuco and Alagôas. Opposite its mouth is a miserable village, which boasts of the last ferry above the Great Rapids. We slept in the bush, and I felt all that depression with which one approaches a long looked-for object, whose fruition appears so fair from afar. At Varzéa Redonda they had compared Paulo Affonso with the Itaparíca, which certainly did not reward 1500 miles of such travel, and all had agreed that the former is grand only between June and September, when the

---

* In 1852 there were 45 houses, with 180 to 200 inhabitants.

† "How the —— do I know?" Emphasis is given by the tone as in that most useful phrase "Pois não."

‡ These places seem generally to have suggested to the older geographers that part of the water disappeared through subterranean passages.

water is at its lowest. I saw no smoke-columns, after hearing that
they were visible from the Serra da Paricornia in the Matinha da
Agua Branca, twenty-four miles off; and after reading Colonel
Accioli, who declares that when condensed by the morning cold,
they may be seen from the Araripe Range, distant thirty leagues.
Nor could I distinguish within two leagues of our destination the
"Zoadão," or thunder, which, they affirm, is audible at the Serra
do Sobrado, thirty-nine miles along the stream.

Apparently I was doomed to a bitter disappointment.

Resuming on the next morning our melancholy way through
the province of Alagôas, we could not but remark the nakedness
of the land. The huts, almost destitute of side-walls or divisions, were mere "tapéras,"* ragged as the population, and from
the bridle-path we could see through them. In the immediate
vicinity of the Great Rapids there is not a hovel, and at the last
house on the left bank, near the Riacho do Corrêa, we asked our
direction from its owner, Manuel Leandro de Resende. He
returned a civil answer, saddled his horse, and accompanied us.
I liked his manner, and engaged him as a guide. Hereabouts
the semi-barbarians are inclined, like the savages of the Congo,
dwelling near the "Yellalahs," to force services upon the
stranger visiting the "Cachoeira." I need hardly say that a
guide, unless he be the rare bird of the right species, often
destroys all the pleasure of the spectacle by his condemnable
attempts to be "agreeable."

---

* "Ce mot seul de Tapéra, qui designe
une maison abandonnée, montre que cet
établissement n'existe plus." (Castelnau,
v. 50.) We shall presently see that this is
not always the rule.

CHAPTER XXIX.

——— wie ein Wassersturz von Fels zu Felsten brauste
Begierig wüthend, nach dem abgrund zu.—*Faust.*

PRESENTLY we heard a deep hollow sound, soft withal, like the rumbling of a distant storm; but it seemed to come from below the earth, as if we trod upon it: after another mile the ground appeared to tremble at the eternal thunder. Sr. Manuel Leandro led us off to the left of whence came the Voice, and began unloading the mules at the usual halting-place. I looked for the promised "Traveller's Bungalow," and saw only the stump of a post, sole remnant of the house run up to receive His Imperial Majesty of the Brazil, who visited the place in October, 1859. The site is a bed of loose sand, which in the height of the floods becomes a torrent. We shall afterwards find where it falls into the main stream. Our rude camp was pitched under the filmy and flickering shade of a tall Carahyba mimosa whose trunk, in places peeled of bark, showed many a name; apparently, however, the great Smithian Gens has been laughed out of cutting and carving its initials and date—here all were Brazilians.

I should advise those who visit Paulo Affonso in the dry season to make at once, with the aid of plan and guide, for the Mãi da Cachoeira, the "Mother of the Rapids," where all the waters that come scouring down with their mighty rush are finally gathered together. To see cataracts aright, it is best, I think—though opinions upon this point differ—to begin with the greatest enjoyment, the liveliest emotion, and not to fritter away one's powers, mental and physical, by working up to the grandest feature. Moreover, this one point displays most forcibly the formation which distinguishes Paulo Affonso from all his great brethren and sisterhood.

Stowing my note and sketch-books in a " Muchíla," or horse's grain-bag, and hanging it over the guide's shoulder, I struck across the left bank of the river ; here a lava-like " Mal-paiz," resembling that of the Itaparíca. The stones, polished as if they were mirrors, or marble-slabs, glittered and reflected the burning sun-beams : in places the ridges are walls of turned and worked rock that looks like mouldings of brass, bronze, or iron. Many of the boulders are monster onyxes, or granite banded and ribboned with quartz : they are of an infinite variety in size and shape, in make and hue, rough and smooth, warm-red, rhubarb-yellow, dull-black and polished jet.

Chemin faisant we crossed an Eastern Channel, at this season almost dry, a thread of water striping the bottom. It forms with the main body a large trapèze-shaped Goat Island, which presents its smaller end down stream. Paulo Affonso differs essentially from Niagara, whose regular supply by the inland seas admits little alteration of weight, or size, or strength of stream, except in the rare winters when it is frozen over. About December, as the floods run high, this tiny creek * will swell to an impassable boiling rapid, ending in a fine fall about the " Vampire's Cave." Upon this " Goat Island," where if there are no goats the walk-ing is fit for them only, are short tracts of loose sand alternating with sheets of granite, and of syenite, with here and there a " courtil " of greener grass. The walk leads to a table of jutting rock on the west side, where we cling to a dry tree-trunk, and peer, fascinated, into the " hell of waters " boiling below.

The Quebrada, or gorge, is here 260 feet deep, and in the nar-rowest part it is choked to a minimum breadth of fifty-one feet. It is filled with what seems not water, but the froth of milk, a dashing and dazzling, a whirling and churning surfaceless mass, which gives a wondrous study of fluid in motion. And the marvellous dis-order is a well-directed anarchy : the course and sway, the wrest-ling and writhing, all tend to set free the prisoner from the prison walls. Ces eaux ! mais ce sont des âmes : it is the spectacle of a host rushing down in " liquid vastness " to victory, the triumph

---

* In M. Halfeld's plan the creek is much larger than it appeared when I visited it : his journey was probably made later in the year.

of motion, of momentum over the immoveable. Here the luminous whiteness of the chaotic foam-crests, hurled in billows and breakers against the blackness of the rock, is burst into flakes and spray, that leap half way up the immuring trough. There the surface reflections dull the dazzling crystal to a thick opaque yellow, and there the shelter of some spur causes a momentary start and recoil to the column, which, at once gathering strength, bounds and springs onwards with a new crush and another roar. The heaped-up centre shows fugitive ovals and progressive circles of a yet more sparkling, glittering, dazzling light, divided by points of comparative repose, like the nodal lines of waves. They struggle and jostle, start asunder, and interlace as they dash with stedfast purpose adown the inclined plane. Now a fierce blast hunts away the thin spray-drift, and puffs it to leeward in rounded clouds, thus enhancing the brillancy of the gorge-sole. Then the steam boils over and canopies the tremendous scene. Then in the stilly air of dull warm grey, the mists surge up, deepening still more, by their veil of ever ascending vapour, the dizzy fall that yawns under our feet.

The general effect of the picture—and the same may be said of all great cataracts—is the "realized" idea of power, of power tremendous, inexorable, irresistible. The eye is spellbound by the contrast of this impetuous motion, this wrathful, maddened haste to escape, with the frail stedfastness of the bits of rainbow, hovering above ; with the "Table Rock" so solid to the tread, and with the placid, settled stillness of the plain and the hillocks, whose eternal homes seem to be here. The fancy is electrified by the aspect of this Durga of Nature, this evil working good, this life-in-death, this creation and construction by destruction. Even so, the wasting storm and hurricane purify the air for life : thus the earthquake and the volcano, while surrounding themselves with ruins, rear up earth, and make it a habitation for higher beings.

The narrowness of the chasm is narrowed to the glance by the tall abruptness, yet a well-cast stone goes but a short way across, before it is neatly stopped by the wind. The guide declared, that no one could throw further than three fathoms, and attributed the fact to enchantment. Magic, I may observe, is in the atmosphere of Paulo Affonso : it is the natural expression of the glory and majesty, the splendour and the glamour of the scene, which Greece would have peopled with shapes of beauty, and which in Germany would be haunted by choirs of flying sylphs and dancing undines. The hollow sound of the weight of whirling water makes it easier to see the lips move than to hear the voice. We looked in vain for the cause: of cataract we saw nothing but a small branch, the Cachoeira do Angiquinho—of the little Angico Acacia—so called from one of the rock islets. It is backed on the right bank by comparatively large trees, and by a patch of vividly green grass and shrubbery, the gift of the spray drifting before the eastern sea-breeze. This pretty gush of water certainly may not account for the muffled thunder which dulls our ears: presently we shall discover whence it comes.

I sat over the "Quebrada" till convinced it was not possible to become "one with the waters:" what at first seemed grand and sublime at last had a feeling of awe too intense to be in any way enjoyable, and I left the place that the confusion and emotion might pass away. The rest of the day was spent at "Carahyba Camp," where the minor cares of life soon asserted their power. The sand raised by the strong and steady trade wind was troublesome, and the surface seething in the sun produced a constant draught: we are now at the very head of the funnel, the vast ventilator which guides the gale to the upper Rio de São Francisco. Far to seaward we could see the clouds arming for rain. At night the sky showed a fast-drifting scud, and an angry blast dispersed the gathering clouds of blood-thirsty musquitos. Our lullaby was the music of Paulo Affonso; the deep, thundering base produced by the longer and less frequent vibrations from the Falls, and from the Rapids the staccato treble of the shorter wave-sounds. Yet it was no unpleasant crash, the deeper tones were essentially melodious, and at times there rose an expression in the minor key, which might be subjected to musical annotation. I well remember not being able to sleep within ear-shot of Niagara, whose mighty orchestra, during the stillness of night, seemed to run through a repertoire of oratorios and operas.

We will now apply ourselves to the prose of the Great Rapids.

The name, as mostly happens in these regions, is a disputed point. Some make "Paulo Affonso" a missioner-shepherd, who was hurled down the abyss by the wolves, his "Red-skin" sheep. Others tell the story of a friar, who was canoeing along the river, when the Indian paddle-men cried, in terror, that they were being sucked into the jaws of the Catadupa: he bade them be of good cheer, and all descended whole.* "Such reverends are now-a-days rare," observed Sr. Manuel Leandro with an unworthy sneer. Similarly in the Province of São Paulo, the Tiété river has a fierce Rapid, known as "Avaremandoura" —Cachoeira do Padre, or the Rapid of the Priest. Here, according to Jesuit legend, Padre Anchieta, one of the multitudinous thaumaturgi of the Brazil, was recovered from the water " some hours afterwards, alive and reading his breviary with a light in his hand." More sober chronicles declare that the poor man was dragged out half drowned.† The gigantic cataract of " Tequendama," we may remember, has also its miracle; it was opened by the great Bochica, god of New Granada, a barbarous land, that had hardly any right to have a god. Others pretend that Paulo and Affonso were brothers, and the first settlers, who gave their names to the place. I would, however, observe, that on the right bank of the stream, opposite the Ilha da Tapéra, one of the many that break the river immediately above the upper break, is a village of fishermen and cultivators, whose name, "Tapéra de Paulo Affonso," shows that it has occupied the site of a ruined settlement, probably made by the colonist who, happier than Father Hennepin, left his mark upon the Great Rapids near which he squatted. The "Tapéristas" are still owners of the right bank : the left belongs to one Nicoláo Cotinguiba,‡ of the Engenho do Pinho, and near "Carahyba Camp" two properties meet. The Cachoeira is in the Freguezia of the Mata da Agua Branca.

The locale of the Paulo Affonso has been very exactly misrepresented by geographers who write geography for the people.§

---

* M. Halfeld (Rel. p. 184) thus gives the legend : "Even they relate that a friar, whilst crossing the river above the Rapid, was sleeping in the canoe which carried him : the pilot, who was an Indian, being unable to manage the craft, and being drawn into the stream, went down. He was never seen again " (a moral, I suppose, pointed against careless pilots, here as common as idle apprentices). "But the friar, who neither woke nor felt the least inconvenience, floated ashore below the Rapids, and was found still sleeping. When aroused by the people, he remembered nothing of what had happened."

† Quadro Historico da Provincia de São Paulo, per J. J. Machado d'Oliveira (p. 58).

‡ P. N. of a place.

§ "The San Francisco River . . . escapes through a break between the Sierras Muribeca and Caryris, between which latter and the Atlantic run the other chains,

This sudden break in the level of the bed, this divide between the Upper and Lower São Francisco, is not formed by a prolongation of the Serra da Borborema, nor by the Chapada das Mangabeiras, nor by Ibyapaba "fim da terra," nor by the Cairirys old or new, nor by the Serra da Borracha, alias Moribéca, so imminent in our maps.* The humbler setting of the gem is a rotting plain brown with stone, scrub, and thicket, out of which rise detached blocks, as the Serra do Retiro, about three leagues to the north-west, and to the west the lumpy Serra do Padre. On the south-western horizon springs, sudden from the flat, a nameless but exceedingly picturesque rangelet of pyramidal hills and peaks, here and there bristling in bare rock, and connected by long blue lines of curtain.

Though our prospect lacks the sublime and glorious natural beauty of Niagara, tempered by the hand of man, and though we find in Paulo Affonso none of the sapphire and emerald tints that charm the glance in the Horseshoe Falls, still it is original and peculiar. In "geological" times, the stream must have spread over the valley; even now, extraordinary floods cover a great portion of it. † Presently the waters, finding a rock of softer texture and more liable to decay, hollowed out the actual "Talhadão," or great fissure, and deepened the glen in the course of ages. We have also here the greatest possible diversity of falling water; it consists, in fact, of a succession of rapids and cauldrons, and a mighty Fall ending in the Mãi da Cachoeira, upon whose terrible tangle of foam we have just looked down. If Niagara be the monarch of cataracts, Paulo Affonso is assuredly a king of rapids; an English traveller who had seen the twain, agreed with me in giving the palm to the latter, as being the more singular and picturesque of the two, which are both so wondrous and so awful. He had not visited

preserving an exact parallelism with it" (p. 141, Physical Geography, from the Encyclopædia Britannica, by Sir John W. F. Herschel, Bart. Edinburgh, Black. 1861). The geography of this most eminent astronomer is frequently at fault: he reminds us of the prophets and the inspired writers of bygone days, who knew everything about heaven, and very little about earth.

* They mostly make the north-eastern extremity of the Moribéca Range hug the

Rapids. There is no such line visible, and the people have forgotten even the name of the old explorer who is mentioned in documents dated 1753-4. Col. Accioli (p. 14) refers to the mountains, and says, I know not upon what authority, that they contain silver and copper. In the Appendix to this volume, the reader will find allusions to Moribéca.

† The fullness of the stream alters, I am told, the shape of the Rapids, but not of the Falls.

the Itaparíca, that foil whose grimness so well sets off its majestic neighbour.

Nature is not in her grandest attire, yet the vestment well suits the shape. Spines predominate. There is the Favelleiro or arboreous Jatropha, with its dark-green oak-like leaves terribly armed; and the Cansanção Maïor (Jatropha urens), a giant nettle, whose white spangles of flowers are scattered in mimicry of snow-flakes upon the sombre verdure. The Cactus is in force; we see the common flat Opuntia, the little Quipá, with its big red fig, the Turk's-head (Cabeça de frade) Melo-cactus, with the crimson fez, whilst amongst the rocks project half-domes of a foot in diameter (C. aphananthemum). Some have flowers quaint as orchids, others are clothed in flue, and the rest are hairy and bald, angular and smooth, giant and dwarf, lorded over by the immense Mandracurú (C. brasiliensis), a tree, but strangely different from all our ideas that make up a tree. The Bromelias are abundant, especially the Carauá, banded like a coral snake, and the Macambira with needle thorns and flower-spikes three feet tall; it is loved by monkeys who, they say, make pic-nics to eat the leaves. The feathery elongated Catingueiro, now tender green, then burnished brown, is remarkable near the dense clustering verdure of the rounded Quixabeira, and the Imbuzeiro with the horizontal boughs, a bush twenty feet high. The Carahyba is the monarch of the bush, and its leek-coloured leafage, hung with long bitter pods, and with trumpets of yellow gold, gains beauty seen near the gay red blossom and the velvety foliage of the "Pinhão bravo," and the whitey-green catkins of the thorny Acacia, known as the Jurema preta. We also remark the black charred bole of the Paó preto, by the side of the sweet-scented Imburana, hung with flakes of light burnished bronze. The scrub is mostly the hard Araça-guava with its twisted wood, and the Bom-nome, whose "good name," I presume, results not from an inedible berry, but because it is found useful for spoons. The cattle straying about the bushes toss their heads, snort, and with raised tails dash through the thicket as we approach; they are sleek, clean-limbed,—much more like the wild animal of the African Gaboon than the European model of bull and cow. Thanks to the drift, they find in the dwarf "courtils" more succulent fodder than usual; they suffer, however, when not penned for the night in the Cahyçára,* from ounces and vampires, and they are sometimes poisoned by the pretty, pink, and innocent-looking blossom, here known as the " Cebolla brava." Finish the picture with clouds of spray and vapour, rising from the abyss, and pouring to leeward an incessant shower of silvery atoms ; with the burnished stones, here singularly gloomy, there mirroring the dazzle of the sun-beam ; and with gay troops of birds, especially the Tanager, the hyacinthine Ararúna, and the red and green parroquet, darting and screaming through the air, whose prevalent hue is a thin warm neutral tint.

My next visit began at the beginning, and thence we followed down the left bank, stepping from slippery stone to stone, and approaching the channel when possible.† Here the São Francisco, running swift and smooth out of the north-west, escapes from the labyrinth of islands and islets, rocks and sands, blocks and walls which squeeze it, and receives on the left a smaller branch, separated from the main by a dark ridge. The two, leaping and coursing down a moderate incline of broken bed, burst into ragged, tossing sheets of foam-crested wave, and tumble down the first or upper break, which is about thirty-two feet high. This kind of " Rideau Fall " is known as the " Vai-Vem de Cima " —the " upper go and come." ‡ The waters are compressed in the central channel by the stone courses rising thirty to fifty feet above them, and are driven into a little cove on the left bank. The mouth of a branch during the floods, now it is a baylet of the softest sand hemmed in by high japanned walls, and here the little waves curl and flow, and ebb again, with all the movement

---

* Also written " Caïsara," and by Sir J. de Alencar " Caiçara :" he derives it from " Cai," burnt wood, and " Cara," a thing possessing. Thus it means the stakes with fired ends thrust into the ground and forming a cattle " kraal," where careful owners shelter their beasts, and where the careless brand them with the iron once or twice a year.

† The formation of Paulo Affonso renders it a Protean feature changing with every month. I visited it in mid-November, when, according to the guide, the water had risen three to four fathoms above the lowest level.

The arithmetic may be briefly taken from M. Halfeld. The left bank of the gorge, called Mãi da Cachoeira, is 365 palms (261 feet, 7 inches) high, and the depth of

the " kieve," or hollow dug by the falls, is 120 palms (86 feet). The narrowest part of the chasm is 72 palms. The first, or Upper Rapid (Vai-Vem de Cima), is 792 palms, 1 inch (= 567 feet, 8 inches) above sea level : and the lowest Rapid (Vai-Vem de Baixo), opposite the Vampire's Cave, is 426 palms, 6 inches (= 305 feet, 9 inches). The united height of Rapids and Fall is 365 palms, 7 inches (= 261 feet, 11 inches).

The Horseshoe Fall of Niagara is 158 (some say 149) feet high, with a width of 1900 feet, and a discharge of 20,000,000 cubic feet per minute. The American Fall is 162—164 feet high, with a breadth of 908 feet. The total breadth of the bed is 3225 feet, and of the water 2808 feet.

‡ Of these Vai-Vens (also written Vaevens) there are, as will be seen, two.

of a tide in miniature. I timed and felt the pulse of the flux and
reflux, but I could detect no regularity in the circulation. The
place tempts to a bath, but strangers must bear in mind that it is
treacherous, and that cattle drinking here have been entangled
in the waters, from which not even Jupiter himself could save
them.

The waters then dashing against the left or south-eastern
boulder-pier, are deflected to the south-west in a vast serpentine
of tossing foam, and form, a few paces lower down, a similar
feature; called by our guide "Half go and come." Here insulated rocks and islands, large and small, disposed in long ridges
and in rounded towers, black, toothed, and channeled, and wilder
far than the Three Sisters or the Bath and Lunar islands of
Niagara, split the hurrying tossing course into five distinct
channels of white surge, topping the yellow turbid flood. The
four to the right topple over at once into the great cauldron.
The fifth runs along the left bank in a colossal flume or launders,
high raised above the rest; meeting a projection of rock at the
south, it is flung round to the west almost at a right angle. Here
the parted waters spring over the ledge, and converge in the
chaudière which collects them for the great fall. When the sun
and moon are at the favourable angle of 35°, they produce admirable arcs and semicircles of rainbows in all their prismatic tintage
from white to red. These attract the eye by standing in a thin
arch of light over the mighty highway of the rushing " burning "
waters ; guides to cataracts, however, always make too much of
the pretty sight.

The third station is reached by a rough thorny descent, which
might easily be improved, and leads to the water's edge, where
charred wood shows that travellers have lately nighted in the
place. Turning to the north-east we see a furious brown rapid
plunging with strange forms, down an incline of forty-nine feet in
half a dozen distinct steps : the flood seems as though it would
sweep us away. At the bottom, close to where we stand, it bends
westward, pauses for a moment upon the billow-fringed lower lip
of the Chaudière that rises snow-white from the straw-coloured
break, and then the low, deep, thundering roar, shaking the
ground and " sui generis " as the rumbling of the earthquake and
the hoarse sumph of the volcano reveal the position of the Great
Cataract. The trend is southerly, and the height is calculated to

be 192 feet. The waters hurl themselves full upon the right-hand precipice of the trough-ravine, surge high up, fall backwards, fling a permanent mistcloud in the air, and like squadrons of white horses, rush off, roaring and with infinite struggle and confusion, down the Mãi da Cachoeira to the south-east. The latter is the grandest point of view which we prospected from the table-rock overhanging the fracture.

Paulo Affonso is always sketched from our third station,[*] where we "realize" an unpleasant peculiarity of his conformation; he has here permitted the eye of man to see the main cataract. A little further down, there is a partial view from above; but the normal central mistcloud curling high and always ascending above the lower lip of the cauldron, veils the depth, and we are not satisfied till we have sighted a Fall from its foot. Now much is left to the imagination, and the mystery is so great as to be highly unsatisfactory. In the depth of the dries it is, they say, possible to climb down a portion of the left wall, and to overlook the cataract. I carefully inquired whether it was visible from the right, or Bahian bank; all assured me that a branch stream allowed no approach to the trough-ravine, and all were agreed that from that side nothing is visible.[†] A moveable suspension-bridge, not, I hope, like that of Montmorenci, could be made to span the chasm; wire-ropes fit to bear cradles, could be passed across; or ladders might be let down and act as the winding staircase which leads to the Horseshoe Fall. At present Paulo Affonso is what Niagara was in the days of Père François Piquet: and we can hardly look forward with pleasure to the time when it will have wooden temples and obelisks, vested interests, 25 cents to pay, and monster-hotels.

The next station is that with which I have advised the stranger to begin. Thence he must retrace his steps, the trough up is too rough and broken to be followed. We again crossed the eastern boulder-channel, and walking to the south-east reached, after a

---

[*] From this point also the photographs are taken, and they afford but a poor idea of the original. M. Halfeld, besides his vignette, gives two lithographs, the first from our third station, and the second from the Paredão, opposite the Angiquinho, a place which we shall presently visit.

[†] Travellers with more leisure than I had, will not take this assertion on trust. The distance from the Porto das Piranhas, where the steamer stops, is only 12 leagues, easily done in two days. And if the "Bahian side" be practicable, it is evidently the place for a flight of steps, cut in the rocky wall of the trough.

few hundred yards, a descent formed by the waters which, in flood-time, sweep over the hollow "Carahyba Camping ground," and course down a stony incline to rejoin the parent flood. We found the bed bone-dry, a slippery surface of bare rock, dark and bright after many a freshet, with here and there steps and deep crevasses. There are stagnant pools and corrie-like holes, green with Confervæ, and rich in landshells. These hollows long preserve the rain-water, and though covered with scum-like aquatic growths, they are during the dries, a great resource for cattle. The hands must be used as well as the feet in descending, and the noonday sun will peel the palms.

The zigzag led down to a Ressáca or bulge in the left bank of the river. Here the torrent is less terrible, but still violent, as it dashes against the south-eastern wall of the trough. The light colour of the precipice, not grown, like the rest of the trough, with moss, Bromelia, and thorn bush, shows that, despite the exceeding hardness of the stone, some part has slipped, and more will slip. Sr. Manuel Leandro assured me that it has not changed since the days of his grandfather and his grandmother.

At the trough-foot we reach a baylet formed by the lower "Go and come" (Vai-Vem de baixo), another back-water of the great rushing tide. People fetching water have fallen into it, but have managed to extricate themselves. No Maid of the Mist, however, will in these ages be able to ascend the line of maëlströms. Now the water recovers from the plunge and dive in the abyss beneath the cascade, it continually rebounds, and as we often noticed in the Cachoeiras of the upper bed, there is no really level surface; the face seems a system of slightly bulging domes.

The little inclined ramp of loose stones at the bottom of the wall is strewed with lumber and with wood brought down by the last floods; its grinding sound and its crash when floods are high, have been compared with the creaking of the ice at the end of a Canadian winter. Light as pumice-stone, the fragments are rounded off and cropped at both ends by the bruising process, and the working takes curious shapes, cheeses and shuttles, ninepins and skittles. Our guide picturesquely called the heaps "Cidade de Madeira," a city of wood, and in them I recognized canoe planks and scantling from the Imperial shed.

The slope ends in a cave opening to the west, and known gene-

rically as the " Casa de Pedra," specifically as the " Furna do
Morcego," or Vampire's Grot. Its appearance is singular. The
entrance, instead of being low, after the fashion of caverns, is a
tall parallelogrammic portal leaning a little to the south. Hence
it has its Saintess (uma Santa), who shows herself at times, and
the people have heard martial music, and singing which did not,
they judged, proceed from mortal wind-pipe. The arch is
formed by a thick table of hard, close-grained granite, spread out
as though it had been lava, and a cleavage-line extends to the
southern corner. Its walls are sandstone, here hard and compact,
there soft and mixed with ochreous iron-stained clay, easily cut
in the days when the now shrunken and sunken stream filled the
trough-ravine with its débâcles.* At present the inundations
extend only half-way up the floor, where an " old man " is in process of degradation. The level sole is strewed with bat's guano,
and with ashes where the people have attempted to smoke out the
blood-suckers. The greatest height is about 90 feet. The eastern
wall overhangs, and the honey-combed upper part shows other
branch caves still forming, whilst the western retreats at the easy
angle of 8°—9°. I saw no bats in this " Cave of the Winds," this
" Devil's Hole ;" but the hour of visit was early afternoon, and
the plague was probably enjoying its nap. The mouth of the
vault has a singular look-out. The fleecy, seething, snow-white
torrent, disposed in vorticose-lozenges and ridges, with its spray
glittering under the sun like myriads of brilliants, strikes upon a
shoulder of polished and intensely black rock, whose parallel and
much inclined bands wall the right side of the crevasse. This
mass deflects the boiling rapid at nearly a right angle, and sends
it roaring between the cliffs of its deep-narrow chasm adown the
abrupt redoublings which end its course in the world of waters to
the east.

* M. Halfeld gives the dimensions 80
palms (57 feet, 4 inches) tall, × 40 palms
(28 feet, 8 inches) wide, and 444 (318 feet,
2 inches) long. The entrance to the smaller
Eastern fissure is 30 palms (21 feet, 6
inches) high, broadening to 60 palms (43
feet) inside. That surveyor elaborately explains the formation of the " Furna." Its
line, he says, presents many veins of calcareous spath, of flesh-coloured felspar, and
of quartz varying from ¼ of an inch to
5 inches in breadth, causing the granite to
lose compactness : moreover, it is sometimes
saturated with muriate of soda, from which
a little salt is made. The granite shows,
it is true, many dykes, some raised above,
others sunk below, and others level with
the surface ; but their thickness is trifling.
The rocks also contain lime, hence the quantity of landshells which, dead at this season,
strew the ground. But the cave is evidently hollowed out in the sandstone grit,
which on the left side of the Lower São
Francisco, forms hills and ridges, and which
further down alternates with or overlies
limestone.

Our last station is upon the "Paredão," lower down than the Vampire's Cave, at the place called "Limpo do Imperador"— the bush having been cleared away for the Imperial visit. There is no shade, and water is far off. A tent and barrels, however, would make all things easy, and a traveller encamped upon this Bellevue would have beneath his eyes by night as well as by day the most beautiful if not the grandest scenery of Paulo Affonso. Here he stands on a level with the stream above the Upper Rapids, and on 300 feet of perpendicular height over the water, which dashes curdling and creaming below. To the westwards the vision strikes full upon the small but graceful Angiquinho branch, which is the American Falls compared with the Horseshoe, and which reminds the traveller of the tall narrow Montmorenci.* This offset is the furthest line on the right side of the river, in which, about the Tapéra of Paulo Affonso, a mass of long islands precedes the narrows and the rapids. It encloses an "Iris Island," a crag which may easily be confounded with the mainland. It is, however, capped with tree and grass, kept green as emerald by the ceaseless drift, and made remarkable by the brown plain forming the distance. Here again the still and silent picture around heightens the effect of the foaming, rushing water. The flood rolls headlong over its own shelf of brown based on jet-black rock, seen in the walls which here jut out and there retire. Dashed to pieces by the drop, it shows about the centre, with the assistance of a projecting rock, at this season clearly visible, a fall within a fall. Puffs of water-douche, looking as if endless mines were being sprung, rise to half its height, and the infinite globules, "spireing up" in shafts, repeat the prismatic glories of solar and lunar rainbows. At its foot, from the spectator's right hand, or from north to south, a section of an arch represents the terminal part of the mysterious cataract, whose upper two-thirds are hidden by a curtain of rock. This, the main stream, impinges almost perpendicularly upon the right-hand wall of the trough-ravine, and the impetus hurls it in rolls and billows high up the face to be thrown back shattered, and to add a confusion more confused to the succeeding torrents. But, subject to the eternal law of gravitation, a sinuous line perforce undulates down the crevasse, which gradually widens, and which puts out buttresses from the right and left.

* The height of Montmorenci is about 250 feet by 50 feet of breadth.

Calmed by the diminished slope, it meets the tall cliffs upon which we stand, and wheeling from north-west to south-west, it eddies down the windings of the ravine, which soon conceal it from the sight. The effect is charming when the moon, rising behind the spectator, pours upon the flashing line of cascade and rapid full in front, a flood of soft and silvery light, while semi-opaque shadows, here purple, there brown, clothe the middle height, and black glooms hang about the ribs, spines, and buttresses of the chasm-foot.

Not the least interesting part of Paulo-Affonso is this terminal ravine, which reminded me of the gorge of Zambezian Mosiwa-tunya, as painted by Mr. Baines. It has given rise to a multitude of wild fables, especially to the legend of the under-ground river, an Alpheus, a Niger, a Nile, that favourite theme with the "old men."* The black sides, footed by boulders which the force of the flood has hurled in heaps, and in places cut by small white streams, preserve their uniformity, and wall in the stream as far as the Porto das Piranhas, forty-two geographical miles below the cataract.† Moreover the elevation profile shows below the actual cataract, a kieve or deep hollow, and a long succession of similar abysses, prolonged to the same point, and gradually diminishing in depth, the effect of a secular filling in. Niagara undermining the soft shales that support a hard structure of lime-stone some ninety feet thick, has eaten back seven miles from the escarpment known as Queenstown Heights. It is supposed to have expended 4000 years in reaching its present position, and to be receding at the rate of a foot per annum. Here we find a similar retreat of the waters. According to the guides, a huge mass of stone above the Chaudière formed an arch under which birds built their nests. This disappeared like the old original " Fall

---

* The Noticias Ultramarinas of 1589 (Chap. 20) makes the Sumidouro, or sink, 80—90 leagues above the Cachoeira, which was then apparently not named. The classical geographical romance soon spread afar. Frei Rio Giuseppe de Santa Theresa (Istoria delle guerre del Brazil) wrote "dopo de aver corso diciotto giornate di paese dentro di cui si nasconde per lo spazio di dodici leghe." Southey (III. i. 44) borrows from the Patriota and Cazal with reasonable correctness, he alludes to the "rapids and falls . . . one of such magnitude that the spray is visible from the mountains, six leagues distant, like the smoke of a conflagration."

† It is found with breaks to the Pão de Assucar, distant sixty-three miles. Of course these are rude estimates, which may have an error of two miles. From Paulo Affonso to Porto das Piranhas, the perpendicular and inclined walls that hem in the fiercely-dashing stream, are often 800 palms (570 feet high). At the Cachoeira da Garganta (of the Gorge), nine miles below the cataract, the breadth of the stream is only 85 palms (61 feet), and the height of the trough is 350 palms (250 feet).

Rock" about ten years ago, and since that time they say the Zoadão, or roar of Paulo Affonso, has not been so loud. Applying, therefore, the rule of the Northern Cataract, we cannot assign to the King of Rapids an age under 2400 years.

      *   *   *   *   *   *

My task was done. I won its reward, and the strength passed away from me. Two days of tedious monotonous riding led to the Porto das Piranhas. The steamer had just left it, but a hospitable reception awaited me at the house of Sr. Ventura José Martins, agent to the Bahian Steam Navigation Company. My companion hurried away to catch the American mail at Pernambuco. I descended the lower Rio de São Francisco more leisurely, under the guidance of Sr. Luis Caetano da Silva Campos of Penedo, whose amiable Senhora made me feel at home. Whilst delayed at "the Rock" I met my excellent friend Dr. A. Moreira de Barros, then President of Alagôas, and visited him at his capital, Maceiô. Thence, with the aid of Mr. Hugh Wilson, I found my way to Aracajú and Bahia, and finally I returned viâ Rio de Janeiro to Santos (São Paulo), alias Wapping in the Far West.

APPENDIX.*

EXTRACT FROM THE REVISTA TRIMENSAL OF THE INSTITUTO HISTORICO E
GEOGRAPHICO BRASILEIRO, RIO DE JANEIRO, JULY 21, 1865.

---

*Historical account of a large, hidden, and very ancient city, without inhabitants,
discovered in the year 1753.*

In America .   .   .   .

in the interior .   .   .   .

adjoining the .   .   .   .

Master of *Can* .   .   .   .

and his followers, having wandered over the desert country (Sertoës) for ten
years in the hopes of discovering the far-famed silver mines of the great explorer Moribeça, which through the fault of a certain governor were not made
public, and to deprive him of this glory he was imprisoned in Bahia till
death, and they remained again to be discovered.   This news reached Rio de
Janeiro in the beginning of the year 1754.

After a long and troublesome peregrination, incited by the insatiable greed
of gold, and almost lost for many years in this vast Desert, we discovered a
chain of mountains so high that they seemed to reach the ethereal regions, and
that they served as a throne for the Wind or for the Stars themselves.   The
glittering thereof struck the beholder from afar, chiefly when the sun shone
upon the crystal of which it was composed, forming a sight so grand and so
agreeable that none could take eyes off these shining lights.   Rain had set in
before we had time to enter (in the itinerary) this crystalline marvel, and we
saw the water running over the bare stone and precipitating itself from the
high rocks, when it appeared to us like snow struck by the solar rays.   The
agreeable prospect of this .   .   .   .   *(uina)* shine

.   .   .   .

.   .   .   .   of the waters and the tranquillity

.   .   .   .   .   of the weather, we resolved to investigate this admirable prodigy
of nature.   Arriving at the foot of the ascent without any hindrance from
forests or rivers, which might have barred our passage, but making a detour
round the mountains, we did not find a free pass to carry out our resolution

---

* Translated by Mrs. Richard Burton, who begs indulgence, on account of this report
having been written in old Portuguese by rude explorers, and therefore very difficult to
render into English.   All the lines that are not filled in, are illegible from the age and
decayed state of the original MS.

to ascend these Alps and Brazilian Pyrenees, and we experienced an inexplicable sadness from this mistake.

We "ranched" ourselves with the design of retracing our steps the following day. A negro, however, going to fetch wood, happened to start a white stag which he saw, and by this chance discovered a road between two mountain chains, which seemed cut asunder by art rather than by nature. With the overjoy of this news we began the ascent, which consisted of loose stones piled up, whence we thought it had once been a paved road broken up by the injuries of time. The ascent occupied three good hours, pleasantly, on account of the crystals, at which we wondered. We halted at the top of the mountain, which commanded an extensive view, and we saw upon a level plain new motives to rouse our admiration.

We discerned about a league and a half from us a large settlement, whose extent convinced us that it must be some city dependent upon the capital of the Brazil. We descended soon to the valley, with the precaution . . . . it might be in such a case, sending explor . . . . gate the quality and . . . . that they should take good notice . . . . chimneys, this being one of the evident signs of settlements.

We waited for the explorers during two days, longing for news, and only waited to hear cocks crow to be certain that it was peopled. At last our men returned, undeceived as regards there being any inhabitants, which puzzled us greatly. An Indian of our company then resolved at all risks, but with precaution, to enter; but he returned much frightened, affirming that he did not find nor could he discover the trail of any human being. This we would not believe, because we had seen the houses, and thus all the explorers took heart to follow the Indian's track.

They returned, confirming the above-mentioned deposition, namely, that there were no inhabitants, and so we determined all to enter this settlement well armed and at dawn, which we did without meeting any one to hinder our way, and without finding any other road save that which led directly to the great settlement. Its entrance is through three arches of great height, and the middle one is the largest, whilst the two side arches are less. Upon the largest and principal we discerned letters, which from their great height could not be copied.

There was one street the breadth of the three arches, with upper storeyed houses on either side; the fronts of carved stone already blackened; so . . . . inscriptions all open . . . . (d)oors are low of ma(ke) . . . . nas noting that by the regularity and symmetry with which they are constructed it appeared to be one long house, being in reality a great many. Some had open terraces, and all without tiles, the roofs being some of burnt bricks and others of freestone slabs.

We went through some of the houses with great fear, and nowhere could we find a vestige of personal goods or furniture which might by their use or fabric throw any light on the nature of the inhabitants. The houses are all dark in the interior; there was scarcely a gleam of light; and as they are vaulted, the voices of those who spoke re-echoed till our own accents frightened us.

Having examined and passed through the long street, we came to a regular square, and in the middle of it was a column of black stone of extraordinary height and size, and upon it was the statue of an average-sized man, with one hand upon his left haunch, and the right arm extended, pointing with his fore-finger to the North Pole. In each corner of the said square was a needle (obelisk ?), in imitation of that used by the Romans, but some had suffered ill-usage and were broken, as if struck by thunder-bolts.

On the right hand of this square was a superb edifice, as it were the principal house of some Lord of the land. There was an enormous saloon in the entrance, and still from fear we did not investigate all the hou(ses) . . . . . . being numerous and the *retret* . . . . *zerão* to form some . . . . *mara* we found o(ne) . . . . mass of extraordinary . . . . (per)sons had difficulty in raising it. . . . . The bats were so many that they attacked the people's faces, and made such a noise that it astonished them. Upon the principal portico of this street was a figure in demi-relief carved out of the same stone, and stripped from the waist upwards, crowned with laurels. It represented a young figure, and beardless. Beneath the shield of the figure were some characters, spoiled by time. How-ever, we made out the following. (See the Plate, inscription No. 1.)

On the left side of the said Square is another edifice, quite ruined ; but from the vestiges remaining, there is no doubt that it was once a temple, for part of its magnificent frontispiece still appears, and some naves and aisles of solid stone. It occupies a large space of ground, and on its ruined walls are seen carvings of superior workmanship, with some figures and pictures inlaid in the stone, with crosses and different emblems, crows and other minutiæ, which would take a long time to describe.

Follow this edifice—large portions of the city totally ruined and buried in large and frightful openings of the earth, and upon all this ground not a blade of grass, tree, or plant was produced by nature, but only heaps of stone and some coarse rough works, by which we judged . . . . *verçao*, because still amongst . . . . *da* of corpses which . . . . is part of this unhappy . . . . *da*, and forsaken perhaps on account of some earthquake.

In front of the said square runs rapidly a mighty and broad river, which had spacious banks and was agreeable to the sight. It might be from 11 to 12 fathoms broad, without considerable turnings, and its banks were free from trees and timber which the inundations usually bring down. We sounded its depth and found in its deepest parts from 13 to 16 fathoms. On the further side of it are most flourishing plains, and with such a variety of flowers that it would appear as if nature was more bountiful to these parts, making them produce a perfect garden of Flora. We admired also some lagoons full of rice, of which we profited, and likewise innumerable flocks of ducks which breed in these fertile plains, and we found no difficulty in killing them without shot, but caught them in our hands.

We marched for three days down the river, and came upon a cataract which made a fearful noise from the force of the water and the obstacles in its bed, so that we thought the mouths of the far-famed Nile could not make more. Below this fall the river so spreads out that it appears to be the great ocean. It is full of peninsulas covered with green turf, with a sprinkling of trees, which make . . . . *davel.* Here we found . . . . in default of it if we . . . . (mu)ch variety of game . . . . . (o)ther many animals bred ; there being no huntsmen to chase and persecute them.

To the east of this waterfall we found several deep cuttings and frightful excavations, and tried its depth with many ropes, which, no matter how long they were, could not touch its bottom. We found also some loose stones, and, on the surface of the land some silver nails, as if they were drawn from mines and left at the moment.

Amongst these caverns we saw one covered with an enormous stone slab, and with the following figures carved on the same stone, which apparently contains some great mystery. (See the inscription No. 2.) Upon the portico of the temple we saw others also, of the following form. (Inscription No. 3.)

About a cannon-shot from the village was a building as it might be of a country-house, with a front 250 paces long. The entrance was by a large portico, and we ascended a staircase of many coloured stones which opened into an immense saloon, and afterwards into 15 small houses, each with a door opening into the said saloon, and each one bore its own water spout . . . . *a* which waters, and adjoining . . . . *mão* in the external courtyard . . . . colonnade in a cir . . . . *ra* squared by art, and hung with the following characters. (See the inscription No. 4.)

After this wonder we descended to the banks of the river hoping to discover gold, and without trouble we found rich " pay-dirt " upon the surface, promising great wealth of gold as well as of silver. We wondered at the inhabitants of this city having left such a place, not having found with all our zeal and diligence one person in these Deserts who could give any account of this deplorable marvel, as to whom this settlement might have belonged. The ruins well showed the size and grandeur which must have been there, and how populous and opulent it had been in the age when it flourished. But now it was inhabited by swallows, bats, rats and foxes, which fattened on the numerous breed of chickens and ducks, and grew bigger than a pointer-dog. The rats had such short legs they did not walk, but hopped like fleas ; nor did they run like those of an inhabited place.

From this spot a companion left us, who, with some others, after 9 days' good march, sighted at the mouth of a large bay formed by a river, a canoe carrying two white persons, with loose black hair, and dressed like Europeans, ... a shot as signal, in order to *ve* . . . . to fly or escape. . . To have . . . . hairy and wild . . . . *ga*, and they all curl up and invest . . . .

One of our companions, called João Antonio, found in the ruins of a house a gold coin, round, and larger than our pieces of 6$400. On one side was the image or figure of a youth on his knees, and on the other side a bow, a crown and an arrow, of which sort (of money) we did not doubt there was plenty in the said settlement or deserted city, because if it had been destroyed by some earthquake, the people would not have had time suddenly to put their treasure in safety. But it would require a strong and powerful arm to examine that pile of ruins, buried for so many years, as we saw.

This intelligence I send to your Excellency from the Desert of Bahia, and from the rivers Para-oaçú (Paraguassú) and Uná. We have resolved not to communicate it to any person, as we think whole towns and villages would be deserted ; but I impart to your Excellency tidings of the mines which we have discovered in remembrance of the much that I owe to you.

Supposing that, of our company, one has gone forth under a different understanding, I beg of your Excellency to drop these miseries, and to come and utilize these riches, and employ industry, and bribe this Indian to lose himself and conduct your Excellency to these treasures, etc. . .

. . . .
. . . . *charão* in the entrances
. . . . *bre* stone slabs . . .
. . . .

(Here follows in the manuscript what is found represented in the plate underneath, No. 5.)

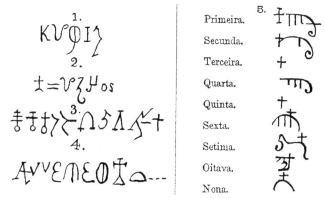

	B.
Primeira.	
Secunda.	
Terceira.	
Quarta.	
Quinta.	
Sexta.	
Setima.	
Oitava.	
Nona.	

Inscripçães encontradas na cidade abandonada de que trata o manuscripto, existente na Bibliothéca Publica do Rio de Janeiro.

Inscriptions found in the abandoned City, of which the MSS. to be seen in the Public Library of Rio de Janeiro, treats.

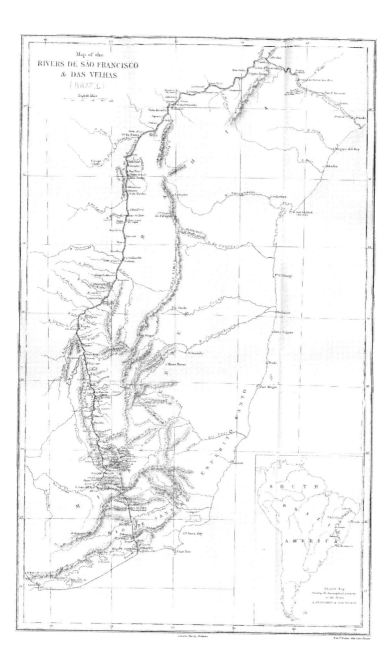

Map of the
RIVERS DE SÃO FRANCISCO
& DAS VELHAS
(BRAZIL)

English Miles

London, Dailey, Brothers.

W.m & Walker Killo Lane, Strand.

INDEX.

THE END.

BRADBURY, EVANS, AND CO., PRINTERS, WHITEFRIARS.

TINSLEYS' MAGAZINE,

An Illustrated Monthly,

Price One Shilling,

CONDUCTED BY EDMUND YATES,

CONTAINS :—

BREAKING A BUTTERFLY ; OR, BLANCHE ELLERSLIE'S ENDING. By the Author of "Guy Livingstone," &c.

A HOUSE OF CARDS. A Novel. By a New Writer.

NOVELS : By EDMUND YATES, Author of "Black Sheep," &c.

WILLIAM HOWARD RUSSELL, LL.D., of the "Times."

ENGLISH PHOTOGRAPHS. By an AMERICAN.

THE LATEST PARIS FASHIONS. (*Illustrated.*)

&c.   &c.   &c.

TINSLEY BROTHERS' NEW WORKS.

BRITISH SENATORS ; or, Sketches Inside and Out-side the House of Commons. By J. EWING RITCHIE, Author of the "Night Side of London," &c. 1 vol.

NOTICE.—A NEW BOOK OF TRAVEL BY CAPTAIN R. F. BURTON.

EXPLORATIONS of the HIGHLANDS of the BRAZIL ; with a full Account of the Gold and Diamond Mines : also, Canoeing down Fifteen Hundred Miles of the Great River São Francisco from Sabará to the Sea. By Captain RICHARD F. BURTON, F.R.G.S., &c. &c. 2 vols. 8vo.

ENGLISH PHOTOGRAPHS. By an AMERICAN. 1 vol.

A WINTER TOUR in SPAIN. By the Author of "Altogether Wrong," "Dacia Singleton," &c. 1 vol. 8vo, with Illustrations of the Alhambra, Escorial, &c., 15s.

THE MARCH TO MAGDALA. By G. A. HENTY, Special Correspondent of "The Standard." In 1 vol. 8vo, 15s.

THE ADVENTURES of a BRIC-A-BRAC HUNTER. By Major BYNG HALL. 1 vol. 7s. 6d.

THE GREAT UNWASHED. 1 vol. uniform with "Some Habits and Customs of the Working Classes."

THE GREAT COUNTRY ; or, Impressions of America. By GEORGE ROSE, M.A. (ARTHUR SKETCHLEY). 1 vol. 8vo.

ESSAYS in DEFENCE of WOMEN. 1 vol. crown 8vo. handsomely bound in cloth, gilt, bevelled boards.

THE SAVAGE CLUB PAPERS. Complete in 1 vol. handsomely bound, cloth, 5s.

TINSLEY BROTHERS, 18, CATHERINE STREET, STRAND.

408755

Made in the USA